FORESTS AND SOCIETY

Sustainability and Life Cycles of Forests in Human Landscapes

FORESTS AND SOCIETY

Sustainability and Life Cycles of Forests in Human Landscapes

Edited by

K.A. Vogt, D.J. Vogt and R.L. Edmonds
College of Forest Resources, University of Washington, USA

J.M. Honea
NOAA Fisheries, Northwest Fisheries Science Center, Seattle, USA

T. Patel-Weynand
International Biological Informatics Program, USGS National Program Office, Reston, USA

R. Sigurdardottir
CAPEIntl Iceland, Iceland

and

M.G. Andreu
School of Forest Resources and Conservation, University of Florida, USA

www.cabi.org

CABI is a trading name of CAB International

CABI Head Office
Nosworthy Way
Wallingford
Oxfordshire OX10 8DE
UK

CABI North American Office
875 Massachusetts Avenue
7th Floor
Cambridge, MA 02139
USA

Tel: +44 (0)1491 832111
Fax: +44 (0)1491 833508
E-mail: cabi@cabi.org
Website: www.cabi.org

Tel: +1 617 395 4056
Fax: +1 617 354 6875
E-mail: cabi-nao@cabi.org

A catalogue record for this book is available from the British Library, London, UK.

Library of Congress Cataloging-in-Publication Data

Forests and society: sustainability and life cycles of forests in human landscapes/
edited by K.A. Vogt . . . [et al.].
  p. cm.
 Includes bibliographical references and index.
 ISBN-13: 978-1-84593-098-1 (alk. paper)
 ISBN-10: 1-84593-098-3 (alk. paper)
 1. Forests and forestry--Social aspects.   2. Forests and forestry--Environmental aspects.
3. Sustainable forestry.   I. Vogt, Kristiina A. II. Title.

 SD387.S55F68 2006
 333.75--dc22

                                                                 2005033354

ISBN-10: 1 84593 098 3
ISBN-13: 978 1 84593 098 1

Typeset by AMA DataSet Ltd, UK.
Printed and bound in the UK by Cromwell Press, Trowbridge.

Mixed Sources
Product group from well-managed
forests and other controlled sources
www.fsc.org Cert no. TT-TOC-2082
FSC © 1996 Forest Stewardship Council

The paper used for the text pages in this book is FSC certified. The FSC (Forest Stewardship Council) is an international network to promote responsible management of the world's forests.

Contents

Contributors

Andreu, Michael G., *School of Forest Resources and Conservation, University of Florida/IFAS, 1200 North Park Road, Plant City, FL 33563, USA; Tel.: 813-707-7330; E-mail: mandreu@ifas.ufl.edu*

Beard, Karen H., *Department of Forest, Range, and Wildlife Sciences, 5230 Old Main Hill, NR 206, Utah State University, Logan, UT 84322, USA; Tel.: 435-797-8220; E-mail: kbeard@cc.usu.edu*

Briggs, David G., *Director, Stand Management Coop and Precision Forestry, College of Forest Resources, Box 352100, University of Washington, Seattle, WA 98195, USA; Tel.: 206-543-1581; E-mail: dbriggs@u.washington.edu*

Brown, Sandra, *Winrock International, 1621 N Kent St., Suite 1200, Arlington, VA 22207, USA; Tel.: 703-525-9430 ext. 678; E-mail: sbrown@winrock.org*

Bushley, Bryan R., *1711 East–West Rd, #574, Honolulu, HI 96848, USA; Tel.: 808-944-6211; E-mail: bushley@hawaii.edu*

Eastin, Ivan L., *Associate Director and Professor, Center for International Trade in Forest Products (CINTRAFOR), University of Washington, Seattle, WA 98195, USA*

Edmonds, Robert L., *Associate Dean for Research, College of Forest Resources, Box 352100, University of Washington, Seattle, WA 98195, USA; Tel.: 206-685-0953; E-mail: bobe@u.washington.edu*

Fanzeres, Anna, *PO Box 125082, Barra de Sao Joao, Rio de Janeiro 28880-000, Brazil; Tel.: (021)242-0725; E-mail: af64@alternativa.com.br*

Gara, Robert I., *College of Forest Resources, Box 352100, University of Washington, Seattle, WA 98195, USA; Tel.: 206-543-2788; E-mail: garar@u.washington.edu*

Gardner, Shelley L., *Center for International Trade in Forest Products (CINTRAFOR), University of Washington, Seattle, WA 98195, USA*

Gordon, John C., *Former Dean and Emeritus Professor, Yale School of Forestry and Environmental Studies, Chairman, Interforest LLC, 27 Evans Road, Holderness, NH 03245, USA; Tel.: 603-536-7571; E-mail: jgordon@iforest.com*

Hecht, Brooke Parry, *Center for Humans and Nature, 2430 North Cannon Drive, Chicago, IL 60614, USA; Tel.: 773-549-0606 ext. 2016; E-mail: brookehecht@humansandnature.org*

Heltne, Paul, *Center for Humans and Nature, 2430 North Cannon Drive, Chicago, IL 60614, USA; Tel.: 773-549-0606 ext. 2016; E-mail: paulheltne@humansandnature.org*

Honea, Jon M., *NOAA Fisheries, Northwest Fisheries Science Center, 2725 Montlake Blvd East, Seattle, WA 98112-2097, USA; Tel.: 206-860-3493; E-mail: jhonea@u. washington.edu*

Innes, John L., *Professor, FRBC Chair of Forest Management, Department of Forest Resources Management, University of British Columbia, Forest Sciences Centre, 2045, 2424 Main Mall, Vancouver, British Columbia, Canada V6T 1Z4; Tel.: 604-822-6761; E-mail: john.innes@ubc.ca*

Lamb, David, *School of Integrative Biology, University of Queensland, Brisbane, 4072, Australia; Tel.: 617-3365-2045; E-mail: d.lamb@botany.uq.edu.au*

Louis, Elizabeth C., *1711 East-West Rd, #574, Honolulu, HI 96848, USA; Tel.: 808-944-6211; E-mail: elouis@hawaii.edu*

Maxwell, Keely B., *Department of Earth and Environment, Franklin and Marshall College, PO Box 3003, Lancaster, PA 17604, USA; Tel.: 717-291-4133; E-mail: keely.maxwell@fandm.edu*

McCool, Stephen F., *Professor, Department of Society and Conservation, University of Montana, Missoula, MT 59812, USA; E-mail: Steve.McCool@cfc.umt.edu*

Medina, Martin, *9107 Bradford Rd, Silver Spring, MD 20901, USA; Tel.: 301-589-7650; E-mail: Medina2525@aol.com*

Minakawa, Noboru, *Center for Tropical Infectious Diseases, Institute of Tropical Medicine, Nagasaki University, 1-12-4 Sakamoto, Nagasaki 852-8523, Japan; Tel. (work): 81-95-849-7868; Fax: 81-95-849-7854; E-mail: minakawa@net.nagasaki-u.ac.jp*

Molina, Randy, *Team Leader Forest Mycology, USDA Forest Service, Pacific Northwest Research Station, Portland Forestry Sciences Laboratory, 620 SW Main St, Suite 400, Portland OR 97205, USA; Tel.: 503-808-3137; E-mail: rmolina@fs.fed.us*

Muller, Gretchen K., *Regional Education Project Manager, National Wildlife Federation, Northwestern Natural Resource Center, 418 First Avenue West, Seattle, WA 98119, USA; Tel.: 206-285-8707 ext. 107; E-mail: Muller@nwf.org*

Ostfeld, Richard S., *Institute of Ecosystem Studies, 65 Sharon Turnpike, Millbrook, NY 12545, USA; E-mail: ostfeldr@ecostudies.org*

Patel-Weynand, Toral, *Currently at: USGS – National Program Office, 12201 Sunrise Valley Drive, Reston, VA 20192, USA; Tel.: 703-648-4217; E-mail: tpatel-weynand@usgs.gov*

Perla, Bianca S., *College of Forest Resources, Box 352100, University of Washington, Seattle, WA 98195, USA; Tel.: 206-616-9661; E-mail: bperla@u.washington.edu*

Reichard, Sarah, *University of Washington Botanic Gardens, College of Forest Resources, Box 352100, University of Washington, Seattle, WA 98195, USA; Tel.: 206-616-5020; E-mail: reichard@u.washington.edu*

Rigdon, Philip H., *Director, Department of Natural Resources, Yakama Nation, PO Box 151, Fort Road, Toppenish, WA 98948, USA; Tel.: 509-865-5121; E-mail: prigdon@yakama.com*

Roads, Patricia A., *10600 Dayton Cincinnati Pike, Miamisburg, OH 45342, USA; Tel.: 973-746-5732; E-mail: OldUnicorn@aol.com*

Sigurdardottir, Ragnhildur, *Director, CAPEIntl Iceland, Stokkseyrarsel, 801 Selfoss, Iceland; Tel.: 354(566)-8778; E-mail: raga@nett.is*

Singh, Surendra Pratap, *Professor and Vice Chancellor, H.N.B., Garhwal University, Srinagar, Uttaranchal, India; Tel.: 91-1346-25216; E-mail: surps@yahoo.com*

Stephanson, Sheri L., *College of Forest Resources, Box 352100, University of Washington, Seattle, WA 98195, USA; E-mail: sheris@myuw.net*

Sturgeon, Janet C., *Simon Fraser University, Geography Department, Burnaby, British Columbia, Canada V5A 1S6; Tel.: 604-291-3716; E-mail: sturgeon@sfu.ca*

Suntana, Asep S., *LEI, Jl. Taman Malabar No. 18, Bogor 16151, Indonesia; Tel.: 62-251-340-744; E-mail: asuntana@lei.or.id; Currently at: 2151 North 92nd Street, #A, Seattle, Washington 98103, USA; Tel.: 206-985-6858; E-mail: asuntana@u. washington.edu*

Thadani, Rajesh, *CEDAR, A-17 Mayfair Gardens, New Delhi, India, 110 016; Tel.: 91(11)51656867/98103-48345; E-mail: thadani_rajesh@hotmail.com*

Upadhye, Ravi, *Deputy Materials Program Leader for Energy and Environment, Chemistry and Engineering Division, Lawrence Livermore National Laboratory, University of California, PO Box 808, L-631, Livermore, CA 94551, USA; Tel.: 925-423-1299; E-mail: upadhye1@llnl.gov*

Vogt, Daniel J., *College of Forest Resources, Box 352100, University of Washington, Seattle, WA 98195, USA; Tel.: 206-685-3292; E-mail: dvogt@u.washington.edu*

Vogt, Kristiina A., *College of Forest Resources, Box 352100, University of Washington, Seattle, WA 98195, USA; Tel.: 206-543-2765; E-mail: kvogt@u.washington.edu*

West, Stephen D., *Professor and Associate Dean, College of Forest Resources, Box 352100, University of Washington, Seattle, WA 98195, USA; Tel.: 206-685-7588; E-mail: sdwest@u.washington.edu*

Wheeler, Robert A., *Alaska Cooperative Extension, University of Alaska Fairbanks, 308 Tanana Loop, Fairbanks, AK 99775, USA; Tel.: 907-474-6356; E-mail: ffraw@uaf.edu*

Zabowski, Darlene, *College of Forest Resources, Box 352100, University of Washington, Seattle, WA 98195, USA; Tel.: 206-685-9550; E-mail: zabow@u.washington.edu*

Foreword

Forests and people have had a mutually dependent but usually uneasy relationship since the ancestors of modern humans evolved to use savannahs, mixtures of grassland and trees. In much of the world, for much of history, clearing forests for agriculture was the major method of adding value to land. Forest-dwelling predators, now highly regarded (at least at a suitable distance) as intriguing and even necessary instances of biodiversity, frequently dined out at the expense of human communities. On the other hand, forests were sources of many useful things, including the hides and fur of the 'diner' species, game to eat, wood for structures, weapons and fuel, medicine and other useful plant products. This duality persists to this day and has left its mark on the professions and organizations that concern themselves with forests and their components. In our urban society, few see forests as a direct threat any more, although some millions of hectares of world forest are still cleared annually for agriculture. The current ambivalence about forests exists between their usefulness for marketable products, mostly wood, and their value as watersheds, reservoirs of biodiversity and places for human recreation. Vogt and her colleagues have undertaken in this book to explain how this tension arose, and to provide the science base for understanding and confronting it in a world that is becoming ever more interconnected, so that forest events in, say, the Pacific North-west of the United States influence forest events in Indonesia, and vice versa.

This seems to me a timely task. On the one hand, the global picture of the forest products industry is changing rapidly. The huge, vertically integrated, forest landowning companies are shedding much of their land. On the rise are 'timber investment management' organizations and real-estate investment trusts that own forests and sell wood but usually have few or no processing facilities. Thus the management of the forest for wood is increasingly taking place in organizations that do not market products to the broad public. At the same time, urban sprawl is converting substantial acreages of forest to ranchettes, effectively removing large areas from timber production and sale. On the other hand, the environmental movement and the non-governmental organizations that fuel and support it have largely won their fight to reduce or eliminate timber cutting on federal forests in the United States. They have also made at least a credible beginning through reserves and certification schemes to protect tropical and subtropical forests. Some (the Nature Conservancy, for example) are beginning to hold and manage forests for a range of values, including timber production. There seems to be a developing consensus that forests are valuable and that their value has many facets, all legitimate if pursued properly. The principal conservation goal is becoming to keep forests as forests, and to recreate them where necessary or useful. Thus, the old duality is beginning to break down, and this calls for a new synthesis to provide science to underlie the new, more integrated view of forests that is emerging.

This book attempts that synthesis on a global scale. My hope is that it is the beginning of a resurgence of thought and writing about forests that question and at least update the older, more polarized views. As democracy and market capitalism sweep the world, many of the old, top-down, government-centric ways of dealing with forests have broken or are breaking down. Books like this one point towards what will eventually be put in their place.

John Gordon
New Hampshire, 2005
Former Dean and Emeritus Professor
Yale University
School of Forestry and Environmental Studies
Chairman, Interforest LLC

Introduction

When writing a book of any kind the author usually tries to accomplish at least one of the following: educate, entertain or enlighten. The individuals involved in the writing of this book have tried to provide all of these.

It is imperative that people from every profession and walk of life become involved in the maintenance of our forests so that we can benefit from them and enjoy them for generations to come. With this goal in mind, as much general non-scientific information as possible has been covered within these pages to enable all individuals to understand the importance of that task. Those areas that have more scientific than general information may, it is hoped, encourage some individuals to become interested enough to explore further.

Forest management, by today's scientific standards, is in its infancy. Theories continue to be explored and data continue to be collected and analysed. This book attempts to demystify some of this complexity so that practical solutions can be pursued. The authors have attempted to more clearly define what some of these issues are.

This book aims to provide a 'road map' to identify what factors need to be considered when making decisions regarding forest uses, when those uses are excessive and when they have become detrimental. It also attempts to document and clarify the complexity of issues involved in managing forests. This book provides a framework for the general public, forest managers and policymakers to understand what factors need to be included when working towards using and protecting the world's forests so that they can be sustained. It also contains information outlining the importance of forests in continuing to provide:

- solutions for our future energy needs that are renewable;
- conservation goals; and
- a sustainable livelihood for people dependent on those resources.

The authors hope this book will not only educate you, entertain you and enlighten you, but will also encourage you to think seriously about these issues. In addition to that, we also hope that you take some action to help relieve these global problems by becoming involved in some way that will improve conditions now and for future generations.

> Never doubt that a small group of thoughtful, committed citizens can change the world. Indeed, it is the only thing that ever has.
>
> (Margaret Mead)

1 Historical Perceptions and Uses of Forests

KRISTIINA A. VOGT, ROBERT I. GARA, JON M. HONEA, DANIEL J. VOGT, TORAL PATEL-WEYNAND, PATRICIA A. ROADS, ANNA FANZERES AND RAGNHILDUR SIGURDARDOTTIR

To understand the present practices being implemented in forest lands and how their resources are used today, it is important to understand the relationships between humankind and forests and how these evolved. That relationship has fluctuated throughout our coexistence and has been recorded in many mythologies, archaeological records and history books. Historical documents have recorded various attitudes to forests, including, among many others:

- Venerating them as sacred because of their beauty and immensity (e.g. magnificent large old trees).
- Fearing them because they appeared dark and concealed animals that could kill or otherwise harm humans.
- Finding them an impediment to agricultural development.
- Seeing them as sources of economic value, which is driving much of the illegal logging of timber that is occurring today.

Despite these different attitudes towards forests, they have supplied humans numerous benefits including food items such as wild game for meat, mushrooms, edible plants and berries, fodder for their animals, medicines (aspirin was extracted from the bark of willows and birch) and wood for heating and cooking.

Wood has also been used for tools, weapons, building materials for houses and fortifications to protect early communities. People use wood to make paper, or as the heating source for smelting metal ores and in making glass, bricks and pottery. Forests of rubber trees provide rubber for automotive parts, vibration dampeners for all types of motors, shock/vibration damping components for earthquake protection in buildings, plus medical and sanitary products.

The uses of forest materials have fluctuated from the beginning of recorded history, but human perceptions regarding forests have generally fallen into three major categories: reverence, fear and exploitation. Each of these will be discussed to some degree, but the primary emphasis will be on exploitation.

The History of Reverence for the Forest

Since ancient times, people have worshipped trees. There are so many articles and books referring to this that it is almost impossible to select just a few to reference in the small space this book provides. The ones selected here serve as examples of the rich variety that have been studied.

Nordic mythology is documented in the *Snorra-Edda*. In 1220, Snorri Sturluson in Iceland wrote the *Snorra-Edda* – a collection of mythological and heroic songs collected and written down in Iceland that were based on oral tradition. *Snorra-Edda* is in fact an Icelandic teach-yourself book in writing poems, but starts with a prologue where the creation of the world is described, along with the roots of religion and the old pagan gods who lived in Aesir. Trees were very prominent in Nordic mythology (Davidson, 1964). The centre of the world was formed from a large ash tree (Davidson, 1964). The great tree at the centre of the world, called Yggdrasil, was so large that its branches reached out over earth and the heavens. This giant ash tree was supported by three roots: one root went into the kingdom of Aesir, where the gods lived, the second root went into the kingdom of the frost giants and the third root was found in the realm of the dead. The saga also records how even in the realm of the gods the persistence of this large ash tree was continuously being threatened by creatures that preyed upon it (e.g. a giant serpent and many other snakes gnawed on the ash tree almost continuously). One of the gods – one-eyed Odin, who rode an eight-footed horse and could practise magic – hung himself with his own spear from Yggdrasil for 9 days to learn songs that had power.

Nordic mythology identifies trees as the original source of humans on earth (Davidson, 1964). According to one story, three Nordic gods (Odin, Vili and Vé) were travelling together on the seashore before humans existed on Earth. As they passed two large wind-thrown trees – an ash and an elm – they decided to create the first man and the first woman out of these trees. The man was created out of the ash and the woman from the elm. Odin gave humans their breath, Vili gave them a soul and the ability to reason while Ve (Vé) gave them warmth and the colour of life. The man was named Askur (Ash) and the woman named Embla (Elm). From this couple the entire human race was created.

Celtic life and culture bring to mind the mysterious priesthood known as the Druids (Matthews and Matthews, 1995; Ellis, 2003). Little written history can be found about Druids except for descriptions found in the writings of Julius Caesar and Pliny the Elder (23–79 CE). Most of this knowledge was based on Caesar's experiences with the Gallic Celts rather than native Britons. According to Pliny:

> The Druids have so high an esteem for the oak, that they do not perform the least religious ceremonies without being adorned with a garland of its leaves. The Druids had no image, but they worshipped a great oak tree as a symbol of Jupiter. Of the Druidical creed it was an article that it was unlawful to build temples to the gods, or to worship them within walls, or under roofs.

Many scholars believe that the word 'Druid' stems either from the old Gaelic word *duir*, which means oak, or the old British word *derwydd*, which means 'oak-seer' (Matthews and Matthews, 1995).

To the ancient Egyptians, the sycamore was the earthly form of the sky goddess, Nut (Mercante, 1978). The sycamore's leaves represented peace both here on earth and in the afterworld.

According to the Buddhist faith, Buddha sat under a bodhi tree (a type of fig) while he meditated and became enlightened (Bo tree, 2005). Even today, bodhi trees are grown near Buddhist monasteries and are considered sacred.

One other god needs to be discussed here since you will see derivatives of his name not only in this book but in forestry in general, i.e. Silvanus, a Roman god of forests, groves and fields, who also presides over boundaries (Cottrell, 1986). Silvanus is very similar to the Greek god Pan, with the same type of mischievous personality. Silvanus evidently liked to scare or play pranks on travellers as they went through the forest. The first fruits of the fields were offered to him as well as meat and wine. He is represented with a pruning knife and a bough from the pine tree. Many forestry terms are derived from his name, such as silva (the forest trees of a certain area), silvical (of or pertaining to forests or forestry), silvicolous (living or growing in woodlands), silvics (the study of forests and their ecology, including the application of soil science, botany, zoology, etc. to forestry) and silviculture (the art of cultivating a forest).

Trees are also used as symbols in many cultures and religions. The flags of many countries have either trees or leaves in their designs. Some examples of the symbolism of trees to different cultures follow:

- In China the mulberry tree symbolizes the cycle of life (Birrell, 1999). Its berries change colour three times while they ripen: white for youth, red for the middle years and black for the wisdom of old age and death.
- In Western countries, the drooping branches of the weeping willow tree represent death and grief, while this same willow tree is a symbol of beauty and grace in Eastern countries.
- Laurel was a symbol of immortality and victory to the ancient Greeks and Romans. The laurel wreath was presented by the Senate to the commanders of the Roman legions that won major wars and to those who excelled in the sciences or the arts. Today, medals presented at the Olympics have laurel leaves as part of their design.
- To Muslim, Christian and Jewish religions, the tree of knowledge symbolizes the temptation of Adam and Eve in the Garden of Eden. The fruit on this tree represents knowledge and God warned them not to eat any. The Devil, in the form of a snake, tempted them into tasting the fruit. When God discovered this, he banished them from Paradise and sent them to live on Earth.
- Ever wonder where kissing under the mistletoe came from? Mistletoe has become a symbol of love and perhaps this is why we kiss under it at Christmas time. In Nordic mythology Baldur, the god of peace, beauty and tranquillity, was greatly loved by almost all of the gods (Hodges, 1974). His mother, Frigg, had all things in the world promise not to hurt him. She spoke to the iron, the fire, water and rocks, she spoke to the trees, the bones, hunger and disease – all of which promised never to hurt Baldur. Mistletoe was the only material on earth she did not bother to ask, since it was so small and insignificant. When Loki, the god of evil, found out about this while masquerading in the form of an old woman, he

arranged the circumstances that resulted in Baldur's death. Loki tricked Baldur's blind brother into throwing an arrow at Baldur made from mistletoe, which killed him. His death ended up being the most significant event leading to Ragnarök, or the apocalypse or the end of the world. Baldur was never brought to life again and remained in Hell, since he did not die in a battle.

The History of Fear and the Forest

At other times and in numerous cultures throughout the world, forests were feared and were places into which people did not venture if it could be avoided. Calvin (2004) suggested that our fear of forests was one of the reasons that our early ancestors moved from living in the forests into the grasslands. This fear either minimized the human impact on forests or resulted in large sections of forest being wantonly destroyed.

These phobias or psychological terms have been created to define this fear of forests. For example:

- Hylophobia is the fear of forests.
- Nyctohylophobia is the fear of dark wooded areas of forest at night.

Some believe that these phobias can be caused by reading the following types of books to young children: Grimms' fairy tales, Hansel and Gretel, Little Red Riding Hood and the Big Bad Wolf and stories about things like vampire bats. Perhaps another story that can induce this phobia would be the American folk tale of Ichabod Crane and the Headless Horseman (Irving, 1980), which is popularly recounted during the American holiday Halloween. Every culture must have at least one or two scary stories that are used to either frighten children for fun or to keep them out of dangerous forests for their own safety.

In some cases, people feared forests because they did not typically travel into a forest so that forests were an unknown entity to them. In Erris (Ireland) in 1847, there were many who had never seen a living tree larger than a shrub. Richard Webb, a representative of the Central Relief Committee of the Society of Friends (Ireland's Potato Famine 1845–1849), was told by the innkeeper at Achill Sound that his 8-year-old daughter was afraid of trees (Woodham-Smith, 1962). He recounted how she feared 'they would fall on her as they waved over her head' (Woodham-Smith, 1962).

Many other reasons to fear forests exist because they directly impact a human's health or survival. During the 1700s, many European colonists moving westward in North America were intimidated by the dark and foreboding forests they encountered. Hostile Indians and wild animals inhabited these areas and confrontations often proved deadly. Malaria was also present in the lowland forests, which either killed people or debilitated them. Quite often, their reaction to this fear was to simply cut down the trees until the area more closely resembled the familiar terrain of European countries they had left (Williams, 1989).

Similarly to the reports of the early historical records from Europe to Africa and Asia, the first colonists in New England were literally afraid of the dense, gloomy forests so full of strange creatures, to say nothing of the 'savages'. William Penn in 1686 stated 'there was a heroic struggle to subdue the sullen and unyielding forest by the hand of man . . . to make it something better than it was . . . openings where God could look down and redeem the struggling inhabitants . . . etc.' (Taylor, 2003). This perspective reinforced the view that forests needed to be removed and that there was a moral rationale that justified this practise. The industrialization or technological development required forest removal and allowed colonialists to justify their ownership of lands already occupied by other people (i.e. the Native Americans) by transforming it. In 1810, Franklin wrote, clearing the land 'tended to measure moral and spiritual progress by progress in converting the wilderness into a paradise of material plenty' (Smyth, 1905–1907). In 1830, Andrew Jackson said:

> What good man would prefer a country covered with forests, and ranged by a few thousand savages to an extensive Republic, studded with cities, towns, and prosperous farms, embellished with all the improvements which art can devise or industry execute . . . and filled with all the blessings of liberty, civilization and religion?
>
> (Jackson, 1830)

Thomas Jefferson wrote in 1780 that the landscape was converted from forests to 'fair cities, substantial villages, extensive fields, an immense country filled with decent houses, good roads, orchards, meadows, and bridges where a hundred years ago it was wild, woody, and uncultivated' (Peden, 1955). These perspectives are commonly held and at the same time mean the demise of forests.

Sometimes, large forest areas around settlements were clear-cut to prevent hostile Indians from using them to attack the early colonists. At other times, forests were cut down and burned because they either restricted the physical expansion of villages/cities or made it difficult to grow agricultural crops needed for survival.

A surprising number of people fear forests today according to the survey conducted by the Harris poll (Taylor, 1999). In that survey, 13% of the people were very afraid of being alone in a forest while 41% were very or somewhat afraid to be in a forest. This fear of being in a forest by oneself was stronger in women (22% very afraid) compared with men (4%). The education level also appeared to determine how much an individual feared a forest, with those with more education having a lower fear of being alone in a forest. For example, 18% of those with secondary school or lower education levels feared the forest. This fear of being alone in a forest decreased to 11% for those with some college education, decreased to 7% among college graduates and was only 4% for those with postgraduate education.

History of Forest Exploitation

Forests and the formation of early civilizations

At the early stages of forest exploitation, societies generally do not worry about timber shortages or that there might not be enough wood to supply the wood products they desire. Large complex communities generally cannot change until forced to, creating a kind of momentum leading to unsustainable rates of exploitation. Sub-communities develop that depend on exploitation, some employed to extract and process the resource and others using the products. The users may not be fully aware of the consequences of resource extraction and the harvesters may be tied to traditional ways of life that appear sustainable simply because they have occurred for so long. The perception of both is that the forests will forever produce an endless supply of resources to meet their needs. This widely held view encourages the exploitation of forests and is a view that resurfaces throughout the history of global societies and forests. It is not until this century that the true vulnerability of forests is being recognized as a global issue.

In previous centuries, when localized forests were over-exploited, political powers extended their reach when possible and even colonized distant regions of the world to maintain their supply of wood. Controlling the wood supply conferred political power on the countries that had the military superiority to protect it. Since wood was needed to build ocean-sailing vessels, control of wood supply also allowed these countries to determine who would trade in the global markets.

To their eventual regret, many early civilizations eliminated their forests. History records how the Egyptian pharaohs launched military and logging campaigns during the 6th millennium BCE into the African continent to search for sources of timber and other non-timber forest products (Winters, 1974). Similarly, the Romans explored the southern region of Europe for forest resources and travelled as far as North Africa from 30 to 476 CE in a search for the timber necessary to maintain their empire. This exploration by the Romans was stimulated by a loss of the wood supply from their own forests in Italy because of excessive exploitation. The Romans became very dependent on acquiring new wood supplies from the regions they conquered (i.e. Africa, Asia Minor, the Iberian Peninsula, Gaul and Britain). When the Roman empire collapsed, the exploitation of these forests did not end because other invaders replaced the Romans in continuing these practices (e.g. tribes from the northern regions of Europe).

During this time, the different groups of invaders exploited forests for firewood and for military purposes such as materials to construct defensive fortifications or timber to build ships needed to control the trade routes (Winters, 1974). It is important to briefly mention some of the civilizations that over-exploited their forests and eventually collapsed because of that misuse. Documented examples of these include: Sumer; Crete; ancient Greece; Cyprus; Rome; Easter Island; the Muslim Mediterranean; the Venetian Republic; the Maya; and Knossos.

This partial listing highlights how the story of forest misuse repeats itself throughout history and how ecological and economic disasters were caused by the destruction of forests and their watersheds. The links between forests and the collapse of the Sumerian and Easter Islander civilizations will be briefly discussed next, but these two examples could as easily describe any of the early civilizations that are listed above. This will be followed by a discussion of the exploitation of forests by the Portuguese and by the English, ending with the exploitation of forests that swept across the Americas.

The Sumerians (4000–2000 BCE)

Mesopotamia (the area of modern-day Iraq and Syria) flourished as an early centre of human civilization 6000 years ago. It is referred to as the 'Fertile Crescent' and was the centre of world agriculture, because the city states that occupied the region were supported by a highly productive agricultural region (Roux, 1992). The civilizations of Mesopotamia were built at the expense of seemingly endless cedar, pine and cypress forests in mountains to the east (today Lebanon) and in the Ammanus Mountains to the north (today Syria). The forests of the Ammanus Mountains also yielded Euphrates poplar, willows and several hardwood species that scholars have not been able to identify. These forests supported a remarkable trade network because wood (especially cedar) was scarce in Mesopotamia.

The first recorded exploitation of forests is documented in an epic mythological story recorded around 4000 years ago on 11 clay tablets entitled *The Epic of Gilgamesh* (Kovacs, 1989). Gilgamesh was a king of Sumer who, with a wild man created by the sky god Anu, had an adventure in cedar forests that resulted in mass cutting of these trees. Gilgamesh and the wild man killed a demon named Humbaba, who was the demon guardian of the cedar trees. After they killed the demon, they proceeded to cut down the trees and to float them back to the great city of Uruk. This epic legend has been suggested to describe the over-exploitation of the cedar forests in what is present-day Lebanon.

Within Mesopotamia, the Sumerians founded many great cities; Ur being one of them (Durant, 1954). Wood was an important material used to build these great cities and parts of many appliances used by the citizens of these cities. The Bronze Age was at its pinnacle, and tools such as axes, hammers, hoes and sickles facilitated common labour. Producing bronze dramatically increased the need for wood to fuel the foundry furnaces. Carpentry shops were common and houses were being built of wood.

In addition to everyday requirements, wood collected in the Sumerians forests was used for the following purposes:

- logs, roof beams;
- levers;
- pegs and rungs of ladders;
- posts and rods for basketry;
- planks and boards;
- boat ribs;
- handles for hoes, ploughs, etc.;
- branches and twigs to make charcoal;
- branch bundles used to reinforce the banks of canals and rivers.

By the third millennium BCE, the growing civilization of the Euphrates and Tigris river basins created a large drain on timber resources to the east and to the north-east. Great battles were fought to possess these resources. The Sumerians ultimately gained control of the forests as well as the log transportation system formed by the Euphrates and Tigris Rivers.

Unfortunately, the loss of forest cover increased upland erosion, filling the waterways with sediment and leaching salt from the salt-rich sedimentary rocks of the north (Jacobsen and Adams, 1958; Hillel, 1991). As downstream rivers became clogged with sediment, debris and wood from transported logs, lowland flooding increased, washing salt over the rich soils that Sumerian agriculture depended on.

Increased salinization was non-reversible and caused progressive declines in crop yields (Jacobsen and Adams, 1958; Hillel, 1991). Harvests of barley averaged

2537 l/ha in 2400 BCE (comparable to modern-day United States harvests). Three hundred years later, yields had dropped by 42%. By 2000 BCE, as barley production collapsed, so did Sumer as it dissipated into mere villages.

Declining food production due to soil salinization was one of the main factors in the collapse of the Sumerian civilization. The loss of forest lands was the primary reason for soil salinization. The centre of development moved north to non-salinized agricultural lands.

Easter Island (approx. 900 CE)

Easter Island (called Rapa Nui by the local inhabitants) lies approximately halfway between the Hawaiian Islands and the Chilean coast of South America (Finney, 2005). It is approximately 15 km wide from north to south and contains roughly 90 km^2 of land. At one time it was a subtropical forest of tall trees and woody bushes similar to other Pacific islands. It was colonized by Polynesian canoe voyagers, originally from Asia, who used islands across the Pacific as stepping stones in their travels (Finney, 2005).

Much of the island's history comes from oral narratives told to Europeans by surviving islanders during the 18th century (Diamond, 2005). In addition, numerous scientists, from archaeologists to palynologists (pollen scientists) to zooarchaeologists, have studied the island since the mid-1950s.

According to local traditions, the leader of the expedition that settled this island was a chief named Hotu Matu'a (McCall, 1995). He sailed in several 'large canoes with his wife, six sons, and extended family' (Diamond, 2005). After their arrival, the island was divided into 11 or 12 pie-shaped wedges, one wedge for each of the tribes or clans. These wedges, though not equal in size, were equal in dividing the island's resources. There was one paramount leader, who integrated the clans religiously, economically and politically.

The immense 8 m high stone statues (called *moai*), made from volcanic rocks, were used in their religion (Diamond, 2005). These statues found on Easter Island have fascinated the world since their discovery. The stone platforms (called *ahu*) that the *moai* were mounted on sheltered a crematorium for the ritual burning of the dead clan leaders. As you will see, these statues greatly contributed to their civilization's collapse.

There does not appear to have been any warring between the clans (Diamond, 2005). There was, however, rivalry, which took the form of competition to see which clan could sculpt and transport the largest statue to their territory. These statues ranged in weight from 10 to 90 tons. After the statues were transported to a chosen location, they were mounted on *ahu*. The method used to move these huge statues from the interior to their coastal positions required the use of timber in numerous ways and in great quantities. Massive sleds were constructed to hold the statues, and then wooden skids (these had the actual appearance and construction of a giant ladder) were built and laid from the quarry to the site selected. These skids may have stretched for kilometres. Fibrous tree bark was used to make the long ropes used by the 50–500 people required to haul these sleds. Once these sleds were pulled across the skids to the chosen site, wooden levers were used to help control the statues as they were erected in place. There are 300 stone platforms and 113 support statues, of which 25 are especially large and elaborate. The amount of wood required to move and erect these statues must have been enormous. In addition, wood was required to cremate the dead, as well as for everyday existence (i.e. cooking fires, construction of houses, etc.).

Palynology (pollen analysis) of sediment cores taken in 1977 and 1983 identified 21 tree species that had existed on Easter Island before 1400 CE (Flenley and King, 1984; Oriac, 1998). Trunk imprints buried in previous lava flows show that Easter Island had a palm tree whose trunk exceeded 2 m (7 feet) in diameter. This size dwarfs the Chilean wine palm, which has been considered the biggest palm in the world. Between 1400 and 1600, 16 tree species – including five found nowhere else in the world – vanished from the island. When the Dutch explorer Roggeveen first landed on Easter Island in 1722, the lack of trees over 3 m (10 feet) in height was noted (Diamond, 2005).

In the 1400s crop yields also started decreasing. Deforestation led to soil erosion during rain and windstorms. Other soil damage occurred as well, including desiccation and nutrient leaching.

By about 1500 CE, most sources of wild food such as porpoises, tuna, land birds and wild fruit had disappeared from the islanders' diet.

Prior to this period, Easter Island was the richest bird-breeding site in all of Polynesia and probably in the whole Pacific (Steadman, 1989). Native bird species included six herons, some chicken-like rails, two parrots and a barred owl. At least 25 seabird species nested on the island. Bird species have become extinct on other Pacific islands, but only Easter Island has ended up with no native land birds at all.

Easter Island's coastal geography prevents fishing along the shores or other island margins with the ocean. Islanders ate the common dolphin that lived far from their shores, but to catch them required harpooning them from big seaworthy canoes. Without timber to replace harpoons or repair the canoes, the islanders were unable to maintain dolphin in their diets.

Although natives and some researchers deny it, scientific evidence indicates that cannibalism occurred sometime during the 1600s (Diamond, 2005). Historical accounts document how people lived under starvation conditions. Surviving islanders' accounts of starvation are graphically depicted by statues (*moai kavakava*) showing starving people with hollow cheeks and protruding ribs. When Captain Cook arrived in 1774 he described the inhabitants as 'small, lean, timid and miserable' (Diamond, 2005). The most derogatory comment one islander could make to an enemy was, 'The flesh of your mother sticks between my teeth' (Metraux, 1957).

In 1680, the peasants overthrew their chiefs and priests, perhaps fed up with the policies requiring them to build the monumental statues that exhausted the island's resources. It appears that, to all intents and purposes, their civilization almost disappeared: the governing body was eliminated and its religion was abandoned. At its peak, it has been estimated that 7000 to 30,000 people lived on Easter Island (Diamond, 2005). This population density was decimated by three epidemics of smallpox in the 19th century, which killed a large proportion of the population. Peruvian slave ships further decimated this population by kidnapping and removing about 1500 inhabitants from Easter Island (Diamond, 2005). By 1872, only 111 native islanders were left.

In 1888, the Chilean government annexed Easter Island and turned it into a sheep ranch.

As for the statues that appear to have greatly contributed to the deforestation of Easter Island, they slowly fell and broke during the next 200 years. In 1774 Captain Cook found that some statues had been thrown down and broken, perhaps some during the 1680 revolt. The last date that Europeans mentioned observing erect undamaged statues lining the coast was 1838, and by 1868 none were reported standing and undamaged (Diamond, 2005).

European colonization and industrialization period (1200–1700 CE)

Between 1200 and 1700, Europeans began an intense period of exploration and long-term colonization throughout the world. During the same time frame, they were also entering a period of industrialization, which was fuelled by forest materials. Locally available supplies of wood were insufficient to provide for the rapidly expanding needs of civilizations as they became industrialized. Due to centuries of severe exploitation of the forests under their control, European countries needed to explore and conquer lands beyond Europe and North Africa. Europeans needed wood to build ships and to heat the foundries that manufactured the weapons (e.g. cannons, guns) that enabled them to compete with other countries for more territory. Many countries found that this created a vicious circle, as shown in Fig. 1.1.

This over-exploitation of forests and their loss from the European landscape can be seen in today's statistics, where only 4% of the global forests are located in Europe. Competition between countries to control this supply and access to these resources was fierce, quite often resulting in military action.

The Portuguese (1200–1700 CE)

Of all the European countries, the Portuguese were the first to pursue maritime expeditions that were totally under the control of the royalty, who provided the financial resources and underwrote the expeditions. They were able to begin their exploration of previously uncharted territories almost a 100 years before other European countries because of their maritime expertise and their advanced technology in shipbuilding

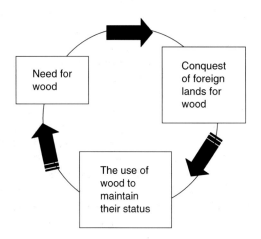

Fig. 1.1.   A vicious circle of wood exploitation.

and navigation (e.g. the Portuguese invented and designed an ocean-travelling ship called the 'caravel').

The caravel was a ship design invented by the Portuguese in the late 1400s and was extensively used until the 1700s. This ship was a three-masted sailing vessel that was lighter than other ships being built at this time but was large enough to carry a sizeable cargo and people (Columbia Encyclopedia, 1995). The caravel was a faster sailing ship and could navigate into the wind, which gave it tremendous advantages compared with other ships. This ship was originally built by the Portuguese to explore the African coastline. Two of the ships (the *Niña* and the *Pinta*) used by Columbus in the late 1400s to travel to the Americas were Portuguese caravels (Columbia Encyclopedia, 1995).

The Portuguese exploited the timber resources from islands such as Madeira for building their ocean-sailing vessels. The deforested island was later replanted with vineyards and today Madeira wine is well known throughout the world.

The English (1450–1547 CE)

During and before the reign of Henry VII (1457–1509), England had to import just about everything – including armaments (Winters, 1974; Williams, 1989). This was a situation Henry VIII (1491–1547) wanted to change as quickly as possible when he inherited the throne. England's navy had been dependent for planking and masts on the Hanseatic League, a union of Germanic and Baltic states. Their dependence on the Hanseatic League, even into the late 1770s, was recorded: 'expansion of English trade necessitates the building of bigger ships, and the Baltic is well equipped with raw materials. For centuries, construction materials such as Danzig plank, Riga masts, and Stockholm tar have been essential to our maritime power' (Anon., 1770, quoted in Winters, 1974).

Henry VIII came into power with a full treasury and many enemies. Because of this, he decreed that England would have an arms industry and a first-class navy. Achievement of these goals required vast amounts of wood for two things: building ships and making charcoal to heat the foundry furnaces. England's iron ore mining and iron foundries expanded and flourished.

At this time, oak was preferred by shipwrights (some beech and elm were also used) and was found in the forests in Sussex, which were rapidly depleted of these tree species. A large number of trees were needed to build or repair a ship. It has been estimated that 1740 oak trees were required to repair four ships, while building a new ship required 2000 oak trees. In England during the time of Henry VIII, forests were intensely exploited in the south-east corner of the country to build his vision of a military power (Winters, 1974).

In response to the rapid disappearance of their forests, Parliament introduced bills to restore these forests by planting trees, to increase the preservation of existing forests, to eliminate the use of hedgerows (vegetation planted along the edges of agricultural fields) for making coal, to prohibit the building of iron mills within 24 miles of London and to stop the construction of new iron mills in Sussex. None of these bills passed, however. The forests of Sussex and elsewhere in England were decimated, but Henry VIII developed a massive navy, which allowed him to colonize and exploit other regions of the world to provide a steady supply of timber for his country.

To maintain their naval superiority over the other European countries, the English would conquer other lands in order to exploit their natural resources, start wars and encourage piracy against ships of other European countries (Neeson, 1991). Those who controlled the access to and uses of forest resources attained considerable global power. Access to forest resources frequently determined the success or failure of many wars. England's naval superiority allowed them to dominate world affairs for some 400 years (Neeson, 1991).

Wooden Ships, Colonization and Controlling Access to Wood

A worldwide problem with ships built in the 17th century was that they only lasted a few decades before they required major repairs or had to be completely replaced. The main problem was shipworms – marine molluscs that make their homes in the holes that they drill in wooden structures. This need to obtain new supplies of timber because of the short lifespan of a ship spurred some of the explorations by European countries. For example, England and other European countries initially targeted and exploited the timber supplies in India because of the large supply of teak (*Tectona grandis*) found there (Winters, 1974). Teak became the preferred shipbuilding material because its resistance to the marine borers was superior to that of other woods. The use of teak significantly increased the life of ships built from it.

Since larger ships were needed to travel across the ocean and a larger navy was necessary to ensure a country's superiority on the seas, the ability to control the supplies of wood for building these vessels was critical. This need is apparent from the decree passed by the English Privy Council of 1631 which stated that trees suitable for building ships were the exclusive property of the Crown. They were to be marked using a broad arrow symbol and were prohibited from being cut for any other purposes (Young, 1979). English foresters enforced the broad-arrow policy.

English naval officers were also charged with searching for, exploiting and controlling timber resources in the newly acquired colonies and in any new region being explored (Winters, 1974). The English broad-arrow policy continued to be enforced in the American colonies when they found that the timber required for the tall masts, planking and resin tars was widely available in North America. This policy of the English contributed to the unrest that grew in the American colonies. Frequently the colonists ignored the broad arrows painted on the trees and cut them for their own use.

The Period of Forest Loss and Exploitation in North America (1770–1960 CE)

The same pattern of forest exploitation and destruction historically documented in Europe also repeated itself in North America. Forest loss began to intensify in New England during the period of 1770–1790, with settlers following a process of clearing the land, farming it and building houses (Williams, 1989). Major periods of exploitative logging occurred in the New England states from 1869 and lasted until 1915. Timber was cut to build the railway system, which would become the transportation network allowing settlers easier access to the western parts of the country. Wood was needed as construction material for the railway tracks and was also used to heat the forges producing their steel rails.

This path of logging moved from the New England states to the Great Lakes region once forests had been cut over in New England (Fig. 1.2). Forest exploitation in New England was being driven by urbanization and industrialization. Clearly by the end of the 19th century, the Great Lakes states were denuded and large parts of their residual forests had been burned. This same process subsequently spread to the south-eastern United States and eventually to the Pacific North-west, each spread followed by a period of over-exploitation of trees in that area. Between 1810 and 1860, the United States was rapidly industrializing and timber exploitation fuelled this growth.

By the mid-1880s, the nation was growing at unprecedented rates and the demand for timber for the construction of buildings was high (Williams, 1989). The huge forest resources originally found in the New England states, the Lakes states and then the South and finally on the west coast supplied and met this demand (Fig. 1.2). There were two periods of intense exploitation of forests on the west coast: 1900–1925 and again in 1930–1960. The decreased cutting of timber between 1925 and 1930 was a result of the worldwide economic depression.

As the United States was becoming industrialized, the changing uses of wood for energy were apparent. For example, four-fifths of the energy use was as fuel wood in the late 1860s. By the mid-1880s, coal was replacing wood. By 1900, fuel wood

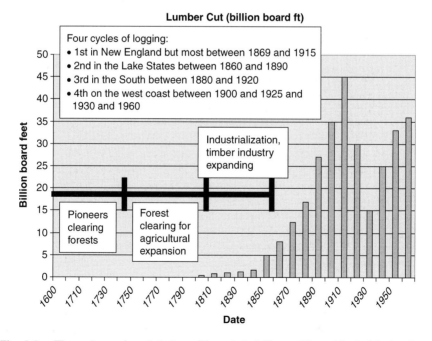

Fig. 1.2. The pattern of exploitation of forests in billions of board feet of timber in the different regions of the United States between 1600 and 1960 (data from Williams, 1989). Only timber harvested in the United States is shown.

contributed only 21% of the United States' energy needs and this decreased to 7.5% by 1920 as coal, oil and natural gas continued to replace wood use for energy (Williams, 1989).

Conservation and Regulation of Forest Uses

Ancient civilizations (2500 BCE–700 CE)

History documents that forests were not always just exploited; there are many instances around the world where forests were managed, preserved or restored. At the same time that some forests were being exploited for firewood, timber for shipbuilding and various other uses, other forests in that country or principality were being set aside and their uses regulated. The reasons are quite varied as to why certain forest areas were set aside and what practises were going to be allowed or restricted in them. A history of protected areas is provided in Case 1.1 – A History of Forest Protected Areas.

Many examples of forest conservation were unintentional benefits caused by the ruling class (e.g. kings, nobles, etc.) setting aside forests as their own private and exclusive hunting preserves (Winters, 1974; McCormick, 1989). Historical records going as far back as 2500 BCE have noted the practise of setting aside forest areas as hunting preserves for the ruling class. This practise of designating forest areas for the exclusive use of the ruling class has been documented in Babylonia (2500 to 1000 BCE), in China during the Chou Empire (1027 to 221 BCE), in Assyria (~700 BCE) and in Persia (350 to 500 BCE) (Winters, 1974). The Indian emperor and religious teacher Ashoka (273–232 BCE) recognized that wildlife needed to be conserved and prohibited forests being burned in the domains under his control (Durant, 1954).

Even though the Roman empire exploited and conquered other regions to obtain the natural resources they needed to maintain their civilization, Pliny the Elder (23–79 CE), a Roman naturalist and encyclopedist, wrote that the mountains of Lebanon, vital to the Sumerians millennia earlier, continued to be an important timber region of the Roman empire. They realized that this timber supply was becoming scarcer. Forty years later, the Roman Emperor Hadrian (76–138 CE) became alarmed by the dwindling timber supply of the Lebanese and Ammanus Mountains and declared a portion of this area to be a 'timber reserve of the Roman empire'.

The practice of setting aside forests for the exclusive use of the ruling class was first recorded in Europe during the Frankish period in the 7th century (Winters, 1974). Many nobles of medieval Europe felt the forests were their private game preserves and should not be ploughed for agriculture or used by the peasants in any way. They were able to brutally enforce these types of decrees because they commanded the military power in the area. These decrees unintentionally limited agricultural expansion and restricted the collection of timber and firewood by the local peasants. The ruling class used their hunting grounds to supply themselves with lumber and firewood for their own exclusive use. These practises resulted in the conservation of forests even though that was not the primary reason for setting aside the forest land.

Even after land ownership shifted from the nobles to the Crown of England after the Norman Conquest and the Battle of Hastings (1066), this pattern of controlling access to forests persisted and kept agricultural expansion from encroaching into these forests. The only difference was that the king now owned and determined forest uses, but this difference was very important. It meant that one person, with one idea, had control over large forest areas as opposed to several individuals, each with differing ideas, controlling small forest areas.

The Middle Ages and monks in Europe begin active forest management (500–1500 CE)

During the Middle Ages in Europe, documentation has revealed that for the first time there was a shift from passive to active management of forests (i.e. forests were actively manipulated) to ensure a steady supply of timber (Winters, 1974; Linnard, 1982; McCormick, 1989; Neeson, 1991; Elliott, 1996). This shift to actively managed forests resulted from the rise in dominance and spread of Christianity throughout Europe and was especially prominent in Germany and France. Monks managed forests and practised agriculture adjacent to new cloisters. The Roman Catholic Church, in particular the Benedictine order, was instrumental in this shift to active management of forests. By the end of the 7th century there were 400 Benedictine monasteries spread over Europe. The spread of these monasteries is an interesting chapter in the story of European forestry.

Monks first began to become involved in active forest management because their orders were deeded tracts of forest land by noble-born landowners to build their new cloisters. In England and Europe small plots of land were deeded to some individuals, but during this period most land was owned either by the Crown or by the nobility. Usually, to acquire land some notable service to the Crown was required. In most cases, the actual people that worked the land were serfs, who did not own land.

It should not be assumed that these 'gifts' to the monasteries were primarily due to religious piety – the nobles wanted these wild lands tamed. At this time, the forest lands had little or no value, so the nobility did not mind giving them to the Church. Many monasteries were built in the untamed forest areas that still existed in Europe and this attracted people who wanted to live close to the new religious centres (Winters, 1974).

The desire of the nobility in many countries to tame and control the wild and unmanaged forests is apparent in historical documents (Winters, 1974). For example, in 1147 the Germanic King Conrad III gave large tracts of forested areas to the Cistercian monks on the condition that they cultivate the land. The Archbishop of Magdeburg encouraged this practise. He gave exemptions to landowners from tithe if they gave their untamed forests and marshes to the Cistercian monks.

So for the next 300 years monks drained swamps, cut forests, farmed the land and attracted settlers – the ultimate desire of the feudal economy of the time. However, as these centres (monasteries, villages) became larger, the forest resources started to become scarce. Since forest materials were an important part of the subsistence economics that existed around these centres, the monks began to establish strict regulations on when, how much and who could have access to them.

During the Middle Ages, the role played by the monks in controlling how forests were to be used was logical since they were the repositories of scientific knowledge and in most cases the only individuals that could read or write. As the forests were cleared by the monks and the villages sprung up around these cloisters, some initial forest management principles emerged. In 1040, monks of the Vallumbrosan Order, an offshoot of the Benedictines,

- preserved forests that were on terrain too tough to farm (places where God would touch their souls);
- encouraged reforestation of cut forested land – prepared the sites for seeding, planted seedlings dug from the forests;
- shaped trees for basketry (called pollarding, which consisted of cutting a tree's top branches back to the trunk, which causes a dense growth of new shoots); and
- left tree stumps of species that sprouted to produce new stems and branches for fuel wood (Winters, 1974; Seip, 1996).

By 1595 forest laws emerged in Europe that stated that commoners:

are hereby forbidden to cut any of the woods (trees) belonging to the abbey before the age of 15 years, seeing the poverty of the soil. They shall regulate their coupes (cutting areas) into 15 equal fellings and they shall leave standing at each felling at least five standards per coupe (management units with an area of XX). They shall allow one-third of their forest area to grow as high forests on the best soil . . . etc.

(Seip, 1996)

Besides establishing forest reserves and providing rudimentary rules on how to manage these resources, the monasteries were the first to establish coppice and pollarding silvicultural systems. The coppice and pollarding silvicultural systems were developed to manage tree sprouts for different products. The shapes of sprouting stems and branches were manipulated while the tree was growing into the shapes desired in the end product (Sharpe *et al.*, 1995). For example, planks for ships needed to have a bend to them so trees were systematically tied down so that they would grow with the right shape and angles.

Monks are accredited with initially establishing the technical practises of sustainable forestry, first in Germany and then in France, that are still the basis of forestry today (Winters, 1974; Neeson, 1991). A key element of sustainable forestry was the introduction of forest inventories and the division of forested areas into management units based on the number of years it takes for a forest to regrow. The amount of wood that was allowed to be collected could not exceed the annual growth in any given area.

The Portuguese audit forest uses: the roots of forest certification (1200–1300 CE)

Since the 13th century, the Portuguese have actively synthesized and managed their forests, using many of the concepts that later became critical elements of forest certification protocols, which were not developed until the early 1990s (see Chapter 7, this volume). Historical records have described how King Don Diniz of Portugal developed and required certain reforestation practises to be implemented (Winters, 1974; Seip, 1996). King Don Diniz of Portugal was an enlightened ruler who was known as '"the Labourer" – administrator, reformer, builder, educator, patron of the arts, and skilled practitioner of literature and love' (Durant, 1957).

King Don Diniz of Portugal even provided incentives to stimulate local communities to pursue practises that would facilitate the natural regeneration of native pine trees in the region of Leira. In addition, he set aside a forest area of 4300 ha (about 10,625 acres) where access was controlled and logging was only allowed if permits had been previously obtained. He also determined who would have access to the forests for firewood and identified which groups would not be restricted in the amount they collected. For example, people who lived in the region were allowed to collect firewood every other weekday, while employees at the local glass factory were allowed to collect firewood with no restrictions.

To ensure that his regulations would be followed, King Don Diniz hired many guards to protect and to monitor this forest reserve. Similarly to forest certification systems used today, the guards had to verify the location where the trees were harvested in a forest and how much timber was being extracted daily. If the guards did not enforce the rules that had been established, they themselves were punished. The main difference between forest certification today and what the early Portuguese practised is that today's audits are voluntary, third-party and non-regulatory (see Chapter 7, this volume).

Most of King Don Diniz's nobles were exempted from following these rules since they returned a profit to the Crown. However, even the nobles could collect only dead wood (no live trees could be harvested) and only that dead wood without mushrooms growing on it. Any timber, other than fuel wood for local community members and the glass factory, being transported out of the forest had to have a note attached to it that specified the volume of wood being taken out and where the trees had been cut. When the timber arrived at the harbour to be transported by ships, shipment documents were issued. These documents had to be presented to an accountant of the Crown, whose job was to compare and to verify that the records provided were in fact correct and matched what was being shipped.

France becomes a dominant sea and economic power in Europe: building a navy while vigilant over forests (1665–1683 CE)

Charles Colbert was the Secretary of State for Foreign Affairs, Controller General of Finance and Secretary for Naval Affairs under King Louis XIV. He carried out a programme of economic reconstruction that helped make France dominant on the sea and an economic power in Europe during his tenure. He was very interested in forests and recognized that, if he was going to make France a future shipbuilding nation with a great navy, the French needed to wisely manage their forests and not waste natural resources (Withed, 2000). Colbert created a department of wood and used it to develop regulations that were intended to reform how forests were used and that would ensure a sufficient supply of wood for shipbuilding for the navy.

In 1669, Colbert passed an ordinance that restricted how natural resources were to be used; the ordinance included water and forests and was entitled *Ordonnance des*

Eaux et Forêts. The goal of this ordinance was 'to abundantly provide its navy . . . universal reformation of all the forests of its kingdom'. This ordinance regulated how the ecclesiastical and lay communities were to use the forests in their region and stated that communities needed to put one-quarter of their wood in reserve to allow the forests to mature. Only a council of the king could authorize cutting any trees in this reserve. This ordinance also provided communities with financial reimbursement if they were forced to reserve a quarter of their wood for the public good (e.g. churches, public buildings, bridges needing wood for rebuilding after a fire). In 1689, rural landowners owning forests were required to afforest at least 5% of their land that had been previously deforested.

Colbert originally planted the forests because he was concerned that France would run out of oak to build ships. Today these oak trees from Colbert's forests are an important source of oak for constructing wine barrels. So, the next time you have a glass of French wine aged in oak barrels, you can credit Colbert.

French introduce scientific knowledge into forest reserves to reduce soil erosion (early to late 1700s CE)

The French were among the first to utilize scientific knowledge in the design of forest reserves and to format this information into policy. Early in the 18th century, the French implemented a series of plans to preserve forest resources in their overseas colonies (Grove, 1992). This policy contrasted greatly with that followed by most of the other colonial powers. The majority of colonial powers exploited forests and, once resources were depleted in any given locale, they simply moved to new areas.

In 1709 on the Island of St Helena, the local French government established the first forest reserve and planted trees in plantations to ensure that future supplies of timber would be available (Grove, 1992). Fifty-five years later, the French established a system of forest reserves on the Caribbean islands they had colonized (i.e. St Vincent, Tobago and Barbados) (Grove, 1997).

Several decades later, the French issued the King's Forest Hill Act of 1791. This Act established protected areas on the island of St Vincent and was the first such legislation for the Americas (Grove, 1997). It mandated the restoration and protection of the dry forests growing on the hill slopes of St Vincent so that they could continue to function as rain collectors.

Between 1730 and 1750, after the Dutch abandoned them, the French took over the Mauritius Islands and began to establish similar types of forest reserves. Their goal was to reduce and prevent further soil erosion by limiting deforestation.

Forestry becomes a systematic science and a profession: links among German forestry, sustainable teak management in India, university training of professional foresters and the United States conservation movement (1850–mid-1900s CE)

During the middle to late 1800s, the next important change occurred when forestry was recognized as a science and a technically based profession. The basic tenets of sustainable forest management were introduced and used to train forestry profession-als in universities.

These changes were driven by several factors. Among them was the appoint-ment of a German forester named Dietrich Brandis (1824–1907) to the newly created position of Inspector General of Forests in India. Brandis's appointment indi-cated that the British intended to manage the teak forests in India instead of just exploiting them (Dawkins and Philip, 1998).

Dietrich Brandis was one of the most respected foresters when Queen Victoria appointed him as the Inspector General of Forests. In that capacity, Brandis made enormous contributions to sustainable forestry in India, Burma, Pakistan and Bangladesh and later in Europe and the United States. He founded the Indian Forestry College at Dehra Dun, India. For the first time in history there existed an institution that would train individuals to become foresters using the principles and science of sustainable forestry. Sustainable forestry at that time meant that the amount of timber to be harvested had to be based on the capacity of the site to annually replace that amount.

Brandis had been given the task of managing the teak forests of India and Burma to maintain a long-term supply of shipworm-resistant teak for the shipbuilding industry in Great Britain. Instead of simply decimating the teak forests in India, Great Britain was determined to use science in managing them.

This is the first time that an effort was made by a colonial power to manage any forest with regard to its long-term sustainability. This measure was necessary because of previous teak exploitation in this region. Teak's economic returns were diminishing as the cost of transporting it from further distances increased. These economic losses created a great urgency in Great Britain to not only manage teak so that it would be a sustainable timber species but to restore areas that had already been heavily exploited.

The idea of planting teak was not new; it had been done since the 10th century in Java. The Dutch, from 1450 onwards, had also been planting teak, and planting intensified after other European powers arrived in the India–Burma region. What was different and innovative about the enterprise in India was the development and implementation of the science to evaluate whether teak production and harvest were sustainable.

Other efforts to regulate teak harvesting had occurred prior to the arrival of the British; however, these rules were not based on a systematic scientific analysis that allowed one to determine the level of timber that could be harvested in a sustainable manner. For example, the Dutch East India Company had established the following timber extraction rules: no cutting below a minimum stem diameter size for har-vested trees; following the cutting of trees, seedlings of the same species had to be planted; and timber areas severely exploited were closed to allow regeneration to occur (Peluso, 1991). These rules were important to ensure that teak regeneration occurred and to make certain that the supply would not totally disappear, but they could not indicate what was or was not a sustainable harvest level.

This is where Brandis's techniques become new and innovative (Dawkins and Philip, 1998). These tools and techniques included:

1. The first use of silviculture in defining how logging operations would be con-ducted and the development of scientifically based logging rules.

2. Development of a forest protection programme based on entomology, pathology and an excellent fire management system.
3. Institution of rules on timber purchase, penalties for violations of regulations and conditions under which land could be cleared.
4. Establishment of large management areas, which Brandis called 'conservancies', where teak forests could be intensively managed in India.

And the most important tool:

5. Development of sustained-yield forest management, where you can only cut an amount of wood equivalent to annual growth of the entire forest. Sustained-yield forest management was based on the calculation of teak biomass and rate of growth so that one would know how much wood was in a forest and one could calculate how much wood could be sustainably harvested every year.

These intensive, hands-on classical forestry concepts permeate not only the beginnings of forestry, but also much later into the 20th century. Sustained-yield forest management is what was taught to generations of forestry students. In India, the British established the modern counterpart to professional institutions, including research and training centres for the management and conservation of tropical forests (Dawkins and Philip, 1998). Unique at this time was the establishment of an infrastructure to study and develop sustainable forest practises. Graduates of these institutes were appointed to oversee timber extraction and to develop conservation measures. The consistent long-term use of these measures and infrastructures was guaranteed by having a succession of scholars holding prominent positions in these institutions. They were responsible for experimenting and establishing field trials that allowed them to develop the silvicultural procedures used in India, including pre-harvest cutting of vines, girdling of trees, thinning, fire management and pest management. These institutional structures and research centres developed for the tropics were the forerunners of the development of future organizations specifically mandated to manage forests.

The downside of the British formalization and systematization of the process to scientifically analyse forest sustainabilty was that it codified the 'empire forestry' model for how to control and manage forest resources. This resulted in the colonial governments controlling the access and management of forests in the countries they colonized – a pattern that was maintained until the mid-1900s. This policy was implemented under the rubric that scientific forest management was under the institutional purview of the government, since it was the government that had paid for the scholarly research in the first place. In addition, and more importantly, they were the only organization with the power to reverse the forest losses that had already occurred. The reality was that it was essential for the government of Great Britain to manage all aspects of a forest resource because these products were necessary for its continued naval superiority and essential to maintaining control of the trade routes.

The maturation of models to conserve forests (1850–1900 CE)

The powers colonizing countries beyond the European borders gave little consideration to conserving forests in their colonies, and much less to restoring an entire forest ecosystem (a system made up of a community of animals, plants and bacteria and its interrelated physical and chemical environment). It was much easier for them to move on to other, still unexploited, forests. However, as the United States acquired independence and eventually recognized the need to conserve its resources, the concept of sustainability that had been introduced and developed by Brandis in India was eventually transferred to North America.

The ideas that began in India evolved in the United States to form the basis of a model to preserve natural areas, such as parks and protected areas, for more reasons than just timber (Case 1.1 – A History of Forest Protected Areas). The model for protected areas matured in the United States in the mid- to late 1800s. This approach was adopted by many other countries and eventually became the standard used throughout the world. This was based on the need to obtain several items from the land: clean water, game animals, pastures, croplands and timber. All of these items were to become part of the park or protected area. As forward thinking as this was, it still did not consider the forest as an ecosystem encompassing a group of species that were dependent on the forest for their survival. The ecosystem-based model did not come into the forestry vocabulary until the 1970s (see Chapter 7, this volume).

These new ideas for conserving forests grew during the Industrial Revolution in Europe and the United States and were probably best vocalized in the United States between 1850 and 1900 (Williams, 1989). Despite the need for natural resources to power industrial development, society was disturbed by the negative and often visible impacts of exploiting forest resources. An increasing number of people wanted to preserve forests and other natural areas for their aesthetic values and as locations that enabled them to 'commune' with nature. In the United States, the public enthusiastically supported establishing forest or nature reserves that would be under the control of the federal government (Winters, 1974; Vogt et al., 1997).

By centralizing the control in the federal government, the United States mimicked the empire forestry model developed by Great Britain during the 1500s to control access to and uses of forests. The major difference between Great Britain and the United States was that forest lands in the United States really belonged to the public and were initially managed in 'protective custody' and eventually as 'stewardship' contracts by the federal agencies.

Approaches to conserving natural areas and forests were systematized in the United States in the 19th century. A brief listing of the acts voted into law during this period in the United States demonstrates how frequently natural areas were being mandated or their use regulated (see Box 1.1).

The views on natural resources or forest conservation in the United States were dominated by individuals like Gifford Pinchot, who advocated sustainable use for the greatest good (Williams, 1989). Since Pinchot acquired most of his training in Germany, France and Switzerland and was very strongly influenced by Brandis's systematic scientific approach on how to conserve and manage forests, the European influence on United States policy was evident. Pinchot greatly influenced President Theodore Roosevelt, who appointed him head of the United States Forest Service when it was first formed in 1905 (Williams, 1989). In this way the United States approach to forest conservation was rooted in German and British scientific forestry. Forests that were set aside during this time had to satisfy a combination of uses, which could vary from cattle or sheep grazing, timber production, drinking-water

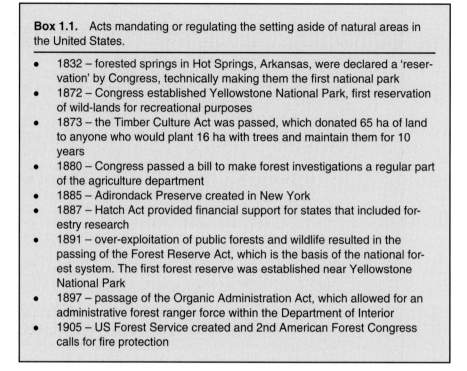

Box 1.1. Acts mandating or regulating the setting aside of natural areas in the United States.

- 1832 – forested springs in Hot Springs, Arkansas, were declared a 'reservation' by Congress, technically making them the first national park
- 1872 – Congress established Yellowstone National Park, first reservation of wild-lands for recreational purposes
- 1873 – the Timber Culture Act was passed, which donated 65 ha of land to anyone who would plant 16 ha with trees and maintain them for 10 years
- 1880 – Congress passed a bill to make forest investigations a regular part of the agriculture department
- 1885 – Adirondack Preserve created in New York
- 1887 – Hatch Act provided financial support for states that included forestry research
- 1891 – over-exploitation of public forests and wildlife resulted in the passing of the Forest Reserve Act, which is the basis of the national forest system. The first forest reserve was established near Yellowstone National Park
- 1897 – passage of the Organic Administration Act, which allowed for an administrative forest ranger force within the Department of Interior
- 1905 – US Forest Service created and 2nd American Forest Congress calls for fire protection

supply to recreational space (Williams, 1989). This model varied from the older European tradition of forest preserves established by royalty in Europe or Asia since in the United States everyone had access, although controlled in many cases, to the resources located in the reserves, not just the royalty.

The role of the United States government in forestry (1900 CE to present)

The role of the United States government in regulating forest activities is apparent from the many acts specifically related to forests. During the early 1900s, several of the acts were related to fire control, giving the Forest Service, created in 1905, a strong mandate to fight forest fires. The number of acts adopted also reflects the importance of forests to American society and the conflicts that have occurred around forest uses and conservation. Key acts are summarized in Box 1.2.

The political importance of controlling timber resources decreased when ships were built using metals instead of wood, and oil and hydropower replaced wood as sources of fuel. However, during the First World War (1914–1918), wood became important again because wood was needed in mining operations. In fact, the British war campaign was almost undermined due to a shortage of timber; however, the USA was able to meet their timber needs from the large area of forests existing within the US borders (Winters, 1974).

A similar situation occurred during the Second World War, when the USA again supplied much of the European timber needs. During the Second World War, Germany tried to control wood production and supply over the entire European continent.

Box 1.2. Major acts passed and enacted in the United States since 1911.

Time	Act and what was enacted
1911	Weeks Act – allowed cooperation in fire fighting between US federal and state forest fire protection agencies; created the National Forest Reservation Commission to create national forests in the eastern part of the US
1924	Clarke–McNary Act augmented the cooperative fire-fighting efforts between the federal, state and private forest owners
1934	Article X of the Lumber Code was enacted to control logging on private lands and was advocated by Pinchot. This law was struck down in less than a year but resulted in stronger cooperation between the US Forest Service and the timber industry
1937	O & C Sustained Yield Act – required management of the national forests in Oregon and California to be managed for sustained yield
1947	Forest Pest Control Act – protects all lands, regardless of ownership, from destructive forest insects and pests. First legislation to direct management across ownership and political boundaries
1960	Multiple Use Sustained Yield Act – created multiple-use planning, which brought in new specialists (e.g. wildlife biologists, recreation specialists)
1962	McIntire–Stennis Act – boosted forest research funds
1964	Classification and Multiple Use Act – required US Forest Service to manage lands for multiple uses, to sustain in perpetuity outputs of various renewable resource commodities and other uses
	Wilderness Act – established National Wilderness Preservation System
1976	National Forest Management Act – required US Forest Service to develop long-range land-use planning, required outside scientific committee not employed by the agency, required preservation and enhancement of plant and animal community diversities equivalent to what is found under natural conditions
1978	National Parks and Recreation Act – required National Park Service to develop long-range land-use or general management plan for their lands
1980	Alaska National Interest Lands Conservation Act – protected 41.2 million ha of Alaskan land as national wilderness, wildlife refuges and parks

They needed access to forest resources to build containers and tools, in the production of bomber aeroplanes and to produce gunpowder (Winters, 1974). Even the Japanese during the Second World War considered access to timber important and launched ~9000 incendiary balloons in an unsuccessful attempt to burn down the western forests of the United States to cut off their timber supplies (Winters, 1974).

Search for Forest Resources and Scarcity after Over-exploitation at Regional Scales

History has recorded how forests were used to aid in the creating of civilizations, but also their collapse when the political power structures mismanaged and exploited forests. Many civilizations have collapsed following periods of intense exploitation, where resources needed to maintain the social infrastructure were depleted and readily available replacements did not exist. Some countries were able to exploit forest resources outside their political boundaries by the use of military force. This military force allowed many nations to become important global powerhouses regulating world economics and important trade routes.

During these same time intervals, local scarcity of forest materials stimulated the pursuit of regulations to control how forests were to be used and managed. Despite the fact that there were many isolated examples of wise forest management and regulations requiring the regeneration of trees in forests, there was no systematic effort to control how forests were used on larger scales.

Since this global supply was deemed endless most governments felt that its management was unnecessary. It was a matter of searching for a wood supply elsewhere if you had depleted the local resources and there appeared to be new lands and new frontiers for conquering and exploiting. This 'limitless supply' view has driven much of the exploitation of forests and their subsequent abandonment once they became over-harvested and unproductive.

However, as numerous civilizations have learned, when the forests are gone cultures decline and in some cases disappear. In the late 1700s, firewood shortages were reported in Philadelphia, Pennsylvania, United States (Williams, 1989). The late 1860s saw a timber deficiency in the United States because of extensive conversion of forest lands into agricultural fields. An important issue addressed at the First American Forest Congress, held in 1881, was the fear of a timber famine (Williams, 1989).

Even though wood was not commonly used to make paper until 1900 (paper was made from old cloth), shortages in the starting materials needed to make paper occurred in the 1860s in the United States. At that time, mummy wrappings from Egypt were imported into the United States for use in making paper (Pollack, 1978). This occurred until concerns were raised about human health problems from the use of wrappings that had previously encased dead bodies. Initially, paper was commonly made from rags since its chemical composition (i.e. cellulose) is one of the basic chemical components of paper (see Chapter 6, this volume).

Global Scarcity Acknowledged in the Twentieth Century but also the Need for Verification of Sustainable Management of Forests

In the late 1970s, the notion of a limitless supply of forest materials was beginning to be recognized as a fallacy by enough people to result in important new changes. There was a growing acceptance that forests are a global resource and there are no new frontiers to explore for untapped supplies. This recognition occurred at the same time that the importance of forests as global resources that provide global environmental services became apparent. Some people, especially environmentalists, argued that the global scope of the challenge and effect of forest resource management required collaborative participation by the international community. Since deforestation in regions like the Amazon impacts climates globally and forests are critical for providing habitat for biodiversity, many international scientific and environmental groups concluded that a global solution was required.

At the same time that this was happening, many local or regional communities in industrialized countries began to pursue policies to decrease or eliminate the cutting of old-growth trees in their areas in order to maintain the habitat of old-growth-dependent species and the high biodiversity found there. These attempts to regulate or eliminate tree cutting have had the unintended consequences of shifting the exploitation of forests from the industrialized countries, which are conserving forests, to less industrialized regions, which are more dependent on forest resources for their survival. Since worldwide use of and demand for forest-based products have not decreased, the exploitation of forests in the less industrialized regions of the globe (which tend to be in poorer tropical regions) has significantly increased during the last decade (Perez-Garcia et al., 1999; FRA, 2000).

While the demand by industrialized countries for wood from tropical regions is increasing, changes have occurred in the less developed countries, where local communities expect and want more control over decisions made on forest resource uses in their area. This change started after the Second World War, when the colonialists began to lose control of the lands that they had conquered starting back in the 16th century. Transfer of control meant that decisions being made on forest uses in Asia, Africa and South America were no longer under the control of the former European colonial countries (Dawkins and Philip, 1998). Despite the European colonial powers having less direct influence on how forests are managed in these countries today, the high consumptive demand by the industrialized countries for forest products is allowing them to indirectly control forest uses since timber is sold on global markets and they have the economic resources to ensure that their desires are met.

Since global markets determine who the suppliers of timber are, the location of timber exploitation will shift with time. For example, the Pacific North-west United States was an important supplier to global markets from the mid-1900s to the 1980s but not today. The shift in timber harvesting away from the regions like the Pacific North-west United States is being driven by lower labour costs or less concern about environmental issues, since tree harvesting has been occurring for several decades already in other regions of the world (e.g. Finland, Chile, New Zealand) (Perez-Garcia et al., 1999). These current suppliers of timber will probably shift with time as other cheaper sources of timber become available. The significant area of remaining forest cover (see Chapter 2, this volume) in South America and Russia suggests that these are going to be likely areas of future timber harvest. Both these areas have significant ecological and political constraints that may reduce the risk of these areas being over-exploited. For example, the vast Siberian forests have slow rates of tree regrowth because of the cold growing conditions and a reoccurring natural fire disturbance cycle that increases the insecurity of timber supply (AFPA, 2004b). On the other hand, history has provided many examples of over-exploitation given analogous ecological and political constraints.

These shifting suppliers of wood to global markets are occurring at the same time as concepts like sustainability have become more important (i.e. managing and using forests so that they will persist into the future in a non-degraded condition). Sustainability expanded to include a broad range of organisms that lived in forests and was accepted and formalized into forest practises. Timber buyers in the early 1990s, especially in the United States and Europe, began to want verification that the timber they were buying was coming from forests managed sustainably (see Chapter 7, this volume). Strategies such as forest certification originated at this time as a process to evaluate the ecological, political, economic and cultural contexts of forest uses (see Chapter 7, this volume). Certified forests have expanded to approximately 2% of the global forests (FRA, 2000).

The pursuit of auditing and certifying forest sustainability is imbalanced globally, with 92% of the certified forests worldwide found in the United States, Finland, Sweden, Norway, Canada, Germany and Poland. The Scandinavian countries are well represented in the list of countries that have had their forests certified. What is obvious from these data is that almost all of the countries where certified forests are found are in the temperate zone rather than the tropics. This imbalance is illustrated in the following data, which present the amount of forest area certified in various regions of the world: Africa = 1.2%; Asia = 0.2%; Oceania = 0.5%; Europe = 57.9%; North and Central America = 38.3%; South America = 1.9% (FRA, 2000). Forest certification was developed to deal with deforestation in the tropics (see Chapter 7, this volume); however, these data suggest that it is not working in the regions of the world where it is probably most needed.

This transition towards sustainable management of forests was easier for European countries to adopt compared with many tropical countries. For example, the rate of change in forest practises was less dramatic and had a lower impact in Europe compared with many of the tropical countries since Europe had:

- a long tradition of forest management, where forests in mountainous regions were managed to protect lower-lying areas from flooding and avalanches during winter months;
- industrialization and urbanization occurring gradually so there was less of an impact of people dependent on forests and who lived in rural areas;
- a smaller role for forests in the economies of most European countries; and
- a lower population of people living in rural areas (Dawkins and Philip, 1998).

It has been easier for many European countries to become certified since there was less conflict in society over the decisions being made in forests because of the points made above. In contrast, developing countries located in the tropics have had and continue to have serious conflicts over how to achieve sustainable livelihoods and sustainable forests in an environment that does not have the capacity to provide for both goals simultaneously.

A Trend towards More Intensive Management and Planting of Exotic Species in Commercially Managed Forests

In the past, when forests were managed at all, it was by encouraging the regeneration of native tree species. During the past few centuries, there has been a major shift in how forests are managed, which has its roots in the development of scientific forestry in Germany. Increasingly, trees that are highly valued and have desired characteristics are being planted worldwide in areas where they are not native. Planting trees in plantations is being pursued as a solution to the decreasing supply of tropical trees and the high rates of deforestation that are occurring in these regions (Panayotou and Ashton, 1992). Many of these plantations are being planted for fuel wood. Panayotou and Ashton (1992) reported that two-fifths of the tree plantations in the humid tropics are being used for fuel wood and other non-industrial uses.

There has been a shift to increase the area of forest plantations dedicated to the practise of monocultures (i.e. single-species plantings). This introduction of non-native tree species on a massive scale is a phenomenon that is changing the global make-up of commercially managed forests. They are converted from an assemblage of numerous different species of trees in one area to a single dominant species. The introduction of non-native species that will use the available nutrient or water resources at a site to grow faster than the native species will improve the timber supply from commercial forests.

The Global Forest Resources Assessment 2000 found that globally 1.5 million ha of natural forests are converted to plantations annually (FRA, 2000). Most of this conversion of natural forests to plantations is occurring in the tropics (see Fig. 3.2). The amount of forest area in plantations is still a small fraction of the total forest area, since 95% of forests are natural and only 5% are in plantations. However, the conversion of natural forests to plantations is a concern when it is happening in the tropics, because this is the region of the world where the highest rates of deforestation are being recorded (comparing data from 1990 to 2000). Deforestation rates of 14.2% have been recorded in the tropics compared with the 0.4% estimated for non-tropical regions (FRA, 2000).

The past offers us information from which we can learn so that we do not exploit ourselves into oblivion. We make up many different and often competing societies, but we are also one species sharing the same world and its resources. History and the trends described above provide us an urgent call to pursue sustainable forestry practices. Sustainable forestry has to become not only a phenomenon of the industrialized countries but must be practised on a global scale. Our dependence on forests is not decreasing so there is a need to adopt new technologies that convert forest materials to useful products in a manner that does not degrade the ecosystems that we depend on to produce those materials. Plantation forestry also has a role to play in the provision of forest products, especially as the land area in forests decreases. High rates of production in plantations should allow greater areas of forests to be set aside for conservation elsewhere.

Forests have been important to people throughout history even though what we require of them has changed. Even today, forest resources continue to be important for the survival and the improvement of living conditions for millions of people in different regions of the world. New uses are being developing today that are based on technologies that provide a higher-quality product from our forests and that allow those same forests to be used to produce energy that does not require burning wood. This technology allows us to use less timber for better products and eliminates the emissions that scientists believe are contributing not only to our global warming problems, but to the depletion of atmospheric ozone. This will be covered in more detail in Chapter 7 of this volume.

It is worth exploring both the history of forest exploitation and also the historical efforts towards protecting forests. Such an analysis highlights the changing roles of forests in human society and provides examples of what we should be doing to prevent repeating the disasters of the past while adopting their successes. The debates on forest uses and conservation are not a thing of the past but are being hotly debated even today.

CASE 1.1. A HISTORY OF FOREST PROTECTED AREAS (BIANCA S. PERLA)

While most people equate the invention of protected areas with the emergence of national parks in the United States in the mid-1800s, this was actually a modern remembering of much earlier practices. Information on early forms of forest protection is sparse and scattered. However, the information we do have suggests that setting aside protected sites is probably the oldest, most elemental form of conservation adopted by human societies. There is evidence that indigenous people on all continents developed intricate social and cultural mechanisms to regulate the use of common property resources, including setting aside sacred places to protect a variety of wildlife, subsistence and religious values (Stevens, 1997).

Early tribal protected areas were created and maintained through oral history traditions, local ecological and cultural practises passed down through generations, and spiritual or religious connections to the natural world that emphasized sustainability. For example, in Europe, around 2500 years ago, Celtic tribes designated certain forest groves as sacred for spiritual and religious practices (Bunnell and Johnson, 1998; Ellis, 2003). Native Americans in California instituted a system of seven cultural constraints or rules that governed plant harvesting in forests and meadows including 'the amount taken does not exceed the ability of the plant population to recover . . . gathering mimics a natural disturbance to maintain or enhance plant production . . . and taboos, codes, or other constraints are put in place to discourage depletion' (Anderson, 1993). Other indigenous tribes throughout the world designated protected areas for many reasons, including conserving certain populations of animals or plants that were otherwise regularly hunted or gathered, protecting individual plants or animals of specific ages or genders, and protecting nesting or breeding sites

of important wildlife or plant populations (Stevens, 1997).

Many traditional societies in the Pacific gave protection to areas they recognized as important to their survival long before European occupation, and some of these practises continue today. For example, the Benuaq Dayak culture of Indonesia protects a common traditional forest area from tree harvest and rice cultivation to provide a continual supply of non-timber forest products for the community. While protection is generally accepted as a cultural norm, there are also serious social and spiritual sanctions in place in case of violation (Sardjono and Samsoedin, 2001). Similarly, in many parts of Africa local communities still adhere to long-term designations of sacred forest groves (IUCN, 1999).

The close connection between culture, nature and survival that led to native conservation practises and designation of sacred places has been hypothesized as enabling many indigenous cultures to support high levels of natural biodiversity in their homelands for many generations (Stevens, 1997). A geographical relationship between native societies and highly diverse natural environments exists in Central America (Chapin, 1992) and has also been proposed by Nietschmann (1992) for areas of South America, Africa and Asia.

Protected areas are also apparent in early records of more formalized and larger societies, such as the Maurya empire (321–185 BC) in India, from where the earliest written record of protected-area legislation and law has been found. The Maurya empire was established after the withdrawal of Alexander the Great from India. Emperor Chandragupta, founder of the Maurya dynasty, obtained power by trading 500 elephants to Alexander's defeated general Nicator. Emperor Asoka (265 BC), grandson of Chandragupta, was the first known monarch to officially set aside areas for the conservation of wildlife, bird and fish species. Protected areas were created after Asoka, regretful of a violent conquest he undertook to seize outlying territory, converted to Buddhism. He created the Law of Piety, a moral code that commanded that respect be given to all forms of life, both human and animal. The law had at least 34 edicts, which were written in Pali calligraphy on stones and pillars at the entrances of the empire so that all who entered were aware of how the empire was governed. Asoka banned hunting in the former royal hunting grounds, prohibited the slaughter of animals for food and even went so far as to provide healing stations on roadways for travellers and their beasts of burden (all the above information is from Draper and Meyer, 1995).

In other places in Asia, areas were set aside and managed specifically for sustainable hunting. In China and Persia, powerful rulers created 'hunting gardens' for their own recreational pleasure (IUCN, 1999). Ecosystem services provided by forests were also recognized centuries ago. Protection of forests specifically for soil quality and water retention has been recorded in Japan as early as 500 years ago (WCMC, 1992).

In Western and Eastern Europe, early evidence of formal protected areas also exists. Germanic laws (around 500 CE) are found that classify forests for a variety of different uses, including grazing, honey and beeswax (Bunnell and Johnson, 1998). In these laws, non-timber products were regularly emphasized as reasons for protecting forest tracts. Around 556 CE, the word *forestis* appears in German law records to describe areas of forest set aside to preserve hunting and fishing values (Bunnell and Kremsater, 1990). In 1457 Romania, King Stephen the Great of Moldovia (eastern Romania) protected large forest patches for the conservation of wild game. These forest patches were called *braniste*, which means a forest with abundant vegetation and low forest cover. It was forbidden to hunt or cut trees in designated *braniste* (Soran *et al.*, 2000). Several forest reserves were designated in Wallachia, Romania during the same period through royal decree of various ruling kings of the 14th and 15th century (Giurescu, 1980, as cited in Soran *et al.*, 2000). In addition, 100 years ago, records in Switzerland and Austria reveal the designation of forest protected areas for conservation of water and soil (Holdgate, 1996).

The creation of national parks and the National Park Service in America did not occur in complete isolation. For example, around the same time that Americans were starting to establish the National Park Service in the United States, a handful of influential scientists and naturalists worked towards the establishment of 36 scientific reserves and successfully passed the first legislation in Romania for the protection of natural monuments. Most influential in the creation and lobbying for these laws and designations was Romanian botanist Alexander Borza, who worked for 6 years to pass the laws that were finally written in 1930. Other instrumental players were a student–teacher pair of scientists, Grigore Antipa (1867–1944) and Ernst Haeckel (1855–1891), and Emil Racovitsa (1868–1947), who was a zoologist (Soran *et al.*, 2000).

While protected areas have existed for a long time and while there were parallel efforts to creating protected areas throughout the globe at the same time that the national parks of the United States were established, the national-park model has been the most widely adopted form of protected area in the world today. When forests and other ecosystems became valued for their biodiversity in the 20th century, concerned conservationists used the United States national-park model to protect critical areas around the globe that would otherwise be depleted or converted. The national-park model favours protecting large expanses of land that are uninhabited by humans and conserving these areas in the national public interest through a federal management entity to preserve scenic wilderness and recreational, historical and natural values (IUCN, 1999).

The push for biodiversity conservation, starting in the mid-1900s, resulted in an increase in the number of parks and preserves worldwide from 2 million km^2 in 1960 to over 18 million km^2 (11.5% of the earth's surface) today (UN, 2003). It is now apparent that using one protected-area model, the United States national-park model, does not always sufficiently address conservation objectives when applied to the highly variable cultural and natural settings throughout the world. More recently, there has been a drive to include protected areas that are managed for subsistence, biodiversity and living cultures as well as scenery, and some of the more varied forms of protected areas that existed throughout history are returning to us today (Holdgate, 1996; IUCN, 1999).

2 Global Societies and Forest Legacies Creating Today's Forest Landscapes

KRISTIINA A. VOGT, JON M. HONEA, DANIEL J. VOGT,
TORAL PATEL-WEYNAND, ROBERT L. EDMONDS,
RAGNHILDUR SIGURDARDOTTIR, DAVID G. BRIGGS AND
MICHAEL G. ANDREU

Forests cannot be managed solely by considering whether the trees need to be harvested, thinned, fertilized or sprayed with insecticides. The forests of today are also managed according to global parameters involving economics, social values relating to forest uses, poverty alleviation, pollution, climate change, energy technologies and informatics or the management of information using computer-based technology. Since some of our current environmental problems are linked to forests, it is critical that policymakers and forest managers understand the complexity of interactions that occur between society and forests. In fact, forests can provide solutions to many of these problems. However, the complexity of forest ecosystems and uncertainty of our understanding of forest processes impede this decision process. Clearly, balancing social values with the need to harvest forest resources for human use presents major challenges to resource managers and policymakers worldwide.

Forests have always played an important role in providing environmental services and natural resources and eliciting human appreciation in societies around the world. Most books have adequately documented the biology, ecology and/or silviculture of forests but insufficient attention has been paid to how society and forests can interact where both economic and non-economic returns are possible and even sustainable. Understanding these interactions should help to reduce the conflicts over how forests are used to provide goods and services. It should also facilitate the implementation of sustainable forestry principles that are based on socially, silviculturally and ecologically sustainable practices.

Many of the highly polarized conflicts over resource uses and conservation in forests are a result of the inability of integrating societal values and economic expectations with the silviculture and ecology of forests. These continuing conflicts have been detrimental to forest-dependent communities who want to improve their economic livelihoods using forest resources (Chapter 3, this volume). In developing countries, international development programmes began in the early 1990s to design projects to simultaneously deal with poverty and the lack of sustainable livelihoods in rural

© CAB International 2007. *Forests and Society: Sustainability and Life Cycles of Forests in Human Landscapes* (eds K.A. Vogt *et al.*)

areas while pursuing conservation goals (i.e. integrated conservation and development projects) (McShane and Wells, 2004; Sayer and Campbell, 2004; Chapter 3, this volume). These programmes are still evolving and have struggled to achieve both goals in the same forest. The metrics used to evaluate whether these programmes are successful are also just being developed and it will be challenging to codify and develop consensus among the different stakeholders (e.g. resource use and conservation communities) because each has a different 'lens' by which success is measured (see Chapter 7, this volume – 'A Challenge: Codification and Consensus in Measuring Sustainable Forestry'). At the same time as these integrated conservation and development projects were being initiated, forest certification was introduced as a process to evaluate forest sustainability when forests were used to collect forest products, to provide environmental services and to obtain conservation goals. This led to the development of third-party forest certification programmes (Chapter 7, this volume) to ensure that credible auditing of resources could be verified (Vogt *et al.*, 1999, 2000).

Forest management has been difficult to implement because of the need to include a complex array of interrelated factors from many disciplines while making the trade-offs required to resolve the conflicts between its economic uses, its societal values and its conservation needs (see Case 2.1 – Nepal, Community Forests and Rural Sustainability; Case 2.2 – The Impact of Indigenous People on Oak–Pine Forests of the Central Himalaya; Case 2.3 – Dead-wood Politics: Fuel Wood, Forests and Society in the Machu Picchu Historic Sanctuary). Forest managers cannot ignore society since it is society itself that determines what management practices are acceptable. Even though ecology and silviculture provide the foundations of forest management, human behaviour modifies and impacts forests globally (see Case 2.4 – Icelanders and their Forest History: a Thousand-year-old Human and Nature History Controlling Resilience and Species Composition in Forests Today). In turn, forests impact society and social behaviour because of the economic return and environmental services they provide.

This chapter will discuss how historically humans have selectively used forests in some biomes compared with others because they provided more resources to clothe, feed and build houses for humans. This reflects the varying capacities of different forest biomes to be hospitable environments for human survival. Those areas most hospitable for human survival in the past are now locations with less forest cover. This selective exploitation of some forests has produced the forest landscape of today and determined who will be the suppliers and consumers of forest products today. Since the demand for wood products continues to increase, countries with less forest area are the consumers of forest products harvested in locations historically more inhospitable to humans. For example, many large timber companies are eyeing forests in Siberia to harvest wood despite the many limitations that exist to sustainably harvest trees in these cold climates. Superimposed upon these patterns are the different consumption categories for wood globally. These categories differ depending on whether a country is industrialized or developing. Most developing countries are using wood mainly as fuel wood so that little is left for producing paper products or industrial uses of wood. When wood is mainly burnt as fuel wood, communities have fewer options for how to manage their forests and acquiring conservation goals is limited by people's need to survive.

Human Uses of Terrestrial Global Systems

Prior to understanding the socialization and globalization of forests, it is important to understand where forests are located around the world and how forest materials are currently used by different societies. This examination of where current forests are located and their primary use reflects the historical decisions made by society.

Forests cover approximately a third of the terrestrial land surface (Fig. 2.1). How much of the terrestrial land surface was in forests prior to active human modification is difficult to estimate because of the lack of historical records on this topic. In the United States, it was estimated that about 45% (770 million ha or 1903 million acres) of the land area was covered by forests in 1600 (Williams, 1989). By 1920, the United States had only 24.7% of this original forest remaining so that almost half of the forest area had been lost between 1600 and 1920, mostly by conversion to agriculture. Since that time, agricultural abandonment has allowed forests to regrow in the United States so that 28% of its land area is now in forests (Fig. 2.2). These patterns highlight how the amount of land in forests is a dynamic process.

If you look at how much of the terrestrial area is forest compared with agriculture today, you can see that humans modify about 10% of the land area to use directly in the production of food. Approximately 66% of the total land area can be used specifically to grow crops or harvest food materials for human food production.

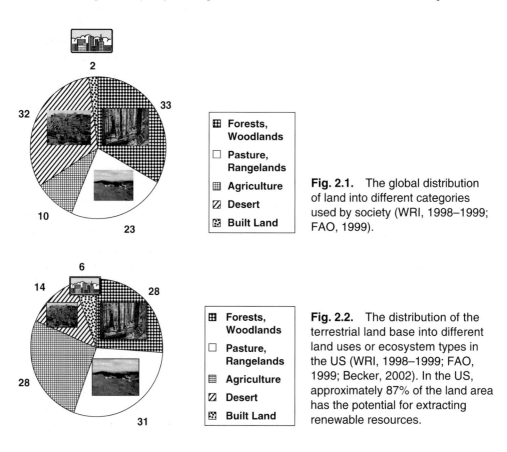

Fig. 2.1. The global distribution of land into different categories used by society (WRI, 1998–1999; FAO, 1999).

Fig. 2.2. The distribution of the terrestrial land base into different land uses or ecosystem types in the US (WRI, 1998–1999; FAO, 1999; Becker, 2002). In the US, approximately 87% of the land area has the potential for extracting renewable resources.

Only 2% of the total (not of the 66%) area is considered built land, which consists of buildings and transportation surfaces.

Another third of the world's land mass is in ice, rock or desert (Fig. 2.1). None of these areas would provide habitation for the easy survival of humans. Some humans have adapted to even these extreme environments (e.g. Aborigines in Australia, Inuit in Alaska and nomads in deserts) even though they are poor locations for human habitation. Survival has dictated that both people and animals usually had to migrate during winter or the drier periods in these regions. Human population densities in these extreme environments are low.

Based on this assessment, about two-thirds of the global terrestrial area is potentially available for habitation and is being used by humans at varying levels. What is interesting at the global level is that humans are able to manage and harvest renewable resources systematically and continuously from a significant portion of the terrestrial land base.

Characteristics of Forest Biomes Determined their Utility for Human Survival in the Past

A forest's location in the world today is a result of how humans used or altered it in the past. Civilizations have developed and flourished around forests. This is especially true for those forests located in areas where the climate was more conducive to human habitation, the soils were productive enough to allow agriculture, sufficient water was available, wildlife was abundant enough to provide the protein people needed, and the forests provided other products that could be used in construction or commerce (Chapter 1, this volume). Some forests have been more heavily used by humans than others, but they have all been used for everything, from simple collecting of firewood to building and maintaining civilizations.

In the past, humans frequently migrated when there was not enough land to produce the food needed for communities to survive. Typically people would transport domesticated animals with them as a food source. For example, when the Vikings first arrived in Iceland more than a thousand years ago, vast areas of grasslands and forests covered Iceland. The Vikings had brought sheep that began to graze in the forests and grasslands (see Case 2.4 – Icelanders and their Forest History: a Thousand-year-old Human and Nature History Controlling Resilience and Species Composition in Forests Today). Humans survived, albeit not at high population densities, because the domesticated animals ate the trees, shrubs and grasses. The humans in turn relied on the domesticated animals for food and clothing. The arrival of the Vikings meant the demise of Icelandic forests. Today Iceland has only 1% forest area remaining. Likewise, the dry, thorn-shrub lands in Africa have only been altered 5% by humans since it is difficult for humans to build a civilization where both humans and animals have to migrate during annual drought cycles.

Forests are found in areas where there is enough water for trees to grow. Access to water has always been an important factor in determining where civilizations have also flourished. Out of all the forest types in the world, the forests with trees that annually drop their leaves (i.e. deciduous) and are located in the temperate zones have been the most conducive for human survival and have been most altered by humans (Table 2.1; see Chapter 4, this volume). These forests have soils that are rich in nutrients and generally are perfectly suited for use as agricultural fields. In addition to providing wood for building homes, deciduous forests also supported wildlife populations large enough to provide food for humans. The coniferous or needle-leaved forests are unlike the deciduous forests, primarily in that their soils are nutrient-poor and rocky and they support a less abundant wildlife population. In the north-west United States, coniferous forests in the wet, rainy, west side were considered to be 'biological deserts' by wildlife biologists in the 1950s because large game animals that could be hunted for food did not occur abundantly in these forests. There was insufficient food for large animals to survive in these forests.

Areas that are prone to drought or where the ground is frozen as solid as a block of ice (i.e. permafrost) are locations with the lowest levels of human use and historical alteration (Table 2.1). Even though some humans have lived in these climatic extremes for months at a time, no large communities have formed and survived for any extended periods in these environments. Examples would be some of the research facilities near either pole. The lack of any resources in these environments requires that all supplies be brought in.

The tundra has only 0.3% of its land area occupied by humans. In this region most animals hibernate or migrate during the winter months and even humans followed the herds or lived in small, scattered communities and survived on sea animals for food (Chernov, 1985). The domestication of the reindeer allowed the Laplanders or Sammi of the northern regions of Scandinavia to permanently live in these cold regions because the reindeer provided all their survival needs (e.g. clothing and balanced nutrition as a food source) (Chernov, 1985). Many have considered these regions to be 'biological deserts' as well, because of the low diversity of animals that could be hunted for food. These ecosystems are not technically 'biological deserts' since there is a high biodiversity of microscopic organisms in the region.

Table 2.1. Biome type (a large, naturally occurring community of flora and fauna occupying a major habitat) and the percentage by which it has been used and altered by humans (the higher the percentage indicates environments that have the larger number of items needed for human survival and the greater the alterations or modifications made by humans) (from Hannah *et al.*, 1995).

Biome type	% Human-dominated
Temperate broadleaved forests	82
Tropical dry forests	46
Temperate grasslands	40
Tropical rainforests	25
Deserts, warm	12
Temperate coniferous forests	12
Deserts, cold	9
Tropical grasslands	5
Tundra and Arctic desert	0.3

Unfortunately, microscopic organisms (i.e. bacteria, protists and fungi) may have a very high biodiversity but their microscopic size prevents them from being a good food source for humans. These ecosystems have barely been disturbed because their environments are too extreme for human survival. In the future, regions of the world that previously experienced low levels of human disturbance may become transformed if technological breakthroughs eliminate the constraints that currently prevent the growth of food crops and allow humans to control their microclimate.

Most people think of the tropical zone as an ideal habitat for humans because of its lack of a cold climate. However, the extremely high temperatures and either high or low rainfall have made it difficult to grow sufficient food crops to support large population densities. Hence, this biome is only 25% dominated by or altered by humans (Table 2.1). The extremely high biodiversity of animal and plant species in the wet tropical forests is also misleading. It suggests that humans should survive well here because so many other species do (Chapter 3, this volume). However, high diversity does not necessarily mean that these areas are capable of supporting high densities of humans. The high plant biodiversity is made up of many plants with a complex mix of toxic chemicals in their tissues that make them inedible for humans. The lush forests suggest high growth rates and a large amount of biomass that can be used as food or for building human habitation. Instead, the soil environment is commonly deficient in the required nutrients needed by vegetation to grow. When vegetation is removed, the soils have a poor capacity to grow food crops. These hot, humid regions are perfect breeding grounds for many diseases that are fatal to humans.

In recent years tragic wars have increased the challenge of sustainable forest resource management in the tropics, especially in Africa. These wars have destroyed forests and resulted in widespread starvation because of the inability of local people to grow food crops. Destruction of forests and agriculture also leads to the rampant killing of wildlife for food (Diamond, 2005; Chapter 5, this volume).

Where are Forests in the World Today?

The global distribution of forests in 2000 reflects the results of historical patterns of forest use and their conversion to other uses. For example, today only 3.2% of the global forest is located in Europe (Fig. 2.3). After over-exploiting local sources of timber, most European countries needed to travel to new lands to access the timber needed to fuel their industrialization, to build the ships for maintaining their military dominance and to secure new trading routes. Except for some of the Scandinavian countries, who continue to be significant suppliers of timber on today's global markets, most of the remaining countries in Europe do not have a sufficient supply of timber to be important competitors in the global timber markets.

Even though Europe has a small fraction of the global forest area, some countries in Europe do have forests comprising a high proportion of their total country land area (e.g. 72% in Finland) (FRA, 2000). This contrasts with other European countries (e.g. Germany, France, Italy), which have about a third of their land in forests, while others are closer to having 10% of their land in forests (e.g. Denmark).

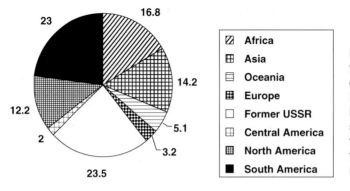

Fig. 2.3. The location of forests in different regions of the world (FRA, 2000; FAO, 2001). The number in the middle of each segment of the pie chart is the percentage of global forests located in a given region.

Today, Russia and South America combined have almost half of the forest area in the world and both locations are predicted to be important suppliers of timber from natural forests for global markets in the future. Siberia's substantial areas of natural forests have led several international timber companies to partner with Russian companies to gain access to this timber. Siberia was not considered an important timber market in the past because it is in an extremely cold region where trees grow slowly, so regeneration of the cut-down forests will take several centuries. Furthermore, reoccurring large-scale fires make it difficult to manage forests in Siberia since there is little security that the forest will exist the next year and will not have already burned down (see Case 5.3 – Wildfire in the Boreal Forests of Alaska). Even though considerable attention is being given to obtaining timber from the former USSR and South America, these supplies are controversial and politically contentious.

North America (i.e. Canada and the United States) has about 12.2% of the global forest area today (Fig. 2.3). This region of the world was an important supplier to the global timber markets less than a decade ago. This region still provides timber for the global markets but is not a dominant global supplier of timber. Timber markets have shifted to other regions of the world where wood is cheaper to buy because labour costs are lower (Perez-Garcia *et al.*, 1999).

Where forests are located around the world does not equate to how much forest material is consumed by citizens in any country. It does tell you who will be the suppliers of wood in the global markets. Since forest products and materials are sold in a global market, there is no need for a country to have forests to be able to acquire and consume forest products.

The types of forest materials consumed and what forest products are made do vary between developing and industrialized countries. In general, more industrialized countries produce more products from forests (e.g. paper products, industrial round wood). Less industrialized countries are limited in how forest materials are used and what products are consumed since most wood is used as fuel wood. This scarcity of wood has resulted in unique ways of obtaining the starting material to make forest products. For example, the low availability of wood to produce paper in Mexico has resulted in the development of an industry driven by the collection of recycled cardboard boxes by scavengers from Mexico, who collect these materials in the United States (see Case 7.7 – Importance of Scavenger Communities to the Paper Industry in Mexico).

Current Global Consumptive Uses of Forest Materials

It is important to understand how forest materials are used by society in different regions of the world (to make paper products, industrial wood products or as fuel wood) because this determines what materials are collected and what other forest products can be realistically collected from the same forest in the future (Table 2.2). In many less industrialized countries like Nepal, more than 90% of the wood is used strictly for fuel wood (e.g. cooking and heating). In fact, more than half of the global population still uses wood mainly for heating. These economies will be unable to be part of the global wood markets for other forest products since they do not have enough wood left to provide for these other uses.

Several trends are apparent from the data presented in Table 2.2:

1. Most of the developing nations use a significant amount of wood as fuel wood or firewood while the industrialized world uses little to none for this purpose.
2. As a country becomes more industrialized, there is a shift from wood to fossil fuels as an energy source.

It is apparent that as a country becomes more industrialized, it begins to use less biomass as fuel wood to produce its energy (Table 2.2). In North America (i.e. the United States and Canada), about 9% of the wood is used for cooking and heating (Table 2.2). Europe's wood consumption has a very similar pattern to that of North America.

Another pattern apparent from these data is that if wood is used primarily as fuel wood for heating or cooking, as occurs in Africa (87% of its total wood consumption) and Central America (73%), little wood remains to produce other products such as paper or lumber (Table 2.2). This pattern contrasts with what is found in the more industrialized countries, where about 10% of the wood is used as fuel wood. This lower dependence on wood for heating and cooking means that three-quarters of the wood can be consumed in round-wood or lumber products such as plywood and a fifth is available to produce paper products.

Table 2.2. World wood consumption data by categories of use for 1996 (from FAO, 1998).

Region	Wood consumption categories Consumption in million m³ (% of total within a region)		
	Fuel wood and charcoal	Non-fibre industrial (plywood, etc.)	Fibre – for paper products
Asia	904.8 (59)	460.0 (32)	141.1 (9)
Africa	519.7 (87)	75.6 (13)	5.1 (1)
South America	192.7 (48)	194.4 (48)	16.8 (4)
North America	93.9 (9)	784.1 (75)	167.1 (16)
Europe	83.3 (12)	513.6 (72)	116.0 (16)
Central America	61.3 (73)	17.5 (21)	5.6 (7)
Oceania	8.7 (43)	31.4 (48)	5.2 (9)

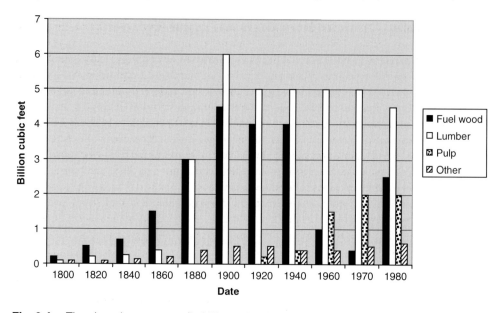

Fig. 2.4. The changing amounts (in billion cubic feet) of wood used for fuel wood, lumber and pulp in the United States from 1800 to 1980 (US Bureau of the Census, 2005).

The changing pattern of wood use is shown in Fig. 2.4 for the United States. These data document wood uses since the 1800s in the United States and demonstrate how the use of wood for lumber dramatically increased during the late 1880s and in the 1960s. The use of wood for pulp increased significantly during the 1960s and 1970s. The decrease in the use of wood as fuel wood is very apparent starting in 1950 when fossil fuels began to replace wood as the major source of energy.

Fuel wood

Many of the tropical regions of the world are subsistence extractive economies (i.e. survival with only the bare necessities of life from collecting forest materials or hunting animals that live in the forests) with high poverty rates. As competition increases for the limited supply of fuel wood in many tropical countries, the poorer members of societies are further marginalized as they are excluded from collecting in lands historically available for their survival (Case 2.1 – Nepal, Community Forests and Rural Sustainability). For example, forests in Nepal are being managed as community forests – a management approach where the community establishes rules on what and how much can be extracted from a forest – as a means of ensuring rural sustainability (see Case 2.1). In some cases people are being denied the opportunity to collect fuel wood from community forests set aside for the exclusive use of the community or from forests that have been converted into national parks (Bushley, 2003).

Since FAO has predicted that fuel-wood demand will be twice what is available by 2025, competition for the limited supply of wood will increase. The decreasing availability of fuel wood is resulting in many communities over-harvesting wood from their forests. Over-harvesting not only changes the species and structure of their forests (see Case 2.2 – The Impact of Indigenous People on Oak–Pine Forests of the Central Himalaya) but is also used to portray these communities as being bad for the environment by national governments (see Case 2.3 – Dead-wood Politics: Fuel Wood, Forests and Society in the Machu Picchu Historic Sanctuary). In the 1980s, negative attitudes about the impact of local communities led to initiatives by conservationists to evict them from their forest homes. These views continue today. In Peru, government officials blamed local communities for over-harvesting wood from the Machu Picchu Historic Sanctuary and wanted to decrease their access to wood. Much of this misinformation was based on a lack of data verifying how much wood local communities were collecting. Fuel-wood collection has become politicized and local people are most affected by this as conflicts occur between indigenous communities, national governments and international organizations (Case 2.3 – Dead-wood Politics: Fuel Wood, Forests and Society in the Machu Picchu Historic Sanctuary).

The regions of the world still having the most available forest area (e.g. South America and Russia) are also those regions where a significant portion of the wood consumption is for fuel wood (Table 2.2). Half of the global forests are located in the regions of the world where a large percentage of the wood is used as fuel wood. Burning wood is not only very energy-inefficient but also unhealthy because of the particulates released during wood combustion. This creates a conflict between the people that harvest trees for their market value to sell in global markets and those that survive by collecting wood primarily for cooking and heating their own homes.

Paper products

The global consumption of paper products has been dramatically increasing during the last decade, suggesting that a greater proportion of future wood supply might be used for producing paper products. Currently 16% of the wood supply in Europe and the United States is used for fibre to produce paper products (Table 2.2). This consumption does not appear to be slowing down even with the high use of computers for digital transfer and storage of information (WRI, 1998–1999). In 1990, the global consumption of paper was 10 million tons and in less than 10 years rose to 140 million tons (FAO, 2001). Of the paper consumed in 2000, 36% of the paper was used for communication (e.g. printing, writing, newsprint), 57% for packing materials (e.g. container board, packaging paper, boxboard) and the remaining 7% was used for miscellaneous items (e.g. tissue, miscellaneous hygienic items, health care, etc.) (AFPA, 2000).

Industrial round wood

Today, approximately 25% of the world's forests are managed for wood production and much of this as single-species-dominated forests (i.e. monoculture forestry). These forests are intensively managed for wood production to produce forest products (e.g. plywood, toilet paper, newspaper, etc.). A subset of these intensively managed forests is in plantations. Even though forest plantations comprise only 5% of the total forest area, they do provide 35% of the round wood or non-fibre industrial wood used globally (FRA, 2000).

A vice-president of an international timber company based in the United States suggested that, if intensive forest management such as monoculture plantations were used in only 5% of the world's forested areas, all of our wood needs could be satisfied from this smaller land base. This intensive management of forests from a small area of land has not occurred. Instead, there have been shifts in which regions of the world are the primary suppliers of wood products. For example, the conflicts over preserving old-growth forests and the species dependent on them for their survival has reduced the area of the Pacific North-west United States that can be used to harvest timber. This region used to be an important supplier of wood but not any longer. Now Chile, New Zealand and Finland are the major suppliers of timber because these regions can supply timber at cheaper prices (Perez-Garcia *et al.*, 1999).

What is the Current Capacity for Sustainable Production of Goods and Services from Forests, Given Past Human Uses?

An important question that needs to be discussed when considering whether any particular forested region is suitable for supporting larger human populations is: 'What is the current capacity for sustainable production of goods and services, given past human uses?' (see Cases 2.1–2.3). Past human land uses have altered the ecosystems sufficiently for different ecological and silvicultural principles to need to be used to manage them sustainably. For example, these previous modifications have changed a particular forest's health, its capacity to sequester carbon and the types and degree of disturbances it can tolerate without irreparable damage (Chapters 5 and 6, this volume). These factors are important since they alter the productive capacity of the soils to grow crops or for forests to regenerate. Today, humid tropical forests do not generally have highly productive soils (which are not related to previous human activities) and are unable to support large human population densities because of the lower growth rates of plants (Chapter 3, this volume).

In contrast to the humid tropics, the biomes found in the United States characteristically have a greater area of highly productive soils. In general, more of the land area in the United States is in a category that easily supports human inhabitation and provides renewable resources that can be grown for economic gain. For example, the United States has almost three times the usable agriculture land compared with the global average (Fig. 2.2). The United States also has a higher proportion of its land in rangelands that provide meat for human consumption. Only 15% of the United States land contains desert environments that have a limited productive capacity to support humans and where the extraction of renewable resources is negligible (Fig. 2.1). The high productive capacity of United States forests can be seen in the high carbon sequestration (isolating a compound so that it is no longer available for particular chemical reactions) rates in tree biomass that are recorded in forests of the Pacific North-west United States (Chapter 6, this volume).

Even though ecology and silviculture are important to sustainable forestry, they do not determine its current capacity for sustainable production of goods and services. Ecology and silviculture mainly determine the constraints on how much forest material can be collected at any given site. Both disciplines also inform society in the decisions that society ultimately makes. Therefore, society determines the criteria used to assess whether sustainable production of goods and services is occurring from forests. Society makes these decisions when they decide which products to buy or which services they would like to obtain from a forest. Societal decisions are influenced by many factors, which can be summarized as:

- local, regional, national and international values for how forest materials should be utilized (e.g. cut trees for timber, set up nature reserves for conservation, etc.);
- how much of the forest materials are used for fuel wood and subsistence survival;
- the ability to utilize new technology to convert wood to new forest products more efficiently;
- how many jobs are created and what the economic return is from using wood;
- the economics of global markets, which determines who are the suppliers and producers of forest products.

The importance of conflicting values in determining the capacity of forests to provide goods and services can be seen in the Pacific North-west United States. This region used to be a dominant supplier of timber to global markets; however, it lost its dominance because of social conflicts that reduced the forest area available for harvesting trees, especially from public lands. The social conflicts included questions such as whether habitat should be set aside for endangered species (e.g. spotted owl, marbled murrelet), whether large trees or trees over 200 years old (i.e. old-growth or late-successional forests) should be harvested and whether communities dependent on timber for their livelihoods should be maintained (Vogt *et al.*, 1997). In this region, a majority of the public did not support rural communities harvesting timber if it meant cutting large trees. The result of this decision, dictated by society, is reflected by the 50% decrease since the late 1980s in the timber supplied to global markets from the Pacific North-west United States.

Since the infrastructure needed to process harvested timber has shrunk dramatically in the Pacific North-west United States, this region will have great difficulty in re-establishing its global dominance again. For example, during the last 12 years, 50% of the mills have closed in the Interior Columbia River Basin and 35% have closed in the Pacific North-west region itself (Andreu *et al.*, 2005). The labour costs to harvest trees in this region are also higher than in other parts of the world, which reduces the chances that it will again be a dominant supplier of timber for global markets.

Another repercussion from the changing dominance of the Pacific North-west in the global timber markets is the shift in timber harvesting to other regions of the world less capable of collecting wood sustainably (e.g. the tropics, Siberia). Since Pacific North-west forests are located in highly productive zones, wood supply is shifting to regions where a greater land area will be needed to collect the same amount of wood originally collected in this region. Another repercussion of shifting timber markets is its encouragement of illegal harvesting or other mismanagement of forests in regions less capable of regrowing trees rapidly (e.g. Siberia, wet tropics).

The highest economic return from forests continues to be timber harvesting and a quarter of this timber harvesting is probably illegal (Chapter 7, this volume – 'Another New Challenge to Sustainable Forestry: Illegal Timber Harvesting'). A recent World Wildlife Fund report (2005) indicated that most of China's soaring demand for timber is being provided by countries where illegal logging is rampant (e.g. Russia, Malaysia and Indonesia). Since China is the second largest market for industrial timber after the United States, the quantity of illegal timber being sold is considerable. The amount of illegal timber being collected in some regions of the world will be covered in Chapter 7 in this volume.

Since the early 1980s, many highly publicized conflicts have occurred over forest uses that have resulted in indigenous peoples, timber companies, governmental agencies and international organizations fighting one another (Chapter 3, this volume). These conflicts highlight the difficulty of integrating social, economic, ecological and silvicultural principles to manage forests sustainably. These conflicts have been especially contentious in the tropics, since this region of the world contains some of the poorest people in the world while also supporting 90% of the world's existing biodiversity (World Bank, 1991; Patel-Weynand, 1997). Moreover, the lack of strong socio-political power permits the over-exploitation of forest resources by those with money as well as when individuals and corporations with sufficient capital for larger-scale operations engage in illegal logging.

Forest Supply Capacity Increases by Linking Forests to Energy Production

The Food and Agriculture Organization (FAO) has predicted that future availability of forest resources will be insufficient to satisfy demand due to the growing rural population. FAO projects that in the future demand for fuel wood and charcoal will increase by 1.1% per year and that demand for industrial round wood will increase by 1.7% per year (FAO, 1999). These trends are explainable by overall population growth, particularly in developing countries, and by growth in demand for wood products in the emerging economies of Asia, particularly in China and India. Both China and India have a small fraction of the global forests within their boundaries (4.2% and 1.7% of the total global forest area, respectively) so they will need to acquire wood from other regions of the world.

Unless a solution is pursued that satisfies this wood need and that is environmentally friendly while providing a sustainable livelihood, it will be difficult for communities dependent on forests to become sustainable. This issue is not unique to the less industrialized countries but is also found in rural areas of the more developed regions of the world (e.g. western United States, due to the importance of jobs related to the extraction and processing of forest resources). Approximately half of the global population is dependent on burning forest materials as fuel to heat their homes and cook their food.

Rural communities around the world face similar problems when they are heavily dependent on forest resources (i.e. American Indians in North America, rural communities in Alaska and the western United States, Bolivia, Mexico, Brazil, Peru,

Belize, Indonesia, etc.). They all require forest materials to generate income and to survive. At the same time, they also face the following realities:

- To generate income from forests requires cutting the larger, choice trees to ensure a high economic return in timber markets.
- A globalization of timber markets that regulates who will supply its timber.
- The historical suppliers of wood to global markets are now important consumers of wood but now satisfy less of their own demand.
- Niche markets requiring eco-labelled products.
- A shrinking area of forests from which to collect materials for survival or for markets.
- Tremendous pressure to set aside forest areas for conservation but without any alternative means to provide a livelihood for local communities dependent on these resources.

Today, the use of forest materials as a renewable resource provides new opportunities for forests to contribute to mitigating climate change when they are used to produce energy that does not adversely affect the environment. To be viable, such alternatives must conserve the biodiversity of the forest while enabling rural communities to remain in forest landscapes.

Within the last few years, using biomass (the total quantity or weight of organisms in a given area or volume) to produce biofuels such as methanol, ethanol or biodiesel has become a hot topic of interest (Vogt *et al.*, 2005; Chapter 7, this volume). In the western United States, wood that was once unmarketable can now be converted to alcohols or biofuels and these products can be sold in global markets. Using wood for these purposes will produce environmental benefits. For example, small-diameter wood can be removed to reduce forest fire risk and wood wastes no longer need to be transferred to landfills, where they contribute to early decommissioning from accepting more wastes. These opportunities to produce new products from wood are a result of the need to:

1. Reduce the risk of fire in the forests of the western United States, parts of Asia and Europe;
2. Use renewable resources to produce products that will help to mitigate climate change resulting from the combustion of fossil fuels (Chapter 7, this volume); and
3. Manage invasive species and restore forests to healthy conditions, where products can be produced that will help pay for eradication and restoration efforts.

Small-diameter forest materials typically do not have any market value. However, the recent technological developments in the energy sector now allow us to generate income from converting these discarded materials to useful products (Chapter 7, this volume). This direction of research provides new solutions to forests that move beyond harvesting trees for lumber, albeit most of the biomass that is currently being used to produce biofuels and alcohols is from agricultural crops and not from forests. In many of the tropical regions, agricultural residues need to be retained in the fields to maintain soil fertility, and food-crop biomass should not be used to produce biofuels (although there is continued expansion of ethanol use in Brazil, where most cars run on ethanol-based fuels). But, in those areas where the availability of forest materials is high and where wood is primarily used as fuel (i.e. burning wood for heat or cooking), forests can and should be used to produce biofuels to alleviate the poverty of the local inhabitants. It becomes a serious problem to find the balance between poverty alleviation and forest survival when an emphasis on either one alone will tilt forest management towards being unsustainable.

CASE 2.1. NEPAL, COMMUNITY FORESTS AND RURAL SUSTAINABILITY (BRYAN R. BUSHLEY AND ELIZABETH C. LOUIS)

Introduction

In modern, industrialized countries, most people tend to think of forestry as growing trees in large government-managed or private forests for timber, paper or other wood products. However, in many developing countries, and some developed ones, local communities play a significant role in forest management, growing timber and other valuable forest products for both subsistence and commercial purposes.

In fact, communities have managed forests for centuries and there are examples of community-based forest management throughout the world. According to 'Who Owns the World's Forests?' (WWF, 2004), communities and indigenous groups currently own or manage an estimated 11% of forests worldwide. In developing countries this figure is 22% of the total forest area versus 3% in developed countries. In some countries, the proportion of community-managed forests is much higher. For instance, in Mexico and Papua New Guinea, indigenous and community groups own about 80 and 90% of forests, respectively (WWF, 2004).

Nepal has been a pioneer in promoting community-based forest management. The national government and international assistance programmes began supporting 'community forestry' in the late 1970s. Today there are over 1400 community forest user groups spread throughout the country, involving more than 1 million households nationwide. Initially, forests were handed over to local communities to foster conservation, enhance protection of badly degraded forests and contribute towards poverty alleviation in rural areas.

However, the results of three decades of community forestry in Nepal have shown mixed results. On the one hand, many communities are doing an exceptional job of managing and protecting their forests for both conservation and future use. Some are also trying to produce and sell non-timber forest products – such as medicinal herbs, fruits or artisan goods – in order to earn an income while minimizing the biophysical impacts on their forests. On the other hand, many people (whether they are members of a community forest user group or not) are still harvesting significant proportions of their primary forest products from the poorly protected national forests. Furthermore, some communities are plagued by internal or external corruption, resulting in inequitable benefits among users.

Community Forestry and Sustainability

The main question posed by this case is 'Can communities in Nepal manage forest resources sustainably?' The answer depends on a wide variety of factors influencing the autonomy and physical interventions of local communities with respect to forests. These will be further explored next.

Who are the communities and what are the forests they manage?

Management of community forests is carried out by forest user groups, which generally consist of households living near the forest. An executive committee comprised of community representatives leads these user groups. Newly formed user groups solicit forest land and, with assistance from the government, create a group constitution and operational plan. Upon governmental approval of the operational plan, the community has the right to manage the forest for a specified period, usually 5 years. However, this right comes with some strings attached. Often heavily degraded land is handed over to communities to manage, and tenure is in the form of use rights only. In other words, communities cannot own these forests outright, but maintain the right to use the resources they provide. Furthermore, restrictions have been set on the use and sale of specific resources and there is always the threat that the government will reclaim the land.

Whether communities can manage their forests sustainably depends largely on the types and quantities of forest resources utilized. In Nepal, communities use forests primarily to meet their subsistence needs. The single most important forest resource is fuel wood, used to heat homes and cook meals. Other important forest products include animal fodder, fertilizer for agriculture, timber for housing construction, thatch material for roofs and plants and herbs for food and medicinal purposes. In many cases, community forests have not been able to provide adequately for the community's needs and this has led to the use and degradation of nearby national forests, especially in the subtropical plains.

Who are the players?

There is a broad spectrum of players who affect the management and sustainability of community forests in Nepal. Perhaps the most obvious players are the user groups themselves. These groups vary widely in their number of members, their proximity to and the size of the forest and their ethnic and social composition. Some groups consist of fewer than 30 households and control little more than a hectare of forest land, while others may contain a few thousand households and manage several hundred hectares. Furthermore, there are some user groups made up exclusively of women or members of minority ethnic groups. Non-sanctioned users (i.e. those not belonging to a user group) also affect community forests by continuing to utilize forests that are no longer officially available to them and thereby engendering local conflict over use rights. Some of these users are indigenous groups who were the original inhabitants and have been marginalized or displaced by settlers.

Nepal's Department of Forests plays a role in affecting community forestry at both the national and local levels. Nationally, it formulates policies related to community forests and user groups, which can have a considerable impact on local autonomy. At the district level, it both regulates the activities of user groups and assists them with the development of management plans, skills and practices. Some user groups see the Department of Forests' role as positive and supportive, while others feel it impedes their development.

In the rural areas, there is a complex patchwork of local-level governance bodies, including village development councils, district development committees and other administrative structures and associations. These bodies often have authority (official or unofficial) over the mandate and activities of community forest user groups. Their impact depends on the nature of their relationship with forest user groups. There are also a host of non-governmental organizations (NGOs) involved in supporting community forestry. Many of them work with established community-based organizations and local networks. Some of this assistance is in the form of direct financial or technical support to user groups, while some is focused on networking and advocacy efforts. Most user groups see these interventions as positive.

Timber traders and smugglers also exert considerable influence on community forestry through their connections with user groups, forestry officials and buyers at the local, national and international levels. Their impact is often seen as negative because they may take advantage of communities, promote corruption and contribute to internal discord among user group members. Nepal's Maoist rebels, who have been waging a civil war against the government since 1996, are also hindering the cause of community forestry by pillaging forests and terrorizing communities.

Impact of Ecological, Political and Socio-economic Factors

Historically, Nepal's forested landscapes have been shaped by a range of ecological, economic and social factors. Many of these continue to influence outcomes for forests and the communities who manage them.

Forest ecology, natural disturbance and vulnerability

Nepal has three distinct ecological zones: the high mountains, the middle hills and the subtropical plains. Each zone has its own distinct forest types and different forest resources. Differences in the ecological and geological characteristics of landscapes play an important role in determining their vulnerability to natural disturbances. Natural disturbances can be exacerbated by anthropogenic disturbances. For example, unsustainable land-use practices combined with heavy rains can increase vulnerability to soil erosion and cause flooding and silting on the plains. Until recently, it was widely accepted that anthropogenic forces were the main contributor to this process, but it has now been acknowledged that natural forces and disturbances have played an equal, if not greater, role.

In general, forests in the middle hills and high mountains of Nepal are less productive than forests in the warmer subtropical plains. Their ecosystems are more fragile and susceptible to the impacts of anthropogenic activity and, as a result, have become seriously degraded. The fertile lowland plains or *Terai* region is home to nearly half of Nepal's population and is strongly affected by demand for timber across the border in northern India. Due to these ecological and demographic distinctions, Nepal's forestry policy maintains a slightly different focus in each of the three principal geographical zones: mountains – highly degraded, focus on soil conservation and regeneration; middle hills – conservation and provision for livelihoods of local communities; subtropical lowlands (*Terai*) – industrial timber, and use and protection by local communities.

Government policy, regulations and interventions in Nepal

During the 1970s, intense pressure and concern from the World Bank and other international NGOs over the alarming rate of loss of Nepal's forests led to the formulation of the first examples of community forestry policy. The belief that community-based management would not only stop the degradation of forests but also help towards the alleviation of rural poverty is one that is valid even today. Nepal is one of the world's poorest nations, with about 80% of the population living in rural areas and relying on forests for fuel wood, fodder and timber.

Over the last three decades, forestry policy has shifted towards increased community involvement in the management of forests. In fact, Nepal is considered one of the world leaders in progressive community forestry policies. However, the transition from top-down management of government forests to a more grass-roots approach involving local communities has been rife with conflict among local users and government officials, primarily due to government interventions in the form of taxes on sale of products from community forests and threats to seize forests if they are not managed in accordance with the operational plans drawn up with the help of forest officials. Many forest user groups feel that these plans are excessively restrictive and do not allow enough flexibility for changes in community needs and priorities.

Agriculture and migration

In some regions, migration and clearing of land for agriculture and grazing have posed a grave threat to forests. Over the last few decades, settlers have continued to move from the middle hills region to the *Terai* region where the flat landscape, better infrastructure, warmer climate and rich soils are more conducive to agricultural production.

Local power structures and dynamics

Community forestry has the potential to bring about social change in rural communities that have been dominated by a caste system believed to be unjust towards the lower castes and untouchables. Since policy mandates equal representation in forest user groups, women and lower-caste members have had increased opportunities to participate in decision-making for the management of their forests. In the past, forest user groups were dominated by elites belonging to the higher castes, but there has been increasing representation of marginalized groups within communities. However, involvement by members of marginalized groups in decision-making and leadership positions is still not representative of their proportions in the communities as a whole.

Commercialization – good or bad for forests and communities?

Commercialization is a viable option for communities. Revenue earned from the sale of forest products can be of great value, as it can be invested in community development projects like building schools, roads, medical facilities, etc. Communities living in the ecologically fragile forests of the hills and high mountain regions of Nepal have had to focus more on conservation, with very little economic activity allowed. In contrast, communities in the productive subtropical plains have more opportunities for commercialization.

Many government officials have been opposed to commercialization because they feel community forests should be used only to meet subsistence needs and that commercialization will lead to unsustainable harvesting practices. However, some NGOs and officials see commercialization as a way for communities to enhance their own livelihoods while promoting the long-term sustainability of the forest, especially through the development of ecologically friendly non-timber forest products and ecotourism initiatives. Commercialization of timber is a very sensitive issue. The high value of and demand for timber can provoke forest degradation, corruption and community conflict.

Corruption, smuggling and political instability

Problems in the management of community forests have been compounded by corruption and smuggling. It is widely believed that many user group leaders are involved with corrupt local politicians and/or merchants to illegally sell timber from community forests. Demand from India for valuable hardwoods from the subtropical forests fuel illegal smuggling, which can be very lucrative for those involved. When this happens, only a few people gain at the cost of the rest of the community and community forests suffer from unsustainable timber harvesting.

Forests in countries with unstable governments or those experiencing civil war often suffer from inadequate protection and management practices. In Nepal the Maoist insurgency, combined with an unstable government, has forced many international NGOs to withdraw support of community forestry programmes because of dangerous and unpredictable working conditions. Government officials are afraid to enter or patrol forests for fear of being attacked and this has led to illegal use of many forests, including community forests.

Conclusions

Many important lessons can be drawn from Nepal's experience with community-based forest management. One such lesson is that participation and collaboration at all levels are key to effective management and positive outcomes for local communities. Another very important realization is that there is no one single solution for all communities and that policies and programmes need to be flexible to accommodate the varying needs of different communities. In addition, in order to be truly effective, community forestry should be incorporated into integrated landscape-level management schemes and larger community development efforts that also promote education, health care, population control and alternative income-generating opportunities.

See Bushley (2003) for additional information.

CASE 2.2. THE IMPACT OF INDIGENOUS PEOPLE ON OAK–PINE FORESTS OF THE CENTRAL HIMALAYA (RAJESH THADANI AND SURENDRA PRATAP SINGH)

The hills of Kumaon and Garhwal, the region that now makes up the Indian state of Uttaranchal, were once known for their dense forest cover. British foresters in the mid-19th century reported these forests to be 'boundless' and 'to all appearances inexhaustible' (Guha, 1989). Vast tracts dominated by oak, cedar and pine covered much of this part of the central Himalaya. It was here that the first silvicultural systems for Indian conditions were developed for managing conifers such as chir pine (*Pinus roxburghii*) – one of the earliest examples of modern silviculture in the world.

These forests have seen many changes as a result of human-induced perturbations. This case study attempts to document how indigenous people have, often unintentionally and in the absence of whole tree removal, changed species composition and stand structure in the central Himalayan oak–pine forests.

The Oak–Pine Ecosystem and its Utility to Local People

At elevations of about 1500–2000 m, a zone that has a mild climate and is favourable for human settlement, forests are dominated by indigenous oak and pine species. Banj (*Quercus leucotrichophora*) is the most common oak, while chir (*P. roxburghii*) is the dominant pine. Both are evergreen species with a 1-year leaf lifespan and can realize net primary productivity of close to 20 tonnes/ha annually in favourable situations (Singh and Singh, 1992). In a given watershed, chir pine generally occupies ridge tops, southern aspects and is more common in wide valleys, while banj dominates the base of slopes, northern aspects and narrow valleys.

Banj oak is of great utility to the local people, mostly small cultivators. The subsistence agro-ecosystem is dependent on nutrient inputs in the form of compost manure made from banj oak leaf litter swept from the forest floor. The high nutrient concentration in mature oak leaves and a low level of nutrient retranslocation from senescing leaves (Ralhan and Singh, 1987) make banj leaf litter particularly suitable for compost manure preparation. In the winters, the leaves of this evergreen oak are a critical source of fodder for cattle; while fuel wood from banj is easy to split and of high calorific value.

According to one estimate, each unit of agronomic yield in this region entails the expenditure of nearly ten units of energy from forests in terms of litter, fodder and fuel wood (Singh and Singh, 1987), and much of this is supported by banj oak. The transfer of organic matter, mostly as banj leaf litter, from the forest floor to agricultural terraces has sustained soil fertility, but has led to the depletion of nutrients from forest soils at a regional scale. Banj oak is also of higher utility than chir pine in providing ecosystem services such as aiding in spring recharge, topsoil conservation and maintaining biodiversity. A higher amount of fine root production and mycorrhizal infestation in banj oak compared with pine (Usman *et al.*, 1999) is in

part responsible for this higher level of ecosystem services from banj forests.

Early impacts of forest people

It has been speculated that sparsely scattered settlements in the past caused local degradation through clearing for agricultural fields and cutting for fuel wood, charcoal and smelting of local deposits of ores. Recent studies (Thadani, 1999) indicate that shifting settlements and agricultural abandonment might have significantly influenced the domination of oak. Early settlements occurred through the clearing of native forests to establish terraced agricultural fields. Banj acorns, which tend to disperse more widely than those of other oaks, often germinate in protected microsites – in rodent burrows along walls of terrace risers, or trapped in vegetation growing on riser walls. Surrounding oak forests or adult trees left undamaged for shade, aesthetic appeal or religious reasons provided a plentiful supply of these propagules. Young seedlings germinating on these terraced risers remained unmolested by farmers, who recognized the benefits of banj saplings in stabilizing terrace walls. Periodic lopping for fodder prevented these saplings from growing large and interfering with agricultural crops. While lopping checks the above-ground growth, saplings gradually develop extensive root systems, which allows them to grow rapidly on release.

Banj oak was thus able to dominate forests arising on agricultural fields, easily outcompeting early successionals that had to grow from seed. Germinating seedlings of other species, in particular pine – which is thought to have a deleterious effect on crops – were normally destroyed by uprooting or perished due to their lower ability to sprout. A study of cultivated and abandoned terraces supports this hypothesis, as the occurrence of banj saplings was unusually high (Thadani, 1999; Quazi *et al.*, 2003).

Historically, agricultural abandonment has occurred due to reasons as wide-ranging as political insecurity, epidemics, large landslips or loss of soil fertility due to erosion. An example of intense agricultural abandonment may have been around 1790, when the Gorkha invasion of Kumaon caused intense political instability, resulting in outmigration. Given the longevity of banj trees, even a fairly low rate of abandonment would ensure the development of areas dominated by almost pure stands of oak across the landscape.

Social movements and peasant protest during the 20th century

The British rule in the central Himalaya started with the annexation of Kumaon in 1815. In the initial years, there was little interference with traditional forest use mechanisms, but the Indian Forest Act of 1878 abrogated many of the rights of the local peasant. In the years after this, and in particular after 1900, the British exerted an increasing amount of control over forests. The hill people resented this control, claiming traditional rights to managing forests for fuel wood, green fodder and various non-timber forest products (NTFPs). Incendiarism became a form of protest and the forest fires of 1916 and 1921 were particularly damaging to the chir pine forests of the region. As a result, in 1921 a committee was set up to look into the grievances of the local populace (the Kumaon Forests Grievances Committee), which recommended reclassification of forests and transfer of control of less economically valuable forests from the Forest Department to local villagers. This led to the eventual formation of *van panchayats*, or village forest governing bodies, in 1930.

The existence of these *van panchayats* as institutions of community forest management attest to the importance of forests for the local people and the extent to which the local populace was willing to battle the colonial administration when their access to forests was threatened. A special, legally recognized executive committee controls many aspects of use and management of village forests. The *van panchayat* system is among the oldest government-recognized institutions of community resource management in the world. About 10,000 such committees exist to date in Uttaranchal and they have had a tremendous impact on forest management.

The will to exert authority over the forest by local people was once again evident in the 1970s. Many of the forest management practices put in place by the British were continued by Indian foresters up to the 1970s, despite the presence of *van panchayats*. These policies paid inadequate attention to the needs of the local people. This led to resistance by the peasants, ultimately culminating in the famous Chipko movement. Chipko was a social movement where local peasants opposed the exploitation of their forests by symbolically hugging their trees. Chipko ultimately succeeded in getting the Indian government to ban green felling in the Himalayan hills in the early 1980s. This was to eventually help influence national forest policy and introduce social forestry, where 'direct economic benefit' is 'subordinated to the principal aim of environmental stability and ecological balance' (National Forest Policy, 1988).

Current Scenario

The preference of broadleaved species, in particular banj oak, over conifers such as chir is clear among local people. This is despite the higher commercial value of the conifers. The current paradigm of social forestry gives low priority to commercial benefits and prioritizes forests for local people. Commercial forestry has been banned in the hills as a result of the Chipko movement. Local institutions have significant control over their forests. The stated objective is to increase the occurrence of oak and prevent the spread of pine. Reafforestation efforts are being undertaken at a scale never seen before, and the focus is almost entirely on planting hardwoods such as oak. Recent satellite data show that the decline in extent of Himalayan forests is almost negligible (Forest Survey of India, 1999).

However, these facts mask a disturbing fact – the change in quality and species composition of the Himalayan forests. Despite programmes to increase the spread of oak, there is a continual decline of oak forests and a spread of pine. Current forest policy may inadvertently have facilitated this process.

The ban on green felling in the hills of Uttaranchal resulted in a cessation of removal of mature pines, whereas the lopping of oak trees and removal of leaf litter continued as before. The removal of leaf litter is encouraged for compost preparation even though this creates microsites not conducive for acorn germination (Thadani, 1999). Lopping of green leaves for fodder is also considered more acceptable than occasional whole tree removal even though this destroys the productive potential of trees and reduces acorn production. These chronic disturbance regimes increase light availability and create bare-soil microsites that encourage pine regeneration. Vast tracts of forests with denuded oak trees have become a common feature across the landscape.

Humans change forest composition, not by logging with chainsaws or heavy machines, but by lopping with sickles and through the grazing of their cattle. Oak forests get whittled away, twig by twig and leaf by leaf. Human intervention is not at the stand level, but the impacts spread from activities focused on individual branches of trees. Branches that are easier to climb or have more leaves are the ones that get cut first. The result of this management is the reverse of what is desired. Banj oak, a species of great socio-economic value, is being replaced by chir pine – a tree that, despite its good timber value, is of little utility in supporting the agriculture-based livelihood system. Also, pines are believed to decrease productivity of surrounding fields and horticultural plantations and decrease spring discharge, thus making them even less desirable.

However, pine seedlings are rarely grazed by cattle, the needles are not used as fodder and the wood has poor burning properties, thereby ensuring that pine seedlings and saplings are much less frequently damaged than oak. When damaged by ground fires or occasional browsing, seedlings of chir pine have the ability to produce new shoots bearing primary needles. Pine has benefited from its ability to colonize bare sites and to tolerate stress and fire and even resprouts when damaged in its seedling stage (Singh *et al.*, 1984; Singh and Singh, 1992).

The Himalayan scenario is thus characterized by a unique association where the more intolerant (early-successional) pines regenerate in stands dominated by the more tolerant oaks even in the absence of catastrophic disturbance.

Conclusions: the Human Domination of Himalayan Landscapes

The change in species composition in the mid-elevational forests of the Himalaya has been more influenced by human use at a subsistence level than by planned silvicultural practices. Abandonment of settlements in the past may have resulted in the growth of almost pure stands of oak, despite the natural propensity of pine to colonize disturbed sites. More recently, despite the desire of local people to encourage oak, the excessive use of this species has resulted in its decline and created conditions that favour the domination of pine – a species of commercial value but little utility as a fodder or fertilizer, and one with low ecosystem values.

Reinstating scientific management that allows for controlled harvesting of pine may be the best solution if greater control can be given to the local people. Managed harvesting of pine could provide livelihoods and income for local communities. A reduced dependence on subsistence agriculture and an increased focus on community-managed forestry could also reduce the pressure on banj oak, allowing for its better conservation. However, establishing sustainable management practices and instituting regulation to prevent over-exploitation will be a challenge.

CASE 2.3. DEAD-WOOD POLITICS: FUEL WOOD, FORESTS AND SOCIETY IN THE MACHU PICCHU HISTORIC SANCTUARY (KEELY B. MAXWELL)

Introduction

In the Andes, a place renowned for pre-Hispanic ruins, vast *punas* and *páramos*, and hillsides dotted with agricultural terraces, there seems little room for forests, especially when visions of Latin American forests are dominated by the tropical moist forest ecosystems downslope towards the Amazon. Yet not only are forest ecosystems present in the Andes, but they contain significant plant diversity, wildlife habitat, water resources and forest products. Despite this importance, there has been little analysis of Andean forest ecology, management strategies and human influence on forest structure and composition.

Anthropological studies in the Andes have likewise for the most part ignored the role of forests in local cultures and economies, even though resources such as fuel wood are critical to rural households. This lack of investigation is surprising, given the increasing national and international concern about forest cover loss in the Andes and the need to ensure sustainable resource flows to impoverished households. When such concern translates into natural resource policy and management interventions, however, success is often hard to come by. Limited knowledge and appreciation of the multiple ways in which Andean forests and society are connected as well as the underlying values that drive conservation efforts impede sustainable resource management.

Using a case study of the Machu Picchu Historic Sanctuary (MPHS) in Peru, this case examines whether protected-area policies on fuel-wood collection are socially and ecologically sustainable. It analyses how underlying narratives about society and forests influence Sanctuary fuel-wood policy, how collection techniques shape forest ecosystems and how fuel wood is embedded within larger social systems of property rights, household organization and social contracts.

Machu Picchu is a particularly relevant case study because of the multiple social actors and narratives that inform resource management. It also sheds light on larger debates. First, ecologists and resource managers disagree over the role of humans in shaping Andean forests and grasslands, often dichotomizing anthropogenic grasslands from natural forests without a complete understanding of the complex disturbance regimes that shape these ecosystems. Secondly, international concern over insufficient fuel-wood supply and whether fuel-wood collection causes deforestation has persisted for 30 years without resolution. Policy suggestions often focus on trying to reduce the kilograms per capita of fuel wood used through improved stoves or plantations. This discourse of energy efficiency neglects the larger social context of fuel-wood management, an understanding necessary for more sustainable management.

Site and Social Actors

The 32,592 ha Machu Picchu Historic Sanctuary was established by the Peruvian government in 1981, and in 1983 was designated a World Heritage Site for natural and cultural heritage. While most people know of Machu Picchu as the archaeological site, the Sanctuary is also host to 40-plus archaeological sites, ecosystems that span from 1800 to 6200 m in altitude, and over 2200 rural and urban residents. More than 360,000 tourists visit Machu Picchu each year, including over 100,000 Inca Trail hikers. A plethora of institutions are involved in its management, including Peruvian state agencies, international organizations, such as the United Nations Educational, Scientific and Cultural Organization (UNESCO), and private tourism operators. These institutions have difficulty defining and working towards common conservation and social sustainability goals, as priorities are often skewed towards increasing Machu Picchu's economic return as a tourist destination.

The ecosystems under discussion for this study lie between 3000 and 3800 m in altitude, and contain at least 48 tree, 143 shrub and 253 herb species. Predominant tree species include unca (*Myrcianthes oreophyla*), chachakomo (*Escallonia resinosa*), chikllur (*Vallea stipularis*), chalanki (*Myrsine latifolia*) and alisso (*Alnus articulata*). Several woody species are multi-stemmed or grow in either shrub or tree form. Forest patches present a variety of stand development and successional stages. High-altitude *punas* occur above 3700 m, with bunch grasses, low-lying herbaceous mats and scattered shrubs and trees. Grasslands, scrub and savannah also dot the landscape at lower altitudes.

These ecosystems lie within one village in the Historic Sanctuary. Approximately 60 households live in this village, with one to ten children per household. Residents are subsistence agriculturalists, growing maize and potatoes and grazing a handful of cattle and pack animals. They burn fuel wood on open mud and stone stoves to cook and heat their kitchens, the room where people gather, chickens roost and guinea pigs nest. Residents earn cash income by selling treats or carrying packs for tourists on the Inca Trail. Until the 1970s, village territory was under the formal ownership of large estates called haciendas, which conducted large-scale cattle grazing, eucalyptus plantations and tree felling for the railway line to the ruins.

Protected-area Fuel-wood Policy

Three different state agencies manage the Sanctuary: one responsible for natural resources, one for cultural resources and the third to direct over-arching policies. The mission of the Historic Sanctuary is to conserve natural ecosystems, archaeological sites and Inca land-use and historic landscapes. The natural-resources agency prohibits rural residents from felling live trees inside the Sanctuary for fuel wood or construction. Legally, then, residents can only collect already dead woody biomass for fuel wood. Locals complain stridently about this policy, and claim that they must travel far to find appropriate wood.

This policy is derived in part from underlying narratives concerning Sanctuary residents, history and ecosystems. Sanctuary managers state that rural residents are recent immigrants now expanding their agricultural territory and deforesting the protected area, with fuel-wood collection being an integral cause of deforestation. However, the role of historical factors in reducing forest cover such as hacienda-directed tree felling is not well taken into account. Nor are local claims to historical residency in the village. This policy also reflects an idealization of forests as free from anthropogenic influence, in contrast to human-created grasslands thought to be shaped by tree felling, livestock grazing and fire. Sanctuary policies attempt to reduce any human impact on forests under the assumption that forests are the natural ecosystems the Sanctuary is supposed to be protecting and that any human activity leads to deforestation and grasslands. This policy equates biodiversity conservation with live tree protection, and does not accord dead wood the same value or recognize ecosystem interconnections.

The historic landscape Sanctuary managers also attempt to protect is an idealized one, with the Inca portrayed as innate conservationists, in contrast to current deforesters. Yet the Inca directed extensive cultivation in the present-day Sanctuary, so current residents are not necessarily expanding agricultural frontiers. Historic residents must also have depended on natural resources such as fuel wood, a land use ignored by Sanctuary managers as a necessary component of Inca landscapes.

How fuel-wood collection shapes forests

While Sanctuary managers associate any human influence on forests, particularly woodcutting, with deforestation, field evidence does not support this assertion. First, local residents collect fuel wood from a variety of ecosystems, not forests alone.

Residents obtain woody biomass from forests, scrub, grasslands and field edges and along irrigation canals. Fuel materials include trees, shrubs, dung, crop residues and construction waste. Secondly, collection techniques do not necessarily involve chopping down entire trees. One technique used is pollarding, or repeatedly cutting branches off a central trunk, leaving behind highly sculpted trees. A modification of pollarding is cutting off one trunk of a multi-stemmed tree or shrub. Following Sanctuary policies, residents may collect standing or fallen dead wood. Some residents prefer collecting dead wood because it can be burned without extensive air-drying. Fuel-wood collection thus impacts tree architecture, ecosystem structure and nutrient cycling, but does not necessarily reduce forest cover or deplete tree populations.

An average household uses 1.8 kg of fuel wood per capita per day. This wood is not burned indiscriminately, but rather is used with stunning precision in cooking meals. Residents articulate specific ideals for fuel wood, citing factors such as aroma, smokiness and piece size. Favoured species include unca, chachakomo and eucalyptus, but preferences differ according to personal taste and food cooked. The top species actually used include not only preferred species but also p'ispita (*Acalypha aronoides*), rayeta (*Azorella* spp.) and q'amasto (*Nicotenia tomentosa*), smaller shrubs that may not burn as well. However, as these shrubs are readily available in field edges and grasslands, their ease of collection may outweigh their less desirable burn characteristics.

Property Rights to Fuel Wood

Rural residents collect fuel wood within village lands, as resource use rights are accorded with village residence. Within a village, ecosystems are communal in that individual households do not hold exclusive rights to specific wood-collecting territories. Families do tend to collect wood from places with which they are familiar from livestock grazing or potato farming in the vicinity. Although trees in the forest do not have individual owners, the labour put into felling a tree confers rights to dead wood. Cut wood left in the forest to dry may be claimed later by the woodcutter alone (one means of bypassing Sanctuary policies as residents then bring home only dead wood).

Planted trees are owned by whoever planted the tree. Rural residents of the Sanctuary plant primarily exotic trees such as eucalyptus, capulí, pine, avocado and peach around their houses or fields, with a few village-owned eucalyptus groves. Some owners sell eucalyptus for fuel wood, particularly to residents of villages in the *puna*, where few trees grow. There is little local knowledge or motivation to plant native trees that 'grow on their own', a situation that may limit plans to plant native trees for fuel wood. Rural residents, former hacienda owners and state agencies wage ongoing legal battles over the ownership of eucalyptus groves. This uncertainty over property rights has created an inability to actually manage the trees.

Household labour, social contracts and fuel wood

Household labour availability determines how much, when and where households collect fuel wood. Male household heads commonly make one to two trips a month, often with a pack animal, to a site that may be 1–2 h away. They take home 20–50 kg of wood, particularly larger pieces, from forests or scrub. Female household heads rarely have time to make such long trips because of cooking, vending and childcare duties. Instead, they collect smaller quantities and pieces while doing other daily chores such as selling soda or watching livestock, collecting from forests, scrub, grasslands and field edges. If male household heads work outside the village, they may not have time for long collection trips, leaving their families to use undesirable species and sizes, depend on child labour or purchase wood.

In Machu Picchu, fuel wood comprises important social contracts at the household and village level. Anyone who eats at the same hearth is expected to contribute to fuel-wood needs. As such, everyone from small children to elderly relatives collects wood, even though they may bring only a few small twigs. Children as young as 4 are sent to collect wood with their older siblings. As children get older, they collect progressively larger pieces and loads, learning how to tie them into bundles to carry on their backs. At a young age, both male and female children are sent. By young adulthood, more male offspring go, as female offspring have other household duties. Adult offspring also bring wood to their elderly

parents who live alone. At a village scale, getting volunteers to contribute fuel wood to village festivals is part of the responsibility of the festival host. Schoolchildren may be required to bring wood to cook school lunch. Fuel wood thus fulfils social roles beyond simply an energy source, and is part of the myriad social networks necessary to survival and community in the rural Andes.

Implications for Ecosystem Theory and Fuel-wood Policy

This case study demonstrates the many links among fuel wood, forests and society in Machu Picchu. It is not simply the case that the demand for fuel wood leads to deforestation. Collection techniques depend on household labour availability. Male and female household heads and children each collect different species and sizes from diverse source ecosystems. Pollarding live and dead wood does impact ecosystem structure and nutrient cycling, but it is erroneous to characterize fuel-wood collection as a primary cause of reduced forest cover in the MPHS. Historic tree harvest for railway ties and fuel probably had a greater long-term impact on forest cover, but is obscured in contemporary discourse that current rural residents are the principal cause of deforestation in Machu Picchu.

Fuel wood contributes more to rural Andean society than joules of energy; it is embedded within household and village social networks. Pigeonholing fuel wood as per capita energy efficiency results in proposals regarding perceived wood deficiencies that try to increase supply through tree plantations or reduce demand through improved stoves. Such interventions would be difficult to implement in Machu Picchu because of the larger social context. Andean residents do not utilize fuel wood so as to maximize kilos per capita, but select species and sizes based on the food being cooked. More energy-efficient, closed stoves would not be suitable for heating kitchens or for meals such as roasted guinea pigs. Tree plantations would have to take into account the property rights of forests and planted trees, as well as a lack of local habit of planting native trees. Moreover, such interventions ignore the role of fuel wood not only as an energy source but also in social contracts at the household and village level.

Fuel-wood policy of the MPHS is neither socially nor ecologically sustainable. It is socially unsustainable because it does not take local wood needs into account. Residents resent Sanctuary managers, who simply restrict wood use without offering alternatives. It is ecologically unsustainable because it does not consider the potential impact of intensive dead-wood collection on forest ecosystem nutrient cycles. Valuing live trees over dead wood ignores the ecological importance of coarse woody debris. These policies are based more on ideals about Sanctuary residents, history and ecosystems than on actual resource use patterns and ecology. Sanctuary fuel-wood policies attempt to create idealized ecosystems – natural forests untouched by humans and historic landscapes of an Inca who left no ecological footprint. However, Andean ecosystems are shaped by multiple anthropogenic and non-anthropogenic disturbances and don't neatly partition into natural forests and anthropogenic grasslands. Both past and present residents of Machu Picchu depend on forest and grassland resources. To work towards socially and ecologically sustainable fuel-wood management in Machu Picchu, Sanctuary managers need to recognize how these narratives dominate resource use policies and shift their goals from live trees in forests to ecosystems and livelihoods.

CASE 2.4. ICELANDERS AND THEIR FOREST HISTORY: A THOUSAND-YEAR-OLD HUMAN AND NATURE HISTORY CONTROLLING RESILIENCE AND SPECIES COMPOSITION IN FORESTS TODAY (BROOKE PARRY HECHT AND PAUL HELTNE)

Introduction to Human Transformation of Iceland

Many histories can be told about a forest. Some can be told through written histories or oral traditions of peoples who have lived there. Some can be told by the evolutionary past or life history of a particular species. There are also histories that can be imprinted into the forest ecosystem itself. A forest's ecological patterns and processes are more than a statement or an account of the current interactions among the different ecosystem components, including the humans who live there. These patterns and processes also recount a history of interrelationships of the living communities and non-living features that have contributed to the development, characteristics and dynamics of the current forest. The ecosystem's history is therefore as determinative of its future possibilities as the forest's genetic material. An ecosystem can tell a story that has its own emergent features, different from, but inclusive of, the genes, organisms, populations and written human accounts of that place.

In this case study, we will share one such ecosystem history from the Icelandic forests (Hecht, 2003). People often remark on the beauty of Iceland; it is a haunting and dramatic landscape. What most people do not realize is that the landscape looks that way largely because of its human and nature history. We know this for a number of reasons, one of which is that the Vikings were prolific and skilled writers who kept detailed records of the changes in their landscape. For example, in *The Book of the Icelanders* (1125 CE), Ari Thorgillsson wrote, 'At that time [first settlement] Iceland was covered with forests between mountains and seashore' (in McGovern *et al.*, 1988). Pollen dating, soil analyses and other written accounts confirm that, at the time of the Viking settlement, Iceland was considerably more vegetated than it is now, with birch woodlands, willows, heather, lichen and moss covering up to 75% of the landscape (McGovern *et al.*, 1988; Finnbogadóttir, 1992; Ogilvie and Barlow, 2000). Archaeological findings reveal birch charcoal in what are today barren, treeless landscapes. Currently, vegetative cover is less than 25%, with patches of mountain birch (*Betula pubescens* Ehrh.) woodlands amounting to only 1% of the landscape (Arnalds, 1987).

Landscape transformation: the arrival of people and grazing animals in Iceland

One of the most important elements of the human–nature history in Iceland was the Viking introduction of grazing animals (e.g. horses, sheep, cattle, goats and pigs) into the landscape at the time of their settlement in Iceland in the early 870s CE. The early settlers believed the Icelandic landscape would be hospitable to the same domesticated animals that flourished in Norway (Diamond, 2005). However, the introduction of these grazers dramatically changed Icelandic birch forest ecosystems. Amorosi *et al.* (1997) has noted,

> While we should not imagine that the terrestrial ecosystems of Iceland or Greenland were static before the Norse arrived, there can be no question that their arrival caused some of the greatest disturbances to flora and soils since the end of the Pleistocene.

The mountain birch that grows in this cold climate is a slow-growing species, no match for the high densities of foraging sheep and other animals the Vikings brought to graze in the Icelandic forests. To simplify the changes that took place, grazers ate birch seedlings and herbaceous species in the forest understorey. Therefore, as trees were harvested by the Viking settlers or as older trees died,

young birch were eaten by grazers and did not survive to replace the mature trees of the forest. In forests that survived, the introduced grazers shifted understorey species composition, reflected by the decreased presence of the grazers' favourite species, most notably the woolly-leaved willow (B.P. Hecht, personal observation). In some heavily grazed landscapes, little to no vegetation persisted at all, leading to widespread erosion (Fig. 2.5). However, as Hecht (2003) showed, even where vegetation does exist, the ecological structures and functions of the birch forests are dramatically altered.

The transformation of the Icelandic landscape, and especially the native birch forests, will be explored by comparing the impact of different levels of grazing on three forest sites today. Let us go up to the treeline in Iceland, where the vegetative communities are already stressed (from climate and other factors), and compare these three very similar sites (Fig. 2.6). At the forest-limit zone, one might note that the birch forest looks very different from forests at lower altitudes, as forests close to the treeline are at their threshold of existence. Therefore, trees at the forest limit have a number

Fig. 2.5. Two ewes with their lambs, resting in an erosion front near Lake Myvatn in northern Iceland. Photo by Brooke Hecht.

SITE 1
⇨ **Protected for 100 years**
⇨ **Eastern Iceland**

SITE 2
⇨ **Grazed**
⇨ **Eastern Iceland**

SITE 3
⇨ **Grazed**
⇨ **Northern Iceland**

Fig. 2.6. The three selected study sites at the mountain birch forest limit in Iceland. Photo by Brooke Hecht.

of different characteristics, including smaller stature and slower growth than trees at lower altitudes and latitudes.

Recording the reality on the ground: using a retrospective analysis

The three forest-limit study sites are similar from a number of perspectives – their soils are remarkably similar, there is one tree species (*B. pubescens*) in the forest overstorey and, while the diversity of plants in the understorey is not identical across sites, many of the same understorey species are present at all three sites. One important difference among these three sites is their grazing history: Site 1 has been protected from sheep grazing for the last 100 years, and Sites 2 and 3 have been grazed since the Vikings introduced sheep and other grazing animals. However, while there is grazing at Sites 2 and 3, the levels of grazing are less intense than they have been in the past. In the recent past, since the mid-1970s, sheep numbers have dropped across Iceland. In 1999, there were 1,073,542 sheep in Iceland (including lambs), down from 2,007,314 sheep in 1978 (Farmers Association, 2000). The other important difference among the three sites is that Site 3 is further north than the other two sites, so it has a shorter growing season and a harsher climate. Thus, these three sites are on a spectrum of stress, which increases from Site 1 to Site 3.

Viewing the structure of these three ecosystems or perhaps by counting the number of species at each site (one standard measure of biodiversity), one might say that these ecosystems are all similar

and healthy. However, by measuring one critical factor, namely the nitrogen content of the leaves, one can see that the ecological patterns and processes of these three ecosystems are quite different. High levels of foliar nitrogen are so important in these ecosystems because it helps maintain photosynthesis at low temperatures. In fact, all over the world, it has been observed that at higher latitudes and elevations, plants of any particular species have higher levels of foliar nitrogen than plants of the same species at lower latitudes and elevations (Köerner, 1989). Consequently, availability of nitrogen in the soil is critical for survival of mountain birch at the forest limit.

Given that the forest-limit elevation is lowest at Site 1 and highest at Site 3, one would expect that foliar nitrogen should increase from Site 1 to Site 3. In fact, the opposite pattern was observed: a decreasing trend of foliar nitrogen from Site 1 to Site 3. Furthermore, Site 3 had a significantly lower nitrogen level than both Sites 1 and 2 (Fig. 2.7).

After imposing an additional experimental disturbance (additions of sugar), the foliar nitrogen levels significantly decreased at both grazed sites, Sites 2 and 3 (Fig. 2.8). Because sugar is 'free' energy for the microbes in the soil, an active microbial population will gain a boost in activity from sugar addition. When microbial metabolic activity increases, microbes take more essential plant nutrients, like nitrogen, from the soil for their growth. If this occurs, less nitrogen is available for the birch to take up for its own growth. Figure 2.8 shows decreases in foliar nitrogen at Sites 2 and 3 following the sugar disturbance, indicating a boost in microbial activity at these sites. In contrast, the lack of significant change in foliar nitrogen levels at Site 1 (the protected site) suggests a less active microbial community. The forest ecosystem at Site 1, and thus the birch, was therefore resistant

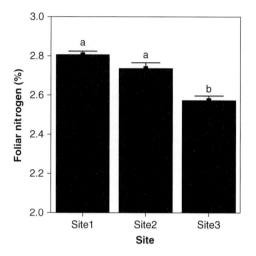

Fig. 2.7. Per cent nitrogen in the birch foliage at the control plots at the forest limit of Sites 1, 2 and 3. Vertical bars represent mean %N + 1.0 SE at the forest limit elevation. Different letters indicate significant differences across sites (*P* ≤ 0.05).

to the sugar disturbance. Furthermore, following the experimental disturbance, nitrogen levels dropped from Site 1 to Site 3 in a stepwise pattern. This nitrogen pattern reveals a decrease in resistance to the imposed disturbance from Site 1 (greatest resistance) to Site 3 (lowest resistance). The pattern of shifting resistance parallels the levels of grazing and climatic stress at the sites, from Site 1 (lowest stress levels) to Site 3 (highest cumulative stress levels).

Sheep and loss of moss layer: suspected link to changing microbial activity

As evidenced by foliar nitrogen levels both before and after the imposed sugar disturbance, sheep grazing is changing how the birch forests of Iceland are able to acquire nitrogen. Overgrazing by sheep is somehow impacting the ability of these forests to obtain sufficient nitrogen to survive and grow. The question is: How? The most obvious impact of intense sheep grazing on the birch forests has already been described – if sheep eat all young birch, there will be no trees to replace the trees as they die or are harvested. However, seedling consumption cannot explain why birch in grazed

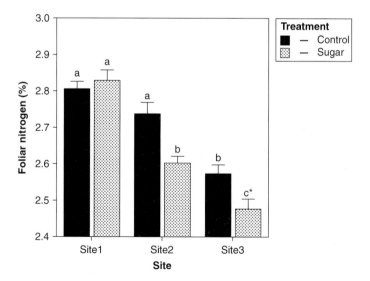

Fig. 2.8. Per cent nitrogen in the birch foliage at the control and sugar plots at the forest limit of Sites 1, 2 and 3. Vertical bars represent mean %N + 1.0 SE at the forest limit elevation. Different letters indicate significant differences across and within both site and treatment, where *P* ≤ 0.05 unless otherwise indicated with an asterisk. *P* = 0.057 for the within-site comparison between sugar and control plots at Site 3.

forests has a decreased potential to acquire nutrients needed for growth and survival. As Hecht (2003) described, sheep alter the moss layers, the soil environment and the soil microbial communities that are involved in the breakdown of organic material. Ultimately, these changes in the soil portion of the ecosystem are the likely reasons that birch in grazed landscapes have greater difficulty acquiring nutrients and thus a decreased resistance to disturbance.

Specifically, sheep impact the soil environment by causing a decrease in the abundance of moss. Moss is important in these boreal forests for maintaining cool soil temperatures. With the decrease or loss of the moss layer at Sites 2 and 3, soil temperatures became higher. There is a strong positive correlation between soil temperature and microbial activity. Microbes become more competitive with plants for essential nutrients like nitrogen when temperatures are higher. Thus, at the grazed sites (2 and 3), where the moss layers were thin or absent, the microbial communities were highly competitive with the birch for nitrogen. Furthermore, following the experimental addition of sugar and intensification of microbial competition, birch at Sites 2 and 3 was even less able to acquire the nitrogen it needed to grow. The diversity of the microbial communities in boreal climatic regions is extremely high, and sheep grazing appears to favour the ability of microbes to become very competitive for limited nutrients as a result of shifting vegetation and soil dynamics.

Importantly, the sheep have changed critical ecosystem dynamics and energy flow among the moss, birch, soil microbes and the sheep themselves. This means that, if sheep were to be removed from a heavily grazed forest, their effects on ecosystem function would not be immediately eliminated but would persist as a legacy of the ecosystem's history.

The most important take-home messages from these data are:

1.   Introduction of grazers decreased the baseline levels of nitrogen available to birch in the grazed ecosystem at Site 3 (Fig. 2.7). This means that the birch forest at Site 3 was closer to its survival threshold.

2.   A history of grazing and climatic stress can change an ecosystem's ability to respond to disturbance, e.g. intensive grazing at Sites 2 and 3 decreased the potential for the ecosystem to resist

additionally imposed disturbances, such as the sugar disturbance (Fig. 2.8).

Conclusions

What kind of forest history is this? Evolutionary history can explain the potential for phenotypic plasticity in foliar nitrogen levels in the birch. However, specific phenotypic expressions correlated with different environments are not necessarily indicative of underlying evolutionary change in the birch; rather the phenotypes here are telling a larger historical narrative about the particular stresses and the abiotic and biotic relationships the ecosystem has experienced. The shifts in nitrogen levels in the birch foliage are stating the cumulative 'memories' of the different stressors and interactions in these ecosystems.

Thus, the history told by shifting nitrogen levels at these different sites is an emergent reflection of a complex series of interactions among microbial populations in the soil, the sheep populations, the climate, the birch populations and the human activities, among other ecosystem components. In other words, changes in many parts of the ecosystem, as well as interactions among different ecosystem parts, can be inferred from the levels of nitrogen in the birch. In particular, the nitrogen levels give insight into how both past and present human land uses have altered fundamental ecological processes.

These shifts in foliar nitrogen levels may be a signal of a swing in evolutionary processes controlling the birch, namely selection intensity. As discussed by Mayr (2004), different evolutionary outcomes follow from two kinds of selection, namely elimination of only the very worst adapted individuals versus survival of only the very fittest individuals or populations. The shift towards a more stringent selection may lead to threshold shifts on the ecosystem scale, such that the current ecosystem collapses or changes to a different ecosystem state.

Clearly, the human activity of grazing sheep in the Icelandic landscape creates more intense selection pressure for the birch. It is evident that such selection pressure on the birch has occurred since the Viking settlement. The distribution of birch is now largely constrained to narrow strips on the mountainsides, as opposed to the more widely distributed birch 'between mountains and sea-shore' recorded by the early settlers and confirmed by archaeological work. The foliar nitrogen patterns revealed by the Hecht (2003) study may be a harbinger of a threshold in birch survival yet to come, offering insight into whether a birch ecosystem might continue to persist at a particular site.

While it may be very clear that the past human–nature relationship at the grazed sites decreased the resistance of the birch to further disturbance, it is important to point out that the resistance of the ecosystem as a whole has also been decreased. Without the birch, there will be no forest, as *B. pubescens* is the only forest-forming species in Iceland. If there are no trees, the ecosystem type has fundamentally changed; there is no forest at all and, in some cases, even the potential for a vegetated ecosystem has also been lost. This is because, in the absence of birch and associated understorey vegetation, soil erosion will occur. Importantly, human activity through the introduction of grazers has been the driving cause behind these shifts in selection pressure and ecosystem structure and function.

While many may count the importance of ecological work in terms of its ability to predict specific ecosystem states, we can understand a great deal about a forest ecosystem through understanding its potential for and timing of threshold shifts, without precise knowledge of which ecosystem state will follow such a shift. As this case study shows, one key element critical to our ability to anticipate threshold shifts is an understanding of heterogeneity in ecological patterns and processes across the landscape. For example, if one were to average the nitrogen values across the three forest sites (in order to report typical foliar nitrogen values for the forests in this region of Iceland), the story that the birch is telling about historical human uses of these different forest sites would be lost.

By not creating such an average, we discover a history that tells us that much of 'nature' in these forests is now an artefact of its human–nature history. Human activity in Icelandic forests not only has significantly impacted the distribution of forests today but has also altered ecological patterns and processes important to the future survival of the birch ecosystems. However, as grazing pressure in Iceland has been reduced, there has been a slow re-expansion of the birch forests (Aradóttir and Eysteinsson, 2005). Thus, if we humans are able to understand both the historical record and the likely future trajectories of complex ecosystem stories, we will be better equipped to make choices about our future relationship with nature before significant ecological thresholds are crossed, both in Iceland and elsewhere.

3 Human Dimensions of Forests: Democratization and Globalization of Forest Uses

KRISTIINA A. VOGT, TORAL PATEL-WEYNAND,
JON M. HONEA, ROBERT I. GARA, DANIEL J. VOGT,
RAGNHILDUR SIGURDARDOTTIR, ANNA FANZERES AND
MICHAEL G. ANDREU

Introduction

Today, greater ranges of people are participating in the decisions that influence how forests are used. This democratization of forest resource management is occurring from local to international levels. At local levels, more and more communities have won or been given the right to participate in the management of forests on which they have historically depended. At much larger international levels, the combination of forest resource limitation and the border-transcending importance of factors such as climate change, pollution and biodiversity conservation have led many communities and nations to influence the use of far-distant forests. This globalization of forest management is occurring purposefully as communities directly attempt to influence resource use in distant regions when it is perceived that some local action may adversely affect them. In addition, globalization is working in less explicitly directed ways and market forces exert their influence on international trade in ways that allow the lifestyle choices of communities in the industrialized world to influence the options available to other communities around the world. This includes attempts to direct local governments in how to best manage local resources around poorer communities.

This chapter will examine how the democratization or the participation of a greater segment of society in decisions being made regarding forests and globalization has changed forest management and uses. These changes have shifted the decision-making from a top-down approach to a bottom-up approach. A top-down approach is when the government or nobles made decisions, as occurred in the past. A bottom-up approach allows the decisions to be made by all individuals in a group or community.

The democratization and, until recently, the globalization of forests have been less of a reality for developing countries (although they have been affected by global-ization). Many of these countries, and particularly their citizens, have had less of a role in the decision-making process regarding the forests within their own borders. Unfortunately, they are still dependent on their country's forests for natural resource extraction to generate income or even to survive. Many of these countries are accustomed to an abundance of natural resources, cheap manual labour, government taxation and land ownership concentrated in the hands of a few wealthy individuals (Luhnow and Lyons, 2005). These factors have not created an environment for the transition of less industrialized countries into contributing members of the industrialized world. In addition, the local population often has no access to the education required to participate in new technologies to produce wood by-products (a by-product is something that is produced as a consequence of the production of some other desired product) used in the industrialized countries (Chapter 7, this volume).

Most indigenous communities in less industrialized countries have not been part of the forest democratization process even though they are being affected by the decisions being made by the more industrialized countries. In fact, international organizations working to solve economic problems in developing countries see forests as the way to alleviate some of their poverty. This, however, is not going to happen if resource uses remain the same as they have traditionally been (i.e. burning or combusting wood) and if trees and forests are not incorporated into integrated country development strategies as effective tools for poverty alleviation (Schmidt *et al.*, 1999).

Today, historical forest uses are occurring at the same time in the developing countries as advanced technologies from the energy sectors (see Chapter 7, this volume) are being adopted in the more industrialized countries. Developing countries continue to be the suppliers of forest materials to the consumers who live in the more industrialized regions of the world. This has resulted in major disputes over who should make economic decisions determining how resources are utilized within a developing country. Conflicts arise over sovereignty issues when some view forests as part of a global commons. When forests are viewed as such, attempts are made to take decisions relating to forests out of the control of national governments and local economic elites. In the countries finding their forests perceived to be part of the global commons, there is a great deal of dissatisfaction in having 'outsiders' making decisions governing forest uses. Individuals living in Brazil ask questions like 'Just because the Amazon forests impact global climates [see Gash *et al.*, 1997], does this give the international community the right to dictate what happens in the forests of the Amazon?'

The democratization of forest use and management has expanded as more people have gained a voice in policy decisions, and differing values are increasingly acknowledged and being included in the debates shaping policy. The discussions over resource uses and conservation have frequently become polarized, as established values held by the more powerful stakeholders resist change. This limits the number of successful cases where solutions included all stakeholders from the local, regional, national and international levels (see Case 3.1 – Debt-for-nature Swaps, Forest Conservation and the Bolivian Landscape).

As countries become highly industrialized or are in transition to industrialization, they no longer rely heavily on forest materials for most of their energy production and instead use fossil fuels. This shift from using forest materials can be seen in the United States, where approximately 4% of current energy production is from biomass (EPA, 2004). Even though forests covered half of Washington State in 2001, less than 1% of its consumed energy was generated from biomass (Duryee, 2004). In contrast, the less industrialized countries typically obtain more than 70% of their energy from combusting wood, a very traditional mode of energy production that has substantial negative impacts on ecosystems (Table 2.2). The resulting decline in productivity and biodiversity undermines sustainable energy production, especially as human populations increase and their impacts worsen – a pattern seen throughout the developing world.

Another pattern that affects forest uses and conservation is the current trend of forest area increasing or remaining stable in the more industrialized countries but decreasing in the developing countries (Table 3.1). Forest loss is highest in areas with the greatest poverty. The relationship between population density and forest loss is effectively demonstrated for the Maya Region (Fig. 3.1). Because of their location in the tropics, these areas are also frequently 'hot spots' of biodiversity (i.e. areas with high concentrations of species). The combination of the high human population

Table 3.1. Population density and proportion of people living in rural areas in 1999 and change in forest cover between 1990 and 2000 (FRA, 2000; FAO, 2001). (Note: the former USSR is included in the Europe data in these databases.)

	Population density (no. of people/km^2)	Proportion of people living in rural areas (%)	Total land area in forests within a region (%)	Change in forest cover (%)
Africa	25.9	63	21.8	−0.78
Asia	117.8	63	17.8	−0.07
Europe	32.2	25.4	46.0	0.08
North/Central America	22.4	26.8	25.7	−0.10
Oceania	3.5	29.8	23.3	−0.18
South America	19.4	20.7	50.5	−0.41
United States	30.2	23	24.7	0.20

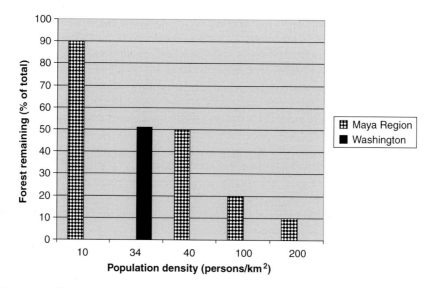

Fig. 3.1. The relationship between population density and the amount of forest remaining in the Maya Region of Central America (data from Meyerson, 2000).

densities, hot spots of biodiversity and high poverty rates results in many conflicts. Unfortunately, the highest poverty levels occur in rural areas, where more people are dependent on the extraction of resources from forests (Table 3.1). The difficulty of establishing viable alternative livelihoods increases the challenge of alleviating poverty and conserving biodiversity in these areas.

The globalization of decisions regarding forest resource use is creating tensions between less developed and more industrialized countries. Developing countries do not want international communities controlling how they use their renewable natural resources. For example, attempts to introduce trade embargoes against Brazilian timber exports inflamed and increased the tensions between the international community and Brazil. Failure of such direct pressure led to a shift in the late 1980s to forest certification, which relies on market forces to achieve the same ends (Vogt *et al.*, 1999).

These conflicts are typically characterized as the more powerful, wealthy North (e.g. United States) trying to impose its will on the less powerful, poorer South (e.g. Brazil). The South is suspicious that the motivations of the North are part of a continued 'imperialist' aggression by the former European colonialists. The industrialized North, on the other hand, frames the issue in terms of the need for more participation in management of resources that impact the 'global commons' (i.e. resources used by and that influence the whole world). Where the South sees threats to its sovereignty, the North sees threats to the global commons. The conflicts between the desires and goals of national conservation organizations and local communities are highlighted in Case 3.2 – Cattle, Wildlife and Fences: Natural Disaster and Man-made Conflict in Northern Botswana. The approaches used by governments towards indigenous communities of shifting cultivators and the implications for whether the shifting cultivators 'destroy' or conserve forests are explored in Case 3.3 – Forest Communities in China and Thailand.

Many of these conflicts are further aggravated by the juxtaposition of different values espoused in less developed versus more industrialized countries. The industrialized countries have focused on stopping deforestation in the Amazon region and conserving its high biodiversity (Myers, 1984), with insufficient consideration given to the communities living in the Amazon and directly depending on its resources for their livelihoods and survival. The values of the international environmental communities were generally forged in nations where large tracts of forest lands had already been set aside for conservation. Within their own boundaries, industrialized countries are more concerned with forest health issues, high fire risks and invasive species (Vogt *et al.*, 2002; Chapters 4 and 5, this volume). This contrasts with what are considered the important issues in a developing country, which are more driven by economic concerns, health problems or human rights issues. At the local and regional level where people require forests to survive, conservation is less relevant to them than knowing where their next meal will come from. In this case, conserving biodiversity means that these people will have to choose between eating and collecting firewood or designating areas as reserves.

International funding organizations have also reflected the different values and conflicts playing out between the developing and industrialized countries. Recently some international environmental groups have come to recognize that successful conservation must take into account local communities that depend on the forests that are the focus of conservation efforts. Accordingly, some groups are working towards the empowerment of indigenous communities with land ownership and better livelihoods (Vogt *et al.*, 1999). However, starting in early 2000, international funding organizations (e.g. the World Bank, European country-level funding organizations) became increasingly less likely to require that their poverty alleviation funds be linked to conservation, de-emphasizing a link that had been a priority in the early 1990s.

The previously acknowledged association between poverty alleviation and conservation was logical since the standard of living of exceptionally poor people has not been improving. This situation was created because alternative economic options were not provided when forests were set aside for conservation, either as community forests or as reserves (see Case 2.1 – Nepal, Community Forests and Rural Sustainability). In some cases, local subsistence extractors have been prohibited from extracting materials from forests that had been traditionally used for harvesting (e.g. in Nepal; Bushley, 2003). In many developing countries, poor people are dependent on extracting resources for survival and for obtaining reliable energy sources (Patel-Weynand, 1997). As forest access is restricted and the total forest area is reduced, people become poorer and more marginalized. This results in decreasing democratization at the local and regional levels.

At the same time that forest area is shrinking for local resource extractors, consumers from the industrialized world are increasing their demand for timber from these same locations. The industrialized world has set aside more of their forests in reserves so that they are not being cut for timber while at the same time shifting timber extraction to the developing world, where much of the forest extraction by local citizens is used as fuel wood for heating and cooking. As more forests are being set aside for preserving habitat or species in the industrialized world, this has the unfortunate effect of shifting the exploitation of forests to the developing world. This shift can be exemplified by the fact that the state of Washington used to be an important global supplier of wood during the 1970s and 1980s but its export has been reduced by more than half as less timber is available for harvest from public lands.

Another emerging pattern is that setting aside forests by the industrialized world has been insufficient to protect forests from being lost at some future date. This is shifting the issues in forest management and conservation from species and old-growth forests (Franklin *et al.*, 1981; Franklin, 1989, 1992) to forest health, fire control and restoration (Chapter 5, this volume). These health issues determine whether forests will persist or be lost by fire and/or insects/pests that kill trees. Therefore, it is no longer just a question of setting aside forests for conservation purposes; the question becomes: Can forests be managed in a way that will protect their health? Leaving forest unmanaged is not a choice. Restoring forests back to conditions that have a healthy complement of natural disturbance processes (e.g. insects, fire, etc.) is crucial (Chapter 5, this volume).

These new issues create new opportunities to resolve forest use and conservation problems in landscapes where people are dependent on forests for their livelihoods. Technology in the energy sector shows great promise in providing a mechanism to deal with forest health issues and their fire risk while providing options for income generation from these same forests (Vogt *et al.*, 2005). Using technology to improve our uses of forests needs to be combined with an understanding of disturbances, conservation biology, silviculture and ecosystem ecology at the landscape levels (Chapters 5 and 7, this volume). Disturbances can reduce the productive capacity of forests so that fewer resources are available for society to obtain. Furthermore, the constraints imposed by ecology (Chapter 4, this volume) and the silvicultural tools available (Chapter 7, this volume) need to be included, because these determine how much humans can manage forests. Humans need to manage forests in order to restore their capacity to provide environmental services, habitats, products and other human values.

The Definition of Best Management Practices Changes as Public Values Change

History has repeatedly demonstrated the exploitation, wasting, replacing or elimination of a forest when society does not value it, is afraid of it, values products other than those collected from forests (e.g. agricultural fields replace forests) or sees the forest as nothing more than an economically valuable product to be exploited (Chapter 1, this volume). Western industrialized forestry has been attributed to the practices of the Roman empire, where forests were removed in order to give room for wide-scale intensive agriculture and to expose communities that could not otherwise be conquered (Drengson and Taylor, 1997). The perception of the Romans was that forests posed a threat to them since their human occupants resisted Roman rule. The forests needed to be cleared to enforce their control over conquered peoples.

What economic value we place on natural landscapes and resources has a significant impact on the decisions that we make as a society and in how we want to use or conserve these resources. Hardin (1968) in an article entitled 'The tragedy of the commons' brought attention to an essay written by a 19th-century amateur mathematician, William Forster Lloyd, on population growth and control. Lloyd drew upon the fate of the common pastures in England, where resource-maximizing cattle herdsmen increased the number of their livestock with the result that the area suffered severe deforestation and land degradation. To the herdsmen, the addition of one animal in the common grazing area resulted in an economic benefit of almost +1, but the negative effects of the additional grazing pressure were shared among all the people using the pasture and were thus only a fraction of −1 for each individual herdsman. Every rational herdsman therefore strived to maximize the number of his flock, resulting in the degradation of the common resources. Much of the English forests were over-exploited and lost when the grazing was communal. This approach needed to change so that individuals bore full responsibility for the impacts of their resource use (Agrawal, 2005).

Industrialization ultimately began to change society's view of forests because people had leisure time and began to travel back to the 'wilderness' to enjoy its aesthetic beauty or the recreational activities it provided. People enjoying the recreational benefits of nature also became more concerned about the destruction of forests that they observed as they travelled through these lands. In contrast, leisure activities are not possible for people surviving at the subsistence level. When most of your time is spent trying to survive, you have a different relationship to nature and different attitudes about its values or its uses (however, see Case 3.4 – Indian Forest: Land in Trust).

Beginning in the mid-1980s to the early 1900s, societies in industrialized countries had more wealth and more of the general public was able to take advantage of recreational activities. Previous to this time, recreation had only been a pastime for the wealthy and not a pursuit of the general public. As more leisure time was available for a greater number of people, different groups were expressing more diverse values relating to the forests. These values varied according to how dependent people were on forest resource extraction for survival (rural versus urban conflicts; developed versus developing country; utilitarian versus aesthetic; local versus global production and sales of resources) and the demographic shifts occurring in society (e.g. fewer people in agriculture) (Box 3.1).

Some values conflict with one another since we can value 'the wild' or non-human or virgin systems at the same time as we value 'pets' or 'collect wild animals' or use their body parts for 'medicinals' (which helps maintain active illegal trade or smuggling). Value conflicts also occur between indigenous communities and outsiders (i.e. immigrants, international organizations) and frequently consist of the introduction of new land-use practices that did not exist before but subsequently become 'valued' by a segment of the community. A good example of this is the value placed on indigenous knowledge and the persistence of indigenous survival activities versus economic activities driven by outside forces or from the national level or from the drive to give global environmental concerns greater importance. For example, scientists suggested the need to stop shifting agriculture (slash/burn agriculture) in the Amazon to increase carbon storage in forests (to help offset the rising carbon dioxide levels) and trying to introduce intensive-agriculture farms at fixed locations (which is not supported by the poor soils that exist in these wet tropical areas). Another example of conflicting values is farmers killing parrots since they eat their agricultural crops while international groups want to protect/conserve these same parrots.

Box 3.1. Typologies of human values.

Kellert (1993) identified a typology of human values for resources. Out of the ten values he identified, seven are provided below as examples of the variety of human values that have been recorded. Some of the values are difficult to measure and therefore the economic value was usually the most common and only metric used by society to express their value for a resource.

- **Aesthetic** – making the world more beautiful or more appealing to the senses and generally more pleasant (e.g. cool hike in mountains)
- **Emotional** – value beyond sensory enjoyment (sense of place), bonds to certain natural areas or plant or animal species, not liking certain animals (snakes, rats)
- **Economic** – linked to tangible products that can be bought or sold (food, timber, energy, etc.), remember need to consider long-term not just short-term benefits
- **Environmental services** – intangible 'services' that allow humans to exist on Earth (e.g. plants produce O_2, microbes purify water, etc.)
- **Ethical** – beliefs
- **Utilitarian** – use resources (farmers)
- **Cultural** – traditions part of a group of people, a community or a country (forests are important for Scandinavians)

In the industrialized countries, highly contentious debates emerged as mainstream societal values shifted towards the environmental side. Conflicts arose due to the lag periods before management agencies began to understand that society's values had changed and it wanted different operational guidelines used to administer forests. In the United States, federal management agencies continued to manage forests using rural and economic values that condoned the continued exploitation of public forest lands for timber (Franklin, 1992). Even though multiple-use legislation was passed in the 1960s, which increased the number of uses that were expected from forests, this legislation did not reflect the conservation or environmental values that were becoming important to some segments of society.

Management agencies were not responsive to changing public values, especially those expressed by people living in urban areas, who wanted more environmental values to be included in management practises. In the western United States, lag periods are apparent where different groups of people (e.g. public, federal resource managers, rural communities) did not recognize society's values shifting from merely obtaining economic return from the forests to also maintaining them as locations to conserve species and provide leisure activities.

This lack of perception resulted in contentious and highly polarized conflicts between all the stakeholder communities with the forests at the centre of these conflicts. At the same time, there was a lag in the scientific knowledge that could be used to support the environmental soundness of decisions. Since ecology and ecosystem sciences were still in their infancy, these new environmental values held by the public could not be supported scientifically. Our understanding of forests as an ecosystem did not really emerge until the 1980s (Vogt *et al.*, 1997). In contrast, silviculture was well developed by the early 20th century (Chapter 1, this volume). The practical tools needed to sustainably manage individual tree species had its roots in India back in the mid- to late 1800s. The management agencies were using the well-developed silvicultural practices to support decisions being made about forests without taking into consideration that silviculture is a science of individual trees and does not consider the conservation needs of species that are not normally hunted for food (Vogt *et al.*, 1997; Chapters 4 and 7, this volume).

Societal values for forest uses can also change dramatically over short time periods and multiple values are present at any given time (Vogt *et al.*, 1997). Operationalization is when ideas become reality and influence some action. Operationalization of attitudes can be summarized as in Table 3.2.

In the United States, the predominant values for how forests should be used or conserved have shifted dramatically over the last 40 years (1960–present) (Sharpe *et al.*, 1995). At all times, a mixture of values existed, with some values being more widely accepted than others. The most recent conflicts over the operationalizing of societal attitudes for forests were especially confrontational in the Pacific North-west United States. The environmental values were quite concerned with forest practices that directly resulted in the loss of habitat or habitat quality important for biodiversity conservation of endangered species. For example:

- Clear-cutting or cutting all trees on large tracts of land, which resulted in the loss of habitat for forest-dependent species.

Table 3.2.   Operationalization of attitudes.

Time	Operationalization of values/attitudes held by industrialized country societies
Pre-1860	• Resource exploitation and use
1870–1904	• Resource exploitation and preservation depending on site
1905–WW II	• Manage by scientific principles or 'wise use'; since Pinchot's definition of 'wise use' includes conservation and providing the greatest good for the greatest number of people over the long term, conservation was by definition being balanced with resource uses (this definition differs from how it is typically defined today, i.e. primary focus: conserve species and habitats; secondary focus: sustainable development for people who survive in these habitats)
WW II–1960	• Resource uses dominant • Conservation less important in most places
1960–1990	• In practice, conservation has priority over resource uses for international organizations funding projects in developing countries • Multiple use becomes important (1960 legislation passed in the United States) but in society multiple use was mainly as recreation • The concept of sustainability accepted by the international communities. The Bruntland Report on Sustainable Development in 1987 called for sustainable development in resource uses so future generations have access to these resources (WCED, 1987). Tools to measure/monitor and audit sustainability had not evolved
Early 1990s	• Ecosystem management adopted – required simultaneous inclusion of economic, social, ecological factors in forest management (Vogt *et al.*, 1997); ecosystem management evolved to adaptive management as it became apparent that insufficient data existed to manage ecosystems • Development of integrated conservation and development projects (ICDP) recognizing conservation needed sustainable development opportunities for local communities or conservation projects were doomed to fail (Sayer and Campbell, 2004) • Sustainable livelihood opportunities based on creating markets for local natural resources become an important research area • Much conflict between conservation and resource uses in the tropics and Pacific North-west United States • Forest persistence driven by human population densities in developing world, with more people subsisting on a smaller land base, especially in the tropics
Post-2000	• Global societies grappling with operationalizing sustainable management of forests that included poverty alleviation, livelihoods of forest-dependent communities, tenure rights of local or indigenous communities for forests

Table 3.2.

• Ecosystem management evolved to environmental sustainability
• Few successful sustainable development projects so continuing conflicts between conservation and resource uses
• Continued persistence of major environmental problems. International funding organizations shifting funding targets to poverty elimination, improving human health, controlling diseases, local economic viability
• International funding organizations providing less funds for conservation

WW II, Second World War.

- Using single species or planting monocultures of one species, usually non-indigenous to a region.
- Road construction – initially roads were built poorly because they were temporary transportation systems, which resulted in decreased habitat quality for salmon in streams. Stream habitat quality was reduced when the roads eroded and increased the amount of sediment in the stream.

Even during the time that environmental values were becoming more dominant, biodiversity conservation was a relatively new concept for the public and still not well understood.

During the last 5 years in the western United States, forest uses and practices have become less contentious in part because forest practices have been modified and adapted in response to concerns expressed by society (see Chapter 7, this volume). New practises emerged that acknowledged ecosystem processes and considered forests in terms of larger systems (called 'new forestry'; Franklin, 1989, 1993). Clear-cutting was no longer acceptable or practised on a large scale. Forests were set aside for biodiversity conservation to maintain old-growth Douglas-fir forests and the species dependent on them.

Conflicts over forest management were also reduced as global timber markets shifted away from the United States Pacific North-west forests as a significant supplier of wood. This shift, however, has not helped to solve the economic problems facing rural communities. In response to the changes brought on by the environmentalists, some timber companies relocated their timber operations to other regions of the world (e.g. the tropics). In general, since timber is a global commodity, timber markets typically shift to regions where it is cheaper for timber processing companies to operate. Several factors stimulated theses moves to the tropics, which can be summarized as:

- fewer regulations exist on how timber should be harvested;
- timber supply is easier to obtain from fast-growing trees in plantations since trees grow two to three times faster in these warmer climates;
- economic returns were more favourable since labour costs and the cost of building the required infrastructures were lower in most developing countries than in the United States.

Despite the Pacific North-west United States not continuing as an important timber supplier in global markets, their forests still need to be managed to obtain the environmental services expected from them. During the last 5 years, there has been growing recognition of the importance of disturbances and the need to manage forests to reduce their risk of loss due to catastrophic disturbances (Edmonds *et al.*, 2005; Chapter 5, this volume). How to manage forests to improve their health and to control their high fire risk are still contentious issues in the western United States. Consensus does not exist on the best management approach to use to reduce the risk of losing forests. For forest health and safety, the following topics have created significant conflict:

- Fire risk.
- Loss of biodiversity.
- Loss of property.

Conflicts in resource uses occur because the issues are complex, many stakeholders with different values are involved and science is unable to provide one simple solution for forest uses and conservation. Science from multiple disciplines (e.g. social sciences, wildlife, recreation, silviculture, ecology, economics and policy) needs to be part of the solution but we are still developing the scientific principles needed to integrate such diverse disciplines. The importance of such a variety of different values increases the difficulty of producing solutions to complex environmental problems. This point was well articulated by the NRC (1996) report:

> Problems in managing and protecting fish populations are in part due to the failure to articulate divergent interests, goals, and values and to address them explicitly . . . From a policy perspective, the salmon problem is one of long-standing and serious conflict in fact, interest, and values.

Science should be used to identify the trade-offs and consequences of decisions being made. Ultimately, public policy decisions should incorporate the science and not be driven by the values of the decision-makers. Most decision-makers have short-term objectives, while the decisions made have far-reaching effects, whose consequences may not be apparent until later. Natural resource management decisions can be based on societal values as long as they are grounded in the sciences, in order to reduce the probability of intended consequences harming the ecosystem at some future time. It is during the last 100 years that the role of society and the need to include society's values for forests have become integral to how decisions are made in forests, especially in industrialized countries. The democratization of forest management decisions will be discussed in the next section.

Democratization of Forest Management and Uses: a Phenomenon in Industrialized Countries

Democratization of forest management and uses initially flourished in industrialized countries. We are not including community forests (started by the international organizations in partnership with local communities in the 1970s) as part of this democratization of forest management and uses. Our rationale for this decision is that, even though community groups set up the rules for forest uses, the forests are still mostly owned by the governments, who have the ultimate power to determine what is done in them (see Case 2.1 – Nepal, Community Forests and Rural Sustainability).

Democratization is defined here as the active participation of the public in policy and management decisions regarding forests. The democratization of forest management and forest uses occurred when the public wanted to be assured that important natural resources (e.g. water, aesthetics, old-growth habitat, wildlife) would be protected before forests were used for consumptive purposes.

Society (especially in the United States) was able to add non-economic intrinsic values (e.g. little, if any, market value that someone would pay for) as a normal part of forest management. The public became more active participants in the decision-making process as new public involvement strategies were needed to include a wide spectrum of social values. In some cases, the public did not trust their government or its agencies to provide the non-economic values they expected from public lands. At the same time, significant conflicts over forest uses persisted because it is impossible for any one forest to indefinitely provide all the services and products expected by every group of concerned citizens.

Even though this process of public participation started in the industrialized regions of the world, it has dramatically changed how forests are managed and used around the globe. For example, once the public became participants in the decision-making process, it was easier for them to think of all forests as part of the 'global commons'. The fact that science could document the global impacts of forests helped to support this thinking (e.g. deforestation occurring in one country can affect climates globally) and made it easier for a global community to justify controlling forest uses everywhere.

Democratization of decision-making in forests: the United States example

How forest decisions became democratized is especially well demonstrated in the United States and will be further discussed next. It is important to recognize that this process occurred without much consideration given to the American Indians who lived on these lands (see Case 3.4 – Indian Forest: Land in Trust). Therefore, the American Indians were not part of the democratization process regarding the decisions being made on forests. The following discussion is therefore based on the changes that occurred in North America dominated mainly by European colonialists.

Changes in land management can be summarized through time as in Table 3.3 (Sharpe *et al.*, 1995).

When the European colonialists first arrived in North America, most resource decisions were not democratic and the public was not concerned with how lands were being managed. Public views began to change between 1782 and 1860 when the federal government was transferring public lands to private ownership (Sharpe *et al.*, 1995). Lands from the federal domain were being disposed of for the following reasons:

- Given away in place of money. The federal government gave lands to the new states joining the union to be managed by them to earn money for maintaining

Table 3.3. Changes in land management.

Management era	Duration
Acquisition of lands with natural resources	1776–1867
Disposal of the federal domain	1782–1860
Reserving lands	1872–now
Custodial management	1897–1950
Intensive management for timber and water	1950–1990
Confrontation and consultation with the public, inclusion of multiple values from public forest lands	1970–now
Ecosystem-based management and adaptive management	1990–now

common schools and institutions of higher education (i.e. universities), building and maintaining government structures (e.g. legislature buildings) and state charitable, penal and reformatory institutions.

- To establish United States government sovereignty over lands by having citizens: (i) settle dispossessed land belonging to American Indians and (ii) establish/control borders against the Spanish, Mexicans, French, British and Canadians.
- To minimize government control over citizens by privatization of land ownership.

This period of land disposition devastated and wasted former public lands, as corrupt individuals and companies exploited natural resources for profit. The public was outraged at the devastation that occurred. However, mismanagement of these lands was common. The public felt that a timber famine was imminent because of the rampant over-exploitation of forests for timber and the high loss of forests due to fires (many set by those exploiting the forests for timber). The public reacted negatively to the following (Williams, 1989):

- Corruption and misallocation of lands – lands were given to the railways cheaply and lands were not being given for their best use, according to the public. For the government, giving land to the railways was important because they built the transportation infrastructure that allowed the west to be settled.
- Deforestation – railways sold land to timber barons, and timber barons had a cut-and-run policy.
- Fires – frequency of large fires had increased dramatically due to drought, sparks from train engines starting fires, slash (waste) from timber harvest creating additional fuel.

Public distrust of government management of resources and a need for oversight of management decisions can be traced back to 1870 and 1890 (Sharpe *et al.*, 1995). This over-exploitation of resources started to change the social climate from supporting resource exploitation and use to wanting to preserve these lands. Despite this public outcry over how resources were being used, the liquidation of resources continued unabated during this time period. The legal structure was used to support this over-exploitation. Starting in 1891–1904, continued societal demand for better management and use of resources resulted in legal mandates to improve and protect federal lands (the Forest Reserves Act of 1891 mandated and halted the sale of public lands).

The continued response of the legal system to societal pressures for better management resulted in a period of custodial management of these public resources (1897–1950) and the passing of laws to support those efforts (O&C Sustained Yield Act of 1937) (Sharpe *et al.*, 1995). Custodial management placed federal agencies in charge of managing resources and lands for the good of the public.

The first agencies to manage public lands were formed in the early 1900s (Forest Service formed in 1905; National Park Service formed in 1916). The early management focus, particularly in the arid West, was primarily fire suppression because of the high number of fires that were occurring. During this period, the public involvement in forest management decisions was minor.

In the United States, the forest policy changes that occurred can be broadly summarized as:

- Forest policy: pre-1960s
  - Development-oriented
  - Professional foresters
  - Iron triangle: Forest Service, Congress, industry
- Forest policy: post-1960s
  - Organized environmental movement
  - New environmental laws
  - Congress controls through prescription
  - Increased judicial scrutiny
  - Increased participation in policymaking
  - Biologists, ecologists, hydrologists, archaeologists and sociologists joined foresters

The continued conflicts over resource uses in the 1990s made public participation an integral part of forest management decisions in the Pacific North-west United States. Much of that information is documented elsewhere but it is worth mentioning two items because of the role they played in legitimizing public involvement in decisions made regarding national forests: the North-west Forest Plan of 1994 and the Healthy Forest Restoration Act of 2003. Both had goals of including all stakeholders in the decision-making process, i.e. attempted to ensure that environmental values could be achieved from forests while also providing sustainable livelihoods for rural communities dependent on forest resources.

The North-west Forest Plan of 1994 included endangered species and involvement of the public in harvest management decisions, and it attempted to provide livelihoods for communities that had obtained all their economic return from timber. The specific points covered in this plan were as follows:

- Covered habitat of the endangered northern spotted owl.
- Addressed conservation needs of the northern spotted owl and then soon-to-be listed salmon species.
- Emphasized involvement of all stakeholders in harvest management.
- Intended to sustain timber-dependent communities by maintaining sustainable wood output.

The Healthy Forest Restoration Act of 2003 is especially relevant to discussions when considering the democratization of forest decisions because it was a step backwards. It reduced public input in forest management decisions. The stimulus for this act was the fact that past fire policy had led to dangerous fuel accumulations in forests, which required managing to reduce fire risk. This act began to more explicitly search for solutions to the economic problems of the timber-dependent communities.

The 1994 plan had not addressed and, in fact, further exacerbated the problems of timber-dependent communities. The North-west Forest Plan of 1994 placed 80% of north-west federal forests off-limits to logging. Only about 40% of the allowable legal sales of timber had occurred because: (i) of a lack of funding to do pre-sale ecological assessments; (ii) lawsuits were holding up timber sales; and (iii) the timber industry was declining in the region because fewer logs were available and the plants processing wood were more automated, requiring fewer employees to run them (job loss). The purpose of the Healthy Forest Restoration Act of 2003 was to:

- Increase thinning.
- Decrease environmental assessment because it was expensive and redundant and wasted time while communities and resources were at risk.
- Decrease public review and appeal because they were expensive and redundant and wasted time while communities and resources were at risk.

On 23 March 2004, changes were also announced to the 1994 North-west Forest Plan. The 1994 plan had involved stakeholders in harvest management and in restoring habitat for endangered species but little had been done to sustain timber-dependent communities. The timber-dependent communities were economically worse off because less public land was available to harvest timber. This loss of timber from public lands resulted in many mills closing because there was no wood to process and many of the jobs associated with the timber industry were lost.

Several factors were identified in 2004 that hindered getting timber to the mills and markets. These factors were changed in order to expedite the process for timber sales from federal lands and as a mechanism to improve the livelihoods of timber-dependent communities:

- The 'survey and manage' requirements:
  Change: No longer necessary to monitor effects on small plants and animals, including rodents, insects and molluscs, if they are not listed as endangered or threatened. This change occurred because: (i) 86% of the 3.2 million ha (8 million acres) of old growth is protected already from logging (projected to grow by 0.4 million ha or 1.0 million acres in 50 years) and (ii) it will save $16 million per year.
- Requirement to monitor harvest effects on individual streams:
  Change: Watershed-scale monitoring was considered adequate and therefore project-scale monitoring was considered unnecessary. This shifted monitoring to a larger scale, i.e. the watershed, so individual stands did not need detailed monitoring data.

These changes reflected a need to balance the ecological risks in forests with the economic and social risks inherent to timber-dependent communities. They also reflect the difficulties of evaluating many ecological problems that require longer time scales to detect change versus the shorter time scales used in political decision-making.

The culmination of including society as part of the decision-making process regarding forest lands and the democratization of forest management occurred with the adoption of ecosystem management as a management paradigm (model). In the United States, federal resource management agencies adopted this approach to management in 1992 (Vogt *et al.*, 1997). In practical terms, ecosystem management is an approach that explicitly includes ecological, societal and economic objectives in natural resource management. The relationship and co-evolvement with regard to the rise of the environmental movement, changing laws and forest management have had effects far beyond the boundaries of the western United States.

The rise of environmentalism and conservation contributing to the democratization of forest uses

Democratization of forest management and uses began to become systemized or formalized with the development of the environmental movement in the mid-1960s. During this time period, societal outrage emerged over the excessive exploitation of forests without consideration of their environmental impacts or the potential loss of species living in these habitats (McCormick, 1989; Vogt *et al.*, 1997). The same outrage mobilized and emerged as the core element of the environmental movement. This movement had its roots in the more industrialized regions of the world and today its values have become codified as part of institutions managing public lands.

In the United States, environmentalists immersed themselves in environmental problems within their own backyard (i.e. Pacific North-west) in addition to those emerging in the tropics. This contrasted with European efforts, where the primary focus was on environmental problems occurring in the tropics. The general public in Europe did not have disagreements with the management of their forests and trusted their forest management agencies (Dawkins and Philip, 1998).

Several factors made it easy for environmental values to crystallize globally during the early 1970s. One of the most important factors was the increasing awareness of the limits to our resource supply and its capacity to feed an ever-expanding global population (e.g. limits to the Earth's carrying capacity). When *The Limits to Growth* report was released in 1972 (Meadows *et al.*, 1972), the authors suggested that environmental problems were due to over-consumption of resources by society, unlimited industrialized growth and inequity in the distribution and consumption of resources. This report generated considerable interest in the general public and emphasized the link between increasing population size and environmental degradation (Meadows *et al.*, 1992).

The environmental movement helped to develop the new discipline of conservation biology (Soulé, 1986; Primack, 1993). Conservation biology has focused on the protection of biodiversity and in developing the science needed to protect individual species. Conservation biology also introduced the idea that forest management needed to include species other than commercial tree species and that a healthy ecosystem would result from maintaining diverse communities of plants and animals.

The environmental movement was also strongly influenced by two acts:

- The Environmental Protection Act of 1969.
- The Endangered Species Act (1973) became law in the United States and further codified the protection of species and habitats endangered by human activities.

Globalization of Forest Management and Uses

Globalization of the science and decisions made in relation to forests occurred in two main ways: (i) international environmental organizations started including environmental values in forest management globally; and (ii) international organizations started monitoring and collecting databases on the amount of deforestation and forest conditions around the world so that policy decisions could be based on scientific data.

The globalization of forest decision-making can be traced back to international organizations that were formed to collect information on forest practises prior to the Second World War (Winters, 1974; McCormick, 1989). The role of international environmental communities can be traced back to the late 1960s and early 1970s, when forests began to be perceived as part of the 'global commons'. The globalization of forest management and uses can be summarized as follows:

- Pre- and post-Second World War: the number of international organizations collecting information on silviculture (to determine where trees would best grow or should be planted), forest condition, forest uses and the amount of forestland in every country in the world increased significantly during and after the Second World War. These organizations have significantly increased the transparency of data used to assess the sustainability of natural resource uses.
- 1960 to present: increased internationalization of global timber markets where developing countries became the suppliers of these resources and the industrialized countries becoming consumers of these products. This focused attention on economic differences between developed and developing countries, and provided solutions for poorer countries.
- 1970s: international organizations attempting to use economic embargoes or incentives to obtain environmental values from forests beyond their jurisdictional boundaries and that they did not own.
- 1990 to present: development of forest certification as an audit system to measure whether sustainable forestry was being practised and whether social, economic and ecological services were being obtained when forests were managed for timber. Forest certification is an auditing system that determines whether sustainable forest management is being practised that includes environmental and societal values (see Chapter 7, this volume). Today, forest certification has become institutionalized around the world and is used globally to measure sustainable management of forests (see Vogt et al., 1999).

International organizations becoming the repository of scientific information on forests and forest management can be traced back to the late 1800s, even though it did not become institutionalized until the mid-1940s (McCormick, 1989). One of the earliest international organizations was the International Union of Forestry Research Organizations (IUFRO), which was established in 1892 by professional foresters to disseminate the science of forests and to help provide local solutions to improving forest growth. Later, in 1932, the Comité International du Bois (CIB) was formed and was based in Vienna, Austria. Before the Second World War, the International Office for the Protection of Nature (IOPN) was formed, which continued to examine wildlife protection efforts occurring in Europe during the war years (McCormick, 1989).

After the Second World War, the Food and Agriculture Organization (FAO) was formed as an umbrella organization in the United Nations (headquartered in Rome, Italy) to provide data for supporting forest resource management. Franklin D. Roosevelt is credited with instigating the formation of the FAO in 1943. Scientists in agriculture and nutrition from 43 countries met when FAO was initially formed. Forests were not part of the original FAO mandate since increasing food production and supply to feed humans around the world was a higher priority (McCormick, 1989). However, the FAO mandate changed when a resolution was passed in 1945 that recommended FAO develop an inventory of global forests; the first inventory was carried out in 1947 and 1948. In 1951, the FAO mandate was expanded to include their collection of global data on forest resources and to conduct a worldwide survey every 5 years (UNESCO/UNEP/FAO, 1978). The original databases collected by FAO were not ecosystem-based and focused on information relevant only to the commercial management of trees. FAO was active in pursuing its goals and transferring its approach to other countries by:

1. Encouraging developing countries to systematize their forest management, using scientific principles to maintain forest capacity to supply timber; and
2. Influencing the dominant economic groups and governments to adopt forest management goals that would maintain the productive capacity of trees for commercial forestry.

In 1951, population growth showed up on the radar screen of FAO when they identified the population growth of poor people as the primary factor causing deforestation (McCormick, 1989). This view implied that indigenous people surviving at subsistence levels were responsible for deforestation. Practices such as slash-and-burn shifting cultivation were identified as being responsible for the loss of soil fertility, soil erosion, river siltation and a reduction in water supply (McCormick, 1989). The FAO came to this conclusion even though these indigenous communities have practised slash-and-burn cultivation for hundreds of years. Slash-and-burn cultivation is a practice used in the tropics where forests are cut down, the branches and stems of the trees are burnt, which acts as a fertilizer, the crops are grown and eventually, after food crop productivity decreases, forests are allowed to grow back.

Interestingly, slash-and-burn cultivation has also been widely used in Eurasian forests and was documented as having been used in Russia since 946 and in Germany since 1290 (Goldammer and Stocks, 2000). In the early 20th century, 50–75% of the Finnish forests were exploited using slash-and-burn cultivation and at the same time about 40% of the cultivated land in Russia was recently burned land. In Europe, slash-and-burn cultivation decreased during the industrial revolution, as it was replaced with more mechanical agriculture practices and the use of commercial fertilizers and pesticides became widespread (Goldammer and Stocks, 2000).

Similarly to the relationships shown in the tropics between increased slash-and-burn cultivation and the growth of the population, slash-and-burn agriculture in the Baltic regions of Europe was also related to demographic changes, rupture of social structure and increased poverty (Goldammer and Stocks, 2000).

Despite examining the causes of deforestation in 1951, the main focus of FAO was on determining the best ways to commercially manage forest resources. As part of these FAO goals, they encouraged the planting of monocultures of plantation forests, using fast-growing trees where the silviculture was well known and using species that had the qualities desired in the finished wood products. This philosophy resulted in substantial areas of natural forests being replaced by exotic or non-indigenous tree species. Planting non-indigenous tree species was logical at the time since little was known about the silviculture or ecology of native forest tree species. It is difficult to manage tree species when the science does not exist for their management.

At the same time that FAO was actively encouraging the management of monoculture plantations, Switzerland, Belgium and Holland led in the formation of an international organization for the conservation of resources. This conservation of resources was supposed to be based on scientific exchange and education in the natural sciences. The United Nations Educational, Scientific and Cultural Organization (UNESCO) – formed in 1946 – was approached to support the development of this new organization. In 1948, UNESCO convened a Congress that launched the International Union for the Protection of Nature (IUPN) and based it in Basle, Switzerland. In 1956, IUPN was renamed the International Union for Conservation of Nature and Natural Resources (IUCN), mainly due to the insistence of the United States. This name change shifted the mandate of IUCN to conserving representative ecosystems with a primary focus on wetlands and a secondary focus on national parks (McCormick, 1989). IUCN continues to exist today and is very influential in determining how environmental issues are dealt with at the international level.

The globalization of forest management and the development of global economic markets really matured after the Second World War. Most countries still managed their resources as if nature were an economic resource that could be endlessly tapped. Continued extraction of natural resources was crucial to the United States view that global trade and aid were a means to bring less developed countries into the global markets and to improve the livelihoods of the world's poor.

In most developing countries, the only commodity they had to sell in the global markets was their natural resources and, since their labour costs were inexpensive, their participation in these markets remained as the suppliers of renewable resources to the more industrialized countries. This role for developing countries in global economic markets does not stimulate the adoption of new technologies nor does it increase the efficiency of producing goods from forests. Therefore, there was no justification for changing how forest resources were used or managed.

The role of international organizations broadened during the 1970s with the birth of the environmental movement. International organizations began to collect information on what materials were collected from forests, production rates of forests, deforestation rates and the health of global forests. Today, similar data are collected but the credibility of these data has greatly improved because of the ability to use remote sensors to monitor forests instead of relying on the data provided by each country (Chapter 7, this volume). Introduction of technology into assessing forest management and deforestation rates has greatly expanded the ability of international communities to monitor forests and their deforestation rates. Many of the data provided by the developing countries were and continue to be questionable and do not reflect the high deforestation rates occurring within their boundaries.

Despite the introduction of environmental values into forest management in the 1970s, the values of the people living in the developed nations still superseded the values and desires of the less developed world. The developing nations still continued to provide the critical raw natural materials needed to fuel the industrialized world. Different goals for forests existed at this time; developing countries were interested in using forests to provide economic return, while the developed countries were more interested in species and biodiversity protection within their own countries. Developed countries paid less attention to people who lived in and were dependent on these forests. This approach resulted from a focus on conservation and the view that people living in forests contributed to deforestation and species losses. An underlying assumption, which continues this pattern of supply and consumption of renewable resources, is the view held by industrialized countries that the modernization of the non-industrialized world would begin to provide market mechanisms to improve the livelihood of people living in these less developed regions of the world.

Beginning in the early 1980s, the general public in Europe and the United States was mobilized by extensive media coverage to pursue the halting of deforestation and forest burning in the tropics. These high rates of deforestation provided a focal point that unified the opinions of the international environmental communities and provided them with a common enemy to condemn.

Timber companies were identified as the main culprits responsible for tropical deforestation. This view is still prevalent today, even though deforestation may not be directly due to logging itself. In 1994, Kobayasha estimated that only 25% of the deforestation of tropical rainforests was a result of logging. Timber companies were blamed for increasing deforestation rates because the timber companies built roads to access wood in remote locations. Once the roads were built, poor peasants would move into these areas and begin to practise their slash-and-burn cultivation. This pattern resulted in more forest area being cleared.

The high deforestation rates of tropical forests brought together scientists from a mixture of different backgrounds to work towards the common goal of stopping deforestation. Some scientists were more concerned with the loss of forest cover in the tropics because of the link between forest loss and changing climates. Other international scientists were driven by conservation concerns. The high biodiversity of species found in the tropics was an important topic of conservation interest to many in the international community and there was concern for the loss of potentially valuable medicinal plants (Myers, 1984). In addition to the environmental and conservation issues, some members of the international community were also concerned with human rights and the rights of indigenous peoples living in these environments.

All these activities pitted the international environmental community against national governments and economic powerhouses in each country. It raised questions about who should make the decision on how forests are used or exploited. This brings us back to the question asked earlier: 'Are forests a "global commons" and no longer strictly the right of individual countries to control and determine how they are used or exploited?' This is a question whose answer is still being debated.

Initially, international communities attempted to use trade boycotts to stop the export of tropical trees and to establish regulatory legislation on trade in tropical woods (Viana *et al.*, 1996). European environmental non-governmental organizations were the most vocal in attempting to halt the total trade of tropical woods (Vogt *et al.*, 1999). Since the boycott of all tropical timber would eliminate its economic value in the global markets, others raised fears that a total boycott would just increase the rate of forest conversion to other uses with higher economic return, such as agriculture and pastures for cattle grazing. As would be expected, governments in the developing world that would be affected by such a boycott reacted very negatively to any attempts to boycott the timber trade in tropical woods. They were extremely concerned that foreign legislation would control resource uses within their country and they would lose their sovereignty. The backlash against trade boycotts eventually led to the development of voluntary systems, which were market-based, to certify that forest resources were being managed in a sustainable manner, i.e. forest certification (Chapter 7, this volume).

The clean development mechanism (CDM), a mechanism to achieve the reductions of greenhouse gas emissions specified in the Kyoto Protocol, has raised new issues regarding land sovereignty of local people in developing countries. CDM projects rely on the decision of the international community that emissions reductions in industrialized countries can be substituted with emissions avoided in developing countries when trees are planted to sequester carbon. Thus, forming new forest plantations in a tropical country, where the trees sequester carbon from the atmosphere into biomass, can be balanced against the emissions from industry in an industrialized country to eliminate the build-up of gases causing the greenhouse effect. Serious criticism has been raised regarding the sovereignty of indigenous peoples in CDM projects. When land is converted to forest plantations or locked up to sequester carbon in tree biomass, this reduces the land area available to local people for the collection of the natural resources needed for survival. The effect is similar to what happened to local people when nature reserves were originally established in tropical forests. These projects preferred planting fast-growing exotic eucalyptus species in monoculture plantations. The introduction of this species has generated some controversy. It has been suggested that it caused a decrease in the available water due to their deep roots accessing groundwater, typically not available to native plants.

CDM projects do have positive benefits such as:

- mitigating the emissions of greenhouse gases into the atmosphere;
- providing ecological benefits in terms of afforestation of degraded lands;
- providing financial benefits: (i) in the form of carbon credits for those who are polluting but have provided resources for afforestation of degraded lands, and (ii) for local communities to replant degraded lands.

Many of the issues being raised in the tropics focused international attention on the need to link social, economic and ecological objectives to resolve these environmental problems. The 1987 Brundtland Report was a significant international document that called for the integration of social, economic and environmental factors in resource uses so that sustainable development would be possible (WCED, 1987). The Brundtland Report considered sustainable development as the only approach that would deal with both global poverty and the continuing degradation of environments in the less developed countries.

Integrating the social, economic and ecological objectives to produce sustainable resource uses is a worthwhile goal even though it is difficult to achieve these objectives when a poorer country pursues rapid economic development or when that country is involved in armed conflict (Chapter 5, this volume). Since timber is a valuable economic resource traded in global markets, social and ecological objectives with regard to those resource uses have taken second place to economic considerations. Timber is used by countries to pay for the costs of fighting in a war, to pay for economic recovery after that war or to pay retributions to the winners of the war. Since timber was one of the few resources with value in global markets, Germany paid some of their retribution payments after the Second World War by cutting down their trees. After the Second World War, there was a major boom in house building that occurred in the United States, which fuelled the demand for wood.

Cambodia is another example in which timber has been used to rebuild a country economically after a war (Le Billon, 2002). Cambodia was ravaged by two decades of warfare starting in the 1970s, which resulted in the genocide of a fifth of its population and a Western-led economic embargo that devastated its economy (Le Billon, 2002). In the early 1990s, the focus was placed on the transformation and selling of timber located in forests in the hostile territory previously controlled by the Khmer Rouge. Timber was the most valuable and internationally accepted trade good and allowed them to take control of economic resources and to fuel economic development (Le Billon, 2002). By the mid-1990s, exploitation of forests provided about 43% of export earnings in Cambodia, which was higher than in any other country at this time (Le Billon, 2002).

Today, the rapid economic expansion of China is replaying the history of what a country goes through when becoming industrialized. China needs wood to fuel its industrialization. Illegal loggers are supplying this demand for wood from other tropical regions of the world (WWF, 2005). China needs to import their wood since it enacted a logging ban on cutting trees in much of the Yellow River and Yangtze River basins after devastating floods related to upstream deforestation in 1998 (AFPA, 2004b). Today, China is one of the major destinations for illegally harvested timber and it has the world's second largest market for timber (WWF, 2005).

Chinese logging companies have, furthermore, become more and more internationalized. In Surinam, Chinese loggers were by far the largest producers of round wood (i.e. whole logs) in 2000–2001 and the export of Surinamese wood to China exceeded by fourfold what was being exported to other destinations (MacKay, 2002). Aided by armed troops from the Surinamese military, these activities have led to a dramatic increase in deforestation rates for this country. Deforestation rates were especially high in areas traditionally used for subsistence by the indigenous Maroon tribes (MacKay, 2002) – the descendants of escaped slaves, who fought for their freedom from slavery and formed autonomous communities within the Surinamese rainforest in the 17th and 18th centuries. The global exploitation of tropical rainforests is being driven by distant countries requiring timber and aided by local governments, often without consulting or obtaining the consent of local indigenous peoples, who depend on these same forests for survival. The end result of this is the further impoverishment of these communities and higher rates of deforestation.

The inclusion of social, economic and environmental factors in sustainable development has been difficult to implement and still remains an elusive goal (Sayer and Campbell, 2004). In the tropics, ecology constraints dictate what can be grown for food and what population density can survive from these lands. This will be discussed next.

Ecological Constraints on Society Dependent on Survival from Resource Extraction from Forests

Humans have been very good at modifying their habitat to make it more liveable, which is why humans can be found living in all parts of the world. The amount of energy needed to survive and to keep the habitat in a condition where humans can live, however, makes some parts of the world difficult for continuous human inhabitation (Chapter 2, this volume). In these environments, the constraints of the natural environment (a big constraint is water availability) limit the ability of humans to live beyond a subsistence level. Our failure to recognize these constraints has caused the collapse of civilizations as these environments were degraded (Chapter 1, this volume – Sumerians).

Today the tropical regions do not support high population densities and those people who survive in forests survive at a subsistence level. These historically lower population densities in the tropics also explain why 52% of the world's forests are found in this climatic zone (FRA, 2000). When forests are poor locations for people to extract resources for survival, there is a greater chance that these forests will not be cut. This contradicts the recent discovery that more than 2500 years ago the Amazon forests did support larger communities of people (Lloyd, 2004). What factors allowed larger populations of people to live in the Amazon in the past are conjecture at this point. Some archaeologists have suggested that it was related to the existence of areas of more highly productive soils, which would allow more people to be fed (Lloyd, 2004). However, the climates were probably also very different 2500 years ago. So there are a number of factors that could be used to explain the larger population densities observed by archaeologists in the Amazon than is currently possible today. What is known today is that these forests do not support high population densities of people whose livelihood is derived from extracting resources from forests.

Historically, people have had more success surviving in forests located in temperate climatic zones than in tropical zones. This is the reason that forests in the temperate climate zones are more altered or no longer exist (Chapter 2, this volume). The current issues in temperate forests listed below are not unique to this climatic zone:

- Increased population growth.
- Increased per capita resource consumption.
- Converting forests to other uses.
- Forest fragmentation – edge effects; reduced connectivity, isolation effects; invasion of open habitat and edge species; roads are corridors for exotic species.
- Fire (see Case 5.3 – Wildfire in the Boreal Forests of Alaska).
- Insect pests (see Case 5.3 – Wildfire in the Boreal Forests of Alaska).

- Exotic organisms – these become pests because we: (i) introduce them across borders to habitats where they were previously not found and often have no native predators; (ii) alter or change habitats so they are less suitable for native species; and (iii) raise our standard of living (e.g. use chemicals to produce blemish-free fruit) (see Case 5.1 – Kudzu).
- Climate change.
- Genetic engineering – to produce either crops resistant to pests, by producing a herbicide when the plant detects the pest, or a plant capable of surviving with less water.
- Engineered wood products (see Case 7.8 – More Efficient Use of Trees to Produce Forest Products).

All of these issues are being dealt with to some extent by all countries and are further examined in other chapters of this book. A more detailed discussion of the tropics will ensue. The tropics face significant challenges (e.g. contain the poorest people surviving at a subsistence level, have the highest deforestation rates, are hot spots of plant and animal diversity) that threaten forest persistence in an environment with significant biological and ecological constraints.

Deforestation rates are higher in the tropics than in the temperate zones (Fig. 3.2). These data also show that the non-tropical areas have more previously unforested

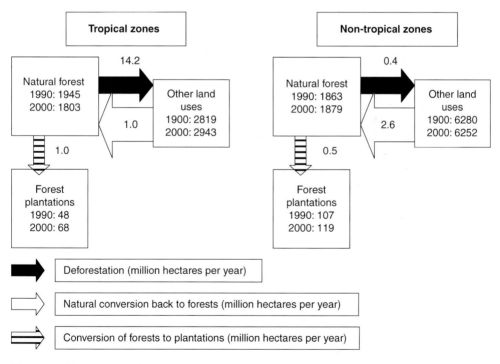

Fig. 3.2. The amount of natural forests and plantations in millions of hectares and the amount of deforestation, natural conversion back to forests from an unforested condition and conversion of forests into plantations in the tropics compared with non-tropical areas (from FRA, 2000). Units are in million hectares for each category of data.

land converting naturally back to forests. Forest land is also shrinking in the tropics while the area in forests is somewhat stable or increasing in the non-tropical areas. Because of these patterns of forest change, it is worth examining the ecological constraints that limit the ability of humans to survive in the tropics – the focus of this discussion will be on the American tropics because of the familiarity of the editors with these forests. The 'old' or highly weathered soils, with low nutrient capacity, in the wet tropics limit its capacity to regrow renewable resources sustainably and without adverse environmental impacts. This is also the region of the world where much of the global forest is located (Chapter 2, this volume). If the demand for wood products continues, it is very likely that much of the timber will come from these locations.

Several factors have been identified with deforestation rates in the tropics. The major factors causing deforestation vary depending on where one is located in the tropics. Deforestation has been blamed on the following:

- The shifting cultivators, because they cut down forests, burn the trees and grow agricultural crops.
- The timber demand from the industrialized world. The amount of demand for timber from the industrialized world varies with time and what other disturbances have occurred. For example, in the early 1990s, approximately 5% of the timber in Brazil was exported, so international buyers were not driving deforestation rates. During this time, most of Brazil's wood was used for fuel wood (85%), while the remainder satisfied local timber requirements (Costanza, 1991). However, after the hurricane season of 2005 and the destruction it caused in southern states of the United States, a considerable portion of Brazil's wood is being exported to the United States and is being consumed by the rebuilding efforts.
- Today, it is recognized that most of the deforestation occurring in the humid tropics is a result of conversion of forests into other land uses, such as agriculture or grazing for domesticated animals. For example, tropical forests have been cut down to plant oil-palm plantations in Asia and cattle ranches in South America.

The ecological constraints of the tropics for humans

Once one understands the ecological and social constraints of tropical environments, it is easier to understand why these environments will not support large human population densities. The characteristics of tropical forest regions can be summarized as follows:

- The only definition that is unique to the tropics, compared with other climatic zones, is the fact that the mean annual temperature is always greater than 24° C in the tropics. The tropics are hot.
- Agriculture is predominantly shifting agriculture because of the poor soils, which are depleted of nutrients quickly, and land has to be left fallow to allow the nutrient reserves to come back. Few agricultural crops are able to grow in the soil (e.g. cassava) because of low nutrient availability (especially phosphorus (P)) and soil chemical toxicity (aluminium (Al) toxicity) (Sanchez, 1976).
- The poorest people live here.

- People mainly live along the *varzea* or riparian zones, where the rivers play important roles in bringing in nutrients; it is easier to survive here since people can practise agriculture and can fish. The most nutrient-rich agricultural soils are found along riverbanks, where nutrient-rich materials are deposited by the river (Pinedo-Vasquez, 1999; Zarin, 1999).
- People are dependent on forests for medicine, building materials, game animals.
- There are locations of high hot spots of biodiversity for particular groups of organisms (e.g. insects, mammals) (Myers, 1984). This high biodiversity explains the significant international interest in this region for conservation purposes. Many of the resources harvested by local communities are the same species valued by international communities for conservation purposes (e.g. parrots, monkeys, crocodiles).
- The high plant diversity found in the Amazon tropics is a result of people's active management of the forests for hundreds of years (Pinedo-Vasquez, 1995; Raffles, 1998).
- Most Amazon *terra firme*, or non-flooded soils adjacent to rivers, are not dominated by widely marketable species of tree and most commercial species are widely dispersed (Pinedo-Vasquez, 1995). Commercial harvesting of trees is limited by the wide dispersion of desirable tree species across large landscapes of forests. It is difficult to find the trees and to transport the trees out of the forest so that it is economical.
- Most of these communities are extractive economies, where they collect natural forest products that lack large consumer markets. Those products that can be sold bring low prices, and income elasticity is caused by the inconsistence of world demand for these products.
- Deforestation occurring in the tropics has significant impacts on global climate (e.g. Amazon) because of its impact on global rainfall patterns (Gash *et al.*, 1997).
- Most indigenous communities do not have land tenure rights to the land on which they live. This lack of land tenure rights means that they have historically had little input on decisions being made on their lands.
- In South America, the original perception was that wet tropical forests were lush and highly productive because of the tremendous growth that the early colonialists saw when they first looked at these forests. However, these forests store most of their nutrients in their biomass and are normally effective at recapturing the nutrients needed for growth. When trees are cut and removed from large areas of the forest, nutrients are removed in the biomass removed from the forests. The nutrient-poor soils are ineffective in regrowing the forests at the same rate because their nutrient supply capacity is too low.

When considering the characteristics of tropical forests provided above, the high number of biological and ecological constraints to human survival as resource extractors becomes evident. If only one factor was used to identify tropical zones, only the high average temperature is unique to the tropics. Some of the factors previously mentioned above are found more commonly in the tropics but they are also general characteristics found in developing countries. For example, the tropics have a high proportion of wood use as fuel wood and a higher concentration of poor people living in the same area where high biodiversity of plant and animals species is found.

The tropics contain the highest biodiversity of plant and animal species but they do not have the highest biodiversities of all species in the world. For example, the tundra has a higher biodiversity of microbes when compared with the tropics. Even though the tropics are poor habitats for humans, approximately 500 million people live in or near forests and depend on it for food, fuel, fodder, timber and income (Patel-Weynand and Vogt, 1999). Most of these people are poor. Of these people, about 200 million people live in clearings or areas where the trees have been removed from the forest.

Shifting agriculture, e.g. slash-and-burn cultivation, is practised in the wet tropics because the soils do not allow for continuous agriculture production (Sanchez, 1976). In shifting agriculture, forests are initially cut in a small patch and then the felled trees are burned. These areas are then planted with an agricultural crop and then allowed to go fallow (no crops) when crop growth is drastically reduced. By allowing agricultural plots to return to a forested condition, farmers allow the soil to replenish its depleted nutrient supply because trees are more efficient at acquiring nutrients from the soil. Trees increase soil nutrient availability when their tissues, enriched with nutrients acquired from the soil, senesce or die and fall on the surface of the soil where they are decomposed by microorganisms.

In the Amazon, shifting agriculture allows agricultural crops to be grown for a few years followed by 10 to 20 years of fallow period (i.e. no crops grown and forest allowed to grow back). The soils closer to the Peruvian border of the Amazon basin have more nutrient-rich soils because weathered rock in the Andes is washed down by rivers and deposited in the flood plains (Pinedo-Vasquez, 1995, 1999; Zarin, 1999). However, the lack of mountain ranges at the mouth of the Amazon River means that little new soil is added here, so these soils are poorer and require longer fallow periods (Pinedo-Vasquez, 1995).

In the Amazon, people live along the riparian zones because these are the most productive areas for growing crops. This is where the best soils are found because the weathered materials washed down from the mountains are rich in nutrients (Pinedo-Vasquez, 1999). It is difficult to grow crops in those parts of the forests that are not enriched by rivers because the soils are so nutrient-poor and have high levels of aluminium, which is toxic to plants (Sanchez, 1976). Aluminium is toxic to most plants and reduces the number of crops that can be grown in a soil when it is present in high concentrations.

The number of food options is limited for people living in wet tropical forests. People cannot go into the forests to 'graze' the native plants that grow there since many/most are not edible. The high biodiversity of species means that there are a lot of insects and mammals (herbivores – eat vegetation and not animals) that eat plants. The plants have evolved to produce chemicals in their tissues so that they are chemically toxic or non-palatable to these herbivores. Interestingly, these same chemicals produced by plants to prevent themselves from being eaten are also what give them their medicinal qualities. Some food plants are found scattered in the tropical forests since the shifting agriculture means that families move around in large tracts of forest and abandon their homes as they move to a more nutrient-rich area. As they leave, home gardens are left behind and many of these plants continue to grow in the forests.

In the Brazilian Amazon, an important food crop can be poisonous to humans if not prepared properly. After the Spanish conquest of this region, cassava or manioc replaced maize as an important food crop. People grow cassava because it is tolerant of the high aluminium levels found in these soils and is high in carbohydrates. Cassava is, however, high in cyanide, which has to be removed before the cassava roots can be eaten. A process for eliminating the cyanide has been developed, which consists of placing cassava roots in water (river or lake) for 1 week. This process allows the cassava roots to ferment and therefore to volatilize (evaporate) the cyanide from the roots. The carbohydrates remain behind in the roots. The presence of cyanide increases the amount of sugar that is stored within the plant so that indigenous communities prefer these plants instead of those that have lower carbohydrate contents (and also lower cyanide).

When the Europeans arrived in the Amazon region, they introduced cattle to flood-plain areas (Raffles, 1999, 2002). Cattle were non-native to South America. This was the most significant impact the Europeans had in South America. Cattle ranching was economically sustainable on the flood plains since the constant addition of alluvium (soil) by the river maintains the growth of vegetation. This is the same reason that so many people lived along river systems in the Amazon since they could grow agricultural crops more continuously. Recently water buffalo are being allowed to graze along the riparian areas and are causing significant degradation to these areas (Raffles, 1999). Their trampling is resulting in large chunks of land being dislodged and washed into the tributaries of the Amazon, continually losing usable agricultural land for the *riverenos*, or river-dwelling people.

A typical family living on a tributary of the Brazilian Amazon is large in size (13+ children is not uncommon), does not have much money to buy goods and usually trades for needed items. The families grow a few crops that are sold in the marketplace to buy clothes and medicine. Everything else is traded and collected from forest or agricultural plots. The resulting lack of opportunity has caused some/many women to turn to prostitution to bring money back to their families.

The human lifespan is short here, only half of what people living in the industrialized world can expect. There are many incurable diseases found in these hot wet tropics and few medicinal cures available at present (Chapter 5, this volume). This inability to cure diseases is ironic because of the high number of plants found growing in this region with secondary chemicals that have been identified as having medicinal value. Another startling statistic is that at least three-quarters of the world's population is dependent on folk medicine – a significant amount of which comes from forests.

International perceptions: a product of fallacies and misunderstanding of tropical forests

The fallacies and misinformation about the tropics held by international groups have determined their values and what decisions they push for these forests. Some of these fallacies are summarized in Box 3.2. Most of them introduce the idea that tropical forests are virgin or untouched by humans. Because of this view, it was believed that it was better to remove the native people from these forests for conservation purposes when humans were found living here.

However, it is clear today that there is no such thing as a virgin or non-human-impacted environment in the Amazon or in other tropical regions (Roosevelt, 1999;

Box 3.2. Fallacies or misinformation that we have about tropical forests.

- Lush, highly productive forests
- Forests, soils 'virgin' – not impacted by human activities
- High plant diversity is natural, not influenced by the indigenous peoples
- Forests fragile, highly susceptible to degradation with any human use
- No fires in wet tropical forests – fire major disturbances in wet tropical forests during El Niño cycles; used to think were too wet and no fires present
- Sustainability of tropical forests is uniquely different from what is found in the temperate zone
- Forests can provide high standard of living for indigenous communities
- Commercialization of non-timber forest products is a viable economic option for local communities and can reduce deforestation rates
- Non-timber forest products are an economically viable alternative to deforestation
- No local-level governance present
- Indigenous communities need to be taught to be environmentally friendly
- Most deforestation driven by demand from the more industrialized world
- Human diseases commonly come from the tropical forests (malaria was introduced into the New World by Europeans and not the other way round)

Raffles, 2002; Willis *et al.*, 2004). Biologists and ecologists traditionally saw the Amazon as being so fragile that even native foragers and horticulturalists would destroy it. This spawned the view that the best tropical forest was one free of humans:

- Because forest people were considered a threat to conservation. In the 1980s and early 1990s, international sustainable development and conservation projects tried to take control away from native people.
- Since biodiversity was viewed as being entirely due to natural factors and needed to exclude people.

When these fallacies are the basis of the solutions developed for these forests by the international communities, it is understandable that so few solutions have worked. These original ideas were extremely incorrect and unfortunately resulted in solutions for the tropical forests that have not worked. For example, when tropical forests are considered 'virgin' or untouched by human hands, the solution is to remove people living in these habitats. Since much of the forest looks like it does today because of humans living here for hundreds of years, it is clear that removing them will alter these landscapes dramatically (Pinedo-Vasquez, 1995). Within the last decade, it has become clear that humans living in the tropics were active managers of their environment and the trees growing in the forest (Moran, 1995; Pinedo-Vasquez, 1995; Roosevelt, 1999; Willis *et al.*, 2004):

- They deforested, introduced new tree species, encouraged the growth of weedy species because they provided products for their survival and they practised active cultivation of species.

- Many trees found in these 'virgin' tropical forests are in fact either domesticated plants, plants that have escaped from human inhabitation, or pre-adapted to human disturbance. Near the villages, people created anthropogenic forests of palm, bamboo and food/fruit-rich trees.
- Large human-produced rubbish dumps have been found from 1000 years ago along the major branches of the river systems and in the interior forest areas.
- Some biodiversity in Amazon forests exists today because of anthropogenic soil enrichment that occurred with rubbish from human settlements, which built up over time and made the soils usable in previously inferior areas for agriculture.

The high biodiversity of species is also misleading since it suggests that there are boundless supplies of trees that can be harvested for commercial uses. Most of the desired, commercial trees species are, however, widely scattered in a natural forest. Therefore locating and collecting these few trees is not economical because of the huge transportation costs. This explains why plantations are commonly found in the tropics (Fig. 3.2) when growing trees for commercial purposes. Plantations can be economically more viable, but are challenging due to the large variety of herbivorous organisms occurring in the tropics. As mentioned above, plantations essentially provide a concentrated source of food for the herbivore specializing in the trees grown there. Most of the other native tree species are poorly known so they are not widely marketable.

Solutions introduced by international organizations for the tropics

In the tropics, many different solutions have been proposed by international organizations who were attempting to simultaneously obtain conservation goals from these forests while providing economic opportunities for communities using local forest resources. Several of the solutions are based on developing economic markets for extracted forest resources or where local communities earn income from servicing ecotourism operations. Increasing the extraction of resources from these forests, however, is not sustainable when the biological and ecological constraints are not explicitly considered prior to the collection of resources.

A few of the approaches that have been pursued are provided below, with a summary of how successful this approach was:

- Economic incentives for products certified to have been collected in sustainably managed forests. Examples of forest products that have been certified are timber and coffee grown in the understorey of a tropical forest.

    Result: Higher economic return has not materialized for the people or communities collecting coffee beans certified to have been collected in a sustainably managed forest nor have higher prices resulted for timber sold with a certification label.
- Developing economic markets for non-timber forest products and other forest products (e.g. Brazil nut, palm heart, *açai*, hardwoods – mahogany, teak).

    Result: Depletion of non-timber forest products has resulted from collecting materials from forests where the resources are not managed to ensure that new plants will replace the ones collected. Furthermore, no management plan determines where non-timber forest products should be collected and the timing of resource collection that considers the ecological constraints of the site. Since no one is overseeing the collection of non-timber forest products, much of it is illegal and over-exploitation is common. Collection of non-timber forest products is currently non-sustainable. Padoch (1992) noted the following results for developing a non-timber forest products project: the general public was marginalized at the expense of a few participants in the programme; resource use rights expropriated by powerful urban dwellers; ruthless exploitation of tribal minorities; no conservation; international community values overrode local and regional values for resources.

- Integrated development and conservation projects were designed to promote the conservation of biodiversity while improving human living standards by ensuring that economic opportunities are provided for communities setting aside forests for conservation. International organizations learned in the early 1990s that conservation projects were unsuccessful if they ignored the survival and livelihood of local people living in the conservation area.

   Results: Most integrated conservation and development projects have been unsuccessful in simultaneously pursuing the two goals of conservation in forests and developing sustainable communities. Today, international funding organizations are shifting to supporting poverty reduction at the expense of conservation projects.

- Ecotourism or an ability of local communities to obtain economic return from tourists interested in visiting and vacationing in a forest or reserve because of its ecological and conservation values. There are several different types of job opportunities associated with ecotourism: tourist guides, managing hotels, running restaurants, fabricating artistic items to sell to tourists, etc.

   Results: Ecotourism has not generated the jobs and income expected by local communities. People from the industrialized world are managing many ecotourism lodges.

- Extractive reserves are legal structures in which the government empowers and entitles local communities to collectively manage their forests and to sell forest products for income. Extractive reserves were formed in Brazil by Brazilian grass-roots organizations, international environmental groups, multilateral funding agencies and local governments. People live in the extractive reserves and manage the forests for multiple purposes, which include agriculture. This is considered an alternative conservation and development approach but where the local communities make the decisions on what to do in the forests.

   Result: When managed by local or regional people who live in the reserves, this has worked quite well and empowered communities to manage their own resources. A good example of a success story is the National Council of Rubber Tappers, which established many reserves in the Brazilian Amazon. Part of their management has consisted of determining what resources and species existed in each extractive reserve, and the development of management plans.

- Collecting plants with pharmaceutical qualities and selling to pharmaceutical companies. Communities are paid for the plants (frequently by a company building the community a schoolhouse). In return, the communities provide the knowledge of which plants might have medicinal value.

   Result: This has not worked well since most pharmaceutical companies artificially produce the compound as soon as it has been isolated from a plant. Pharmaceutical companies have to produce the compounds artificially since the plant content of the compound is highly variable and the content can also vary depending on where the plant is growing. So, once a compound has been identified, continuing to collect this plant from the forest has little value and provides little economic return to the community.

- Clean development mechanisms (CDMs) of the Kyoto Protocol and carbon sequestration in forests where carbon credits are provided for growing trees (see Chapter 6, this volume for more detail on CDM and the Kyoto Protocol). Local communities can receive carbon credits for planting trees, especially in areas previously not forested, which varies depending on how much carbon is sequestered. Polluting the environment results in CO_2 being added to the atmosphere. Growing trees sequesters CO_2 and, since trees grow faster in the tropics and land is cheaper, tropical tree plantations offset CO_2 accumulation and reduce greenhouse gas risks. First started in 1988, when a coal-fired power plant in Connecticut agreed with the government of Guatemala and others to plant 15 million trees, which they calculated were necessary to offset the estimated amount of CO_2 emissions expected from the plant over the utility's 40-year lifespan (Goodland *et al.*, 1991).

  Result: This has worked for organizations that have planted trees and the linking of the tropics to the carbon-trading markets has resulted in higher prices per ton of carbon sequestered in a forest. How many indigenous communities have a better livelihood is less clear. Communities may lose forest areas to these projects that were previously used for subsistence or resource allocation. When forests are used to sequester carbon, these forests are not supposed to be used to collect wood for fuel wood or for other uses. When exotic fast-growing species (e.g. eucalyptus) are planted, there has been concern about whether it is depleting the water resources of an area and there is also concern about the planting of an exotic species that may become invasive.

- Trade embargoes to stop the buying of timber being harvested from forests so that the loss of markets might decrease the cutting of trees. This was originally used in an attempt to control deforestation in the Brazilian Amazon by eliminating global markets for wood from these forests when the deforestation rates were very high.

  Result: This approach did not work because Brazil was outraged at other countries making decisions regarding their forests. This raised serious questions of sovereignty. The countries attempting to implement a trade embargo backed down.

- Land tenure to indigenous communities so that they have more control over the selling and marketing of forest resources from their forests.

  Result: Some indigenous communities have received land tenure but most governments are not interested in providing them the amount of land that is needed for forest-dependent communities to survive from collecting forest materials. A family living from the forest needs larger tracts of land since the productive capacity of the soil is lower in the tropics.

Extractive economies present some unique challenges, since market prices ultimately determine how well they work and the scale of operation is not very conducive for a market economy. When market prices were high for specific rainforest commodities and/or government subsidies were available, extractive economies have been successful.

However, most of the Amazon *terra firme* (land areas not flooded by a river) forests are not dominated by widely marketable species and the labour costs for collecting commercial species are high because they are dispersed in the forest. There also does not appear to be large consumer markets for most natural forest products and other global providers compete to supply these consumer markets (Browder, 1992). An example of this is certified coffee grown in the understorey of a natural forest in Central America. Coffee prices have drastically dropped in the American tropics since the World Bank supported coffee growing in the Ivory Coast during the last decade. The supply of coffee beans in the global market was too high so the coffee farmers in Central America did not make enough money to cover their costs to grow the coffee.

It is important to recognize that extractive economies have a role to play in tropical forests but should perhaps not be considered as the primary solution for sustainable development. When it is made the primary mode of economic development, the ecological constraints make this practise unsustainable. Padoch (1992) recorded how the rise and fall of the river isolates communities for months at a time because of the dangerous waters, and how the markets move to accommodate the rise of the river, which can annually change 9 m in height. The environment dictates, therefore, when some communities can participate in markets.

Padoch (1992) suggested that continuing to maintain extractive economies to supplement incomes might be a strategy worth pursuing. This means that extraction of resources should not be the primary income generator in any community. What are some alternatives? Extractive economies appear to work well when they function as a buffer for local communities during times of unemployment (Padoch, 1992). For example, 'Marketing, especially small-scale wholesaling of forest and agricultural products, is a refuge of many urban unemployed as well as farmers, fishermen, and forest collectors down on their luck' (Padoch, 1992).

Forest product uses in the tropics: are they sustainable?

Forest product use in the tropics can be sustainable if resources are harvested considering the ecological and social constraints that exist here. It is important to recognize the ecological roles of the species in the forest since there is a strong likelihood that over-harvesting will occur when an economic demand is created. The loss of some species may decrease the growth rates of the entire forest (e.g. palms maintain the calcium concentrations in the soil and therefore its nutrient status) or impact species dependent on particular plant species for food (e.g. figs flower almost continuously and therefore provide food for birds during periods of less flowering). For example, constructing roofs made of palm thatch on tourism lodgings has resulted in the over-harvesting and illegal collection of palm fronds from conservation areas in Belize (O'Hara, 1999). This has resulted in the extensive loss of palm in some forest areas.

Another factor that needs to be considered is that most of the products that would be sold in the markets also have ecological roles in maintaining the nutrient status of the forests (O'Hara, 1999). The collection of materials from forests to feed animals or as fertilizer for agricultural fields is not ideal because these materials do maintain forest growth for the long term. This is the reason that shifting agriculture is practised here. The poor soils mean that collecting live plant material from a forest will further reduce its nutrient status. It is the decomposition of this leafy material that restores the nutrient status of the soils and without it the soil loses its ability to grow anything.

In many tropical regions, foliage from forests is already collected as fodder for animals or is added to agricultural fields as a fertilizer to enrich these soils for growing crops. A considerable amount of forest material is removed from forests for enriching agricultural fields or to feed domesticated animals. In Nepal, 40% of the annual feed used to raise buffalo is leaves collected from forests, while it is 25% of the feed for a cow (Patel-Weynand and Vogt, 1999). In dry tropical forests, livestock do not survive without forest grazing but the removal of these materials from the forests means that they are no longer contributing to revitalizing the nutrient status of the soil. Suggesting the removal of new non-timber forest products to sell in markets does not consider the role that these materials have in the tropical ecosystem.

Many examples exist where trees have been cut in mountainous regions and have devastated many low-lying areas when flood events occur. Continual removal of trees will eventually decrease the productive capacity and increase the erodibility (erosion capacity) of these soils even further. This very scenario was recorded in China during the 1980–1990 time frame, with devastating floods as late as 1998 (AFPA, 2004b; Diamond, 2005).

Many of the people who collect and make products from non-timber forest materials (e.g. palms, cinnamon, medicinal plants) are now facing the need to respond to global markets and to understand how global markets work. Since extractive economies lack large locally or regionally based consumer markets, transporting these materials to markets is difficult and results in too many middlemen profiting from these products. This means that the person who made the product ends up making very little money. Other issues that the producer of the non-timber forest product has to worry about are whether the materials will spoil or degrade before getting to the markets. Many non-timber forest products will not spoil or degrade but still have to satisfy a global consumer market requiring uniformity in product quality. In the Philippines, Case 7.6 documents how non-timber forest products are providing a sustainable livelihood for villagers but global markets are shifting to regions where it is possible to construct these products more cheaply (e.g. China).

The problem with most of the solutions suggested for the tropics is that they do not add value to the forests in non-traditional ways. They are a purely extractive mode of resource use, where the economic value is determined by international markets (which are usually low) and where substitutes are readily available. Despite this, forests contribute almost three times as much to the economies of developing countries in comparison with what they contribute to the economies of developed countries.

Another problem is that these forest resources are also an important food source for local people. If selling to regional markets and urban dwellers provides higher economic return, the already limited food options are further reduced for people surviving in the forests. An example is *açai*, a fruit of the palm that is high in protein and is an important food source for a family in the Amazon. How well these families survive is dependent on how well they manage their palm plants. Now *açai* is sold as a drink to the 'exercise generation' because of its high protein content. When the palms are over-harvested, the local people are left with no economic product and the loss of an important food source.

CASE 3.1. DEBT-FOR-NATURE SWAPS, FOREST CONSERVATION AND THE BOLIVIAN LANDSCAPE (GRETCHEN K. MULLER)

Introduction

The high deforestation rates in the tropics emerged as one of the most important environmental crises in the 1980s and they continue to play a prominent role in the agenda of international conservation organizations. Multiple factors complicate the formulation of solutions and the developing of mechanisms to mitigate the high deforestation rates. Conservationists began generating economic incentives to reduce these rates by developing initiatives such as forest certification, marketing of non-forest timber products, ecotourism and debt-for-nature swaps (DNS). Although these initiatives have made important contributions to conservation policy, their efficacy in terms of long-term ecological, economic and social sustainability remains unclear. In order to reduce deforestation rates and optimize conservation funds, it is critical that current environmental mechanisms are evaluated and further refined when they are shown to not achieve their goals.

This case study evaluates one such conservation mechanism, debt-for-nature swaps, by tracing the DNS mechanism from its origins within the international conservation arena to the national and local level. These groups are distinct and yet interrelated and the long-term success of the DNS approach is dependent upon the involvement of the three parties. This study focuses on the local-level implications at the site of the world's first DNS in the Beni Biosphere Reserve, Bolivia. After a series of interviews with residents within four communities both inside and within close proximity to the reserve, several themes emerged.

The research indicates that the establishment of the reserve and the money leveraged to enforce the reserve's management plan have protected the livelihood of the Chimane indigenous communities and economic conditions are linked to a community's proximity to the reserve. It was also determined that forest composition has changed, as has community proximity to the interior forest and the types of wildlife that are commonly seen now versus 16 years ago (prior to DNS). Additionally, community participation in the redelineation of the reserve boundary and buffer zone was low. These findings will be discussed in greater detail,

as will some of the unintended consequences that emerged across the national landscape during and after the negotiation of the debt swap. The DNS was one of the smallest in Bolivia; however, much was learned in terms of the importance of maintaining a transparent negotiation process, enlisting government and public support and incorporating local communities in decision-making.

Debt-for-nature Swap Emerges as Conservation Tool

Dr Thomas Lovejoy proposed the debt-for-nature swap concept to leverage money to support conservation efforts while simultaneously reducing hard-currency debt. This idea evolved from the observation that much of the world's biological diversity was harboured in many of the same countries that face the greatest financial strain from foreign debt burdens. Bolivia's drive to exploit natural resources is believed to be linked to its economic crisis, which has accelerated the fragmentation of pristine forests over the past decade. This finding, coupled with Bolivia's biological and cultural diversity, provided an ideal landscape to negotiate the world's first debt-for-nature swap. The swap was negotiated in 1987 and leveraged funding for the Beni Biosphere Reserve, a 135,000 ha reserve of Amazonian lowlands in northern Bolivia.

The Beni Biosphere Reserve (BBR) received its first source of long-term funding from the world's first DNS, and therefore provides the longest time frame within which to analyse the local-level perceptions of the social, ecological and economic conditions prior to and since the DNS negotiation. It is important to note that the financial support generated from the swap is one of many funding sources received to meet the reserve's operating costs and this research does not attempt to make direct links between the impact of the DNS and local-level perceptions. However, delving into these perceptions provides relevant information in determining the overall efficacy of the reserve and presents the conservation community with information that can aid in the design and implementation of future DNS and the programmes they fund.

Tracing the Debt-for-nature Swap across the Bolivian Landscape

Residents from the communities of El Cedral, San Antonio, El Totaizal and Puerto Mendez were interviewed to gain insight into perceptions of the ecological and community viability of the landscape both within and around the reserve before the inception of the DNS and today. These communities were selected based on their distinct characteristics such as proximity to the reserve and ethnic make-up. El Cedral is a Chimane indigenous community and is located within the reserve boundary. San Antonio is the closest Chimane community to the reserve and is located in the buffer zone. Puerto Mendez is far away from the reserve but located within close proximity to the buffer zone and is also Chimane. El Totaizal is a 'Mestizo-campesina' community located within the buffer zone. The information obtained from the respondents in each community contributes to an increased awareness about the local landscape.

The Beni region has been influenced by many factors, such as the establishment of the Beni Biosphere Reserve, the redelineation of the reserve boundary and the influence of logging and ranching interests in the area. The creation of the Association for Chimane Indigenous Communities and Chimane Indigenous Territories has also influenced the local landscape. One of the indicators used to determine the impact that the establishment of the Beni Biosphere Reserve has had on the livelihoods of the Chimane was identifying the dietary practises of the residents within each participating

community and whether these practises have changed since the establishment of the park.

Current literature about the dietary practices of the Chimane states that they rely on hunting and agriculture. Their traditional primary source of protein comes from animal wildlife. The results in Table 3.4 reiterate this and demonstrate the link between traditional livelihood and forested areas.

Table 3.4 shows that, among the families interviewed in El Cedral, all stated that their primary source of protein derived from bush meat. As this community is located within the reserve, they have direct access to the interior forest and thus its wildlife. The results from the communities of San Antonio and El Totaizal depict a mixed diet of bush meat and beef, with 100% of the respondents in San Antonio stating that their diet consists of more beef than was previously consumed. Among the respondents in El Totaizal, 70% stated that they consume more beef than before as animal wildlife is further away. The responses from Puerto Mendez revealed that bush meat is no longer consumed because, as all of the respondents stated, the animals are located too far away from their community to allow for hunting. Since this community is located further away from the reserve than the other three communities, assumed links can be made as to the percentage of bush meat consumed and access to the reserve. The extensive deforestation that occurred after the redelineation of the BBR and buffer zone left certain Chimane communities far from forested areas and this appears to have resulted in dramatic changes in lifestyle characteristics such as diet.

Table 3.4. Dietary practices.

Community	Primary source of protein			Has this changed since establishment of park?
	Bush meat	Mixed bush meat and beef	Mixed fish and beef	
El Cedral	100%			100% same but hunting easier before
El Totaizal (Mestizo-campesina)		100%		100% eat more beef
San Antonio		100%		70% eat more beef because animals are further away
Puerto Mendez			100%	100% hunting far away

It was also determined that much has changed in terms of forest composition, community proximity to the interior forest and the types of wildlife that are commonly seen now versus 16 years ago (prior to DNS). Local perceptions on the difficulty of hunting, distance travelled for wood, types of forest products used, soil conditions and degree of wildlife change demonstrate that a perceived change in the forest boundary and changes in the forest composition have occurred regardless of the communities' proximity to the forest boundary. However, more dramatic changes were reported in those communities that are further from the forest. This suggests that population pressures coupled with regional land-use activities such as ranching and logging have caused changes in the forest. These results provide the reserve with early warning signs that the reserve is not isolated from many of the political, social and ecological challenges that plague the region. Furthermore, the steady encroachment of private logging and ranching interests into surrounded areas is creating an isolated patch of forest amidst a deforested and economically impoverished landscape. As the reserve becomes increasingly isolated from larger tracts of forest, the vitality of the communities that rely on a subsistence economy for their livelihoods will also become increasingly threatened.

A redelineation of the park boundary and its buffer zone occurred in the 1990s. This altered the regional context surrounding the park and is thought to have impacted the livelihoods of nearby communities. As the redelineation was to affect the protected status of a significant portion of the landscape as well as the communities that would no longer be within the protected-area boundary, it seems that community involvement in the process would have been invaluable. Based on the findings of this research, community participation in the redelineation of the reserve boundary and buffer zone was low. Several unintended consequences arose from their exclusion in terms of pushing indigenous rights into the political arena.

Two Emerging Movements: Indigenous Rights and Conservation

The Beni province is home to a large group of Chimane indigenous communities, many of which were living within and around the BBR prior to its delineation as a reserve. Although the park is viewed as having positively impacted local indigenous communities by maintaining their livelihoods, there were a series of conflicts between the Chimane and the logging and ranching interests in the 1990s as a result of land-use practices within the region. In response to the patchwork of varied and conflicting land uses within the region, a committee was created with the intention of constructing a regional land-use plan. The committee, referred to as the Interinstitutional Technical Commission, consisted of owners of timber concessions, resource users, the Regional Forestry Service, the National Environmental League, Bolivian scientists and national and international conservationists. The only groups not represented on this committee were the indigenous communities.

The committee made a series of decisions that impacted land uses within the region and included a confirmation of the redelineation of the park boundary as well as granting logging concessions to timber companies in a portion of forested area surrounding the reserve. The lack of participation on the part of indigenous groups living within the area and the logging of 650,000 ha triggered the Chimane Indian 'March for Dignity and Territory' in 1990. Indigenous leaders mobilized hundreds of indigenous people on a 500-mile trek from the Amazon to the Andes over a course of 33 days. In response to the indigenous movement, President Paz Zamoro issued the executive decrees that changed the socio-economic and physical composition of the region. The march is credited for placing indigenous rights on the environmental agenda in Bolivia and forging an alliance between two emerging national movements.

Conservation Evolves to Include Community

Deforestation and biodiversity loss continue to pose significant international conservation challenges. Few other natural resource conflicts have been faced with the political, ecological and social complexities that surround these issues. To prevent further mismanagement of the world's forests, innovative mechanisms, such as DNS, which generate funding for conservation programmes that enhance local economies must be further refined. While the international donor community shifts its attention away from biodiversity to focus its efforts on poverty alleviation and ecosecurity, it is imperative that viable tools like DNS are not lost.

The debt-for-nature swap mechanism is well suited to address both social and ecological concerns across a variety of landscapes. It is imperative, however, that future DNS negotiators better understand the contextual landscapes of the proposed conservation programmes prior to their design and implementation, as the human dimensional aspect of conservation cannot be ignored if conservation efforts are to succeed. It is equally important that these mechanisms fund programmes that operate under a strategy to uplift the socio-economic status of resource-dependent communities while reducing stress on the resources to be conserved. In order to remain a viable option for generating conservation funds in the future, the DNS approach must continue to evolve simultaneously with the increased knowledge and experiences gained from research, evaluations and lessons learned from collective experiences in the conservation community.

CASE 3.2. CATTLE, WILDLIFE AND FENCES: NATURAL DISASTER AND MAN-MADE CONFLICT IN NORTHERN BOTSWANA (SHERI L. STEPHANSON)

In 1996, contagious bovine pleuropneumonia (CBPP), or cattle lung disease, raged through northern Botswana and threatened to destroy an important economic asset and significant cultural symbol. The national government eradicated the disease by killing 320,000 head of cattle throughout Ngamiland district (Mullins *et al.*, 2000). To prevent it from spreading further, they rapidly constructed a series of cordon fences. While effective in controlling the disease, the fences also impeded migration for numerous species relying on the bounty of the Okavango Delta. The economic impact devastated the region's cattle owners, for whom cattle is key to wealth accumulation, and disrupted associated socioeconomic relationships. Concerned about protecting the rich biodiversity of the Okavango Delta, an international conservation non-governmental organization (NGO) inserted itself into this context to advocate for the realignment or removal of the fences. Operating within the parameters of Botswana's Community-based Natural Resources Management (CBNRM) programme, it negotiated with local stakeholders to develop options for sustainable resource use in the northern sand veld.

Negotiating successful agreements rests on the legitimacy of both the process and the facilitator of the process. Social scientists contend that if stakeholder interests are represented in a legitimate negotiating process, the resulting decision will be more sustainable in the long term (Brechin *et al.*, 2002). Establishing legitimacy requires a process of fair and transparent dialogue, in which stakeholders trust that their interests are represented (Appelstrand, 2002). In conditions where economic and political instability prevail, trust can be tenuous, and establishing it can be a continuous process. An understanding of these social dynamics must accompany any effort to successfully build and maintain trust. Failure to understand them can contribute to unintended conflict among stakeholders and undermine progress towards sustainable natural resources management. This case study illustrates how one such failure occurred, jeopardizing the NGO's reputation and adding to historically embedded social conflict.

The Okavango Delta: Where Nature and Society Converge

The Okavango Delta is the world's largest Ramsar wetland site measuring 6,864,000 ha and home to several of Botswana's ethnic minorities, which have survived in the region for centuries. The permanent swamp, seasonally flooded wetland and dry land provide habitat for many large mammals, including 14% of Africa's elephant population, zebra, lion and wild dog. Over 600 species of birds migrate through the region and 100 species of fish live in its waters. Several distinct ethnic groups, each with its own history and livelihood mechanisms, also call the Delta home.

Ethnic groups

Three ethnic groups play an important role in this case. The Bayei and Hambukushu raise cattle, farm and fish. In Botswana, cattle play a prominent role,

in both social and economic terms, as cattle ownership indicates wealth, status and livelihood security.
The third group, the Basarwa (Basarwa are alterna-
tively referred to as San or Bushmen, or by names
that identify a specific tribe, such as the Bukakwhe
depicted in this case), are southern Africa's aboriginal people, who migrate seasonally to hunt and
forage in small family groups. All three of these
groups belong to Botswana's ethnic minority,
but the Basarwa fall at the bottom of the power
hierarchy and suffer from societal marginalization
(Taylor, 2000). Throughout the latter half of the
twentieth century, the government of Botswana
actively encouraged them to settle by offering to
provide health and education services in newly
established villages. Once they settled in villages,
the Basarwa struggled to survive in an economically
depressed region that offered few employment
opportunities for unskilled workers. Alienated from
their traditional way of life, many Basarwa became
reliant on government welfare programmes to meet
their daily needs.

Land tenure and resource management

Three categories of ownership comprise Botswana's
land tenure system: customary, or communal land
with usufruct rights, freehold, or private ownership,
and state land. State land makes up 23% of
Botswana's territory and, in rural areas, some of this
land is zoned as wildlife management areas (WMA)
(Mathuba, 2003). WMA zoning allows for wildlife
utilization and some subsistence activities – in the
Okavango Delta, wildlife utilization generally consists of commercial hunting and photographic safaris.
These zones are further subdivided into controlled
hunting areas (CHA), which are geographically
bounded units that the Department of Wildlife and
National Parks (DWNP) uses to allocate wildlife
utilization quotas.

This tenure system stems from a series of land-
use policies that evolved throughout the 1980s and
1990s and also form the basis of the Community-
based Natural Resources Management (CBNRM)
programme. In the 1990s Botswana received
United States Agency for International Development (USAID) funding and technical assistance to
develop a framework for community-driven sustainable development and wildlife utilization. The
CBNRM programme prescribes the creation of

community trusts as a primary mechanism for
implementing the programme. A community trust is
a legal institution entitled to lease a CHA from the
DWNP and derive economic benefits from sustainable activities, such as ecotourism and safaris.

Multiple Stakeholders and Sources of Tension

The Okavango Community Trust (OCT) was
created in 1995 and encompasses five villages
that border the northern edge of the Delta.
Hambukushu and Bayei live in four of the five
villages, whereas the Basarwa comprise a majority
of the fifth village. All members of the five villages
belong to the trust and elect board members to represent them on the OCT board, which manages
operations and distributes revenue. As described
above, these ethnic groups fall within varying levels
of a historically embedded social hierarchy, which
contributed to some conflict among the members of
the OCT from its inception.

Around the same time that the OCT began
operating, contagious bovine pleuropneumonia
broke out and forced a profound shift away from
cattle and towards wage labour and reliance on
cash to meet livelihood needs. The government
offered cash compensation to cattle owners and
many of them chose not to reinvest in cattle. Shops
selling household commodities, such as toothpaste
and sugar, sprang up to receive the cash. The
age-old system of *Mafisa* began to break down
(Hoon, 2004). Under *Mafisa*, cattle owners left their
herd with a caretaker, who took milk in exchange
for caring for the cattle. Since most Basarwa did
not own cattle, they were the caretakers for cattle
belonging to their higher-status neighbours. This
patron–client relationship was prominent in
Ngamiland until the cattle disease event and the
creation of the OCT. This relationship breakdown
created a power vacuum that the OCT board
members sought to fill in their new roles.

These economic stresses probably exacerbated
existing tensions both within the OCT and between its Basarwa members and the government
of Botswana. Having experienced increasing
marginalization from their traditional way of life,
people in Gudigwa, the Basarwa village, viewed
the OCT's creation as another step towards stripping them of their heritage. For example, prior to

the OCT's creation, the government had allocated a special game licence to Gudigwa for subsistence hunting. Concurrently with the creation of the OCT, the government revoked this licence, leading people to believe that the government took their special hunting rights away and gave them to the OCT (Taylor, 2000). This added to their feeling that government actions were undermining their livelihood and that the Basarwa continued to be marginalized under the new OCT structure.

In 1997, a US-based international NGO stepped into this social landscape seeking support for its campaign to relocate the cordon fences. The NGO had been working in other parts of Ngamiland, but was new to the OCT villages and did not conduct a regional socio-economic assessment before engaging in a dialogue with the people of Gudigwa. The NGO staff chose Gudigwa because they believed that the government would grant it a concession for a key piece of land through which the fences cut. And, based on their status as a hunter-gatherer society, the NGO viewed the people of Gudigwa as natural allies in its struggle to conserve wildlife migratory routes. Gudigwa viewed the international NGO as a strategic ally in its struggle to access economic resources. Because it wanted to build trust and, ultimately, support for its agenda, the NGO agreed to invest financial and technical resources into improving the development status of the village.

Unintended consequences

Over time, this alliance between one OCT village and the international NGO added to the existing tension within the OCT board. The tension culminated in a quiet conflict – one that used rumour and obstruction, rather than violence or confrontation, as its primary tools. Even though the NGO contributed to this conflict only indirectly, through its support of Gudigwa, the result of the conflict directly impacted the NGO's ongoing operations and threatened to undermine its long-term objectives in the region.

As people in the other four villages learned of the resources flowing into Gudigwa, rumours spread accusing the NGO of being a 'Bushman organization' that was unfairly favouring Gudigwa over its 'partners' in the OCT. Underlying the rumours was a perception that the international

NGO had leveraged its wealth and power to thrust Gudigwa forward in the new struggle for socio-economic dominance. Given the Basarwa's low status in the social hierarchy, a 'Bushman' ascendancy threatened to change the prior balance of power in the region and, hence, the device of rumour was employed to quell this ascendancy.

As it came under fire from those with a stake in the success of the OCT enterprise, the NGO attempted to mitigate the conflict and restore its image in the region. It quickly shifted its focus away from Gudigwa, held a series of community meetings and partnered with an international development NGO to meet the development needs of all the OCT villages. In the short term, this rapid shift diverted limited funds and staff away from planned activities in Gudigwa and strained its relationship with the members of that village. In the long term, it helped to expand the project's work to a scale more suited to the organization's objectives and its effort to engage the OCT as a body helped to restore trust in the organization. However, had it assessed the regional socio-economic and political context upfront, the NGO might have avoided the conflict altogether.

Lessons Learned

This case study demonstrates the dynamic nature of social relationships that ultimately shape the outcome of resource management efforts. Throughout history, local socio-economic systems developed in response to the unique ecological and social conditions of their place. These localities face challenges when a transition to the global market economy brings accompanying changes in power relations, decision-making and livelihood strategies for the local people. External agents, such as international NGOs, contribute to the complexity of relationships in a landscape of competing interests.

Achieving conservation goals requires skilfully negotiating competing interests and reaching a legitimate agreement on how to manage natural resources. Interests can vary from something as tangible as the right to cut down trees to such intangible forces as the struggle for power exemplified in this case. Successfully negotiating a space for conservation in a social landscape rests on establishing trust and legitimacy among the stakeholders. External groups must gather and analyse as much information as possible in order to understand the social setting it is stepping into and work to avoid fuelling existing conflicts or generating new ones. The unintended consequences, such as

those described in this case, threaten to undermine an organization's work. This, in turn, jeopardizes the effort to institutionalize sustainable natural resources management.

CASE 3.3. FOREST COMMUNITIES IN CHINA AND THAILAND (JANET C. STURGEON)

Governments throughout the tropics have a hard time dealing appropriately with shifting cultivators. State planners often see shifting cultivators as 'forest destroyers' using 'backward' techniques that need to be 'modernized'. State approaches have run the gamut from forced migration of shifting cultivators to lowlands (Indonesia) to excluding upland farmers from the forest (Thailand) to sedentarizing their land uses (China, India, Philippines). Few governments have seen the value of shifting cultivation. The case study presented here offers a comparison of the outcomes for Akha shifting cultivators in China and Thailand – two dramatically different state regimes.

Akha Farmers in China and Thailand

Upland farmers who call themselves Akha originated in China and gradually spread across adjacent hilly parts of Burma, Thailand, Laos and Vietnam. A comparison of Akha livelihoods and forest management in China and Thailand done in 1996–1997 assessed how Akha and their forests had fared under these two governments. The results showed that where state policies gave property rights to Akha and government agents took their cultivation practices into account, as in China, Akha livelihoods and the forests around them were in reasonably good condition. Where policies excluded Akha from property rights and state agents ignored Akha customary land uses, as in Thailand, Akha livelihoods and forests were in a poor and declining condition.

Each of the two Akha villages, one in China and one in Thailand, is located on the national periphery, right on the border with Burma (see Fig. 3.3). Each village rests on a ridge top (1600 m in China; 1000 m in Thailand) amidst a subtropical evergreen oak forest. In China, Akha have

lived in Mengsong for some 250 years, while, in Thailand, Akha have occupied Akhapu for just over 80 years.

The comparison involves relevant state policies under the modernizing states of China and Thailand since the 1950s. In the early 1950s, land use in Mengsong (Fig. 3.4) and Akhapu (Fig. 3.5) was similar enough to be described together. In both villages, farmers kept an area of protected forest right around the village to ward off evil spirits. In this site and in separate watershed protection and cemetery forests, customary rules prohibited cutting or collecting anything. Surrounding the protected forest was a larger wooded area, where farmers could cut trees for housing, fences and other subsistence uses. Beyond the forest was an extensive area for shifting cultivation (Figs 3.6 and 3.7). Here households felled secondary forest for shifting-cultivation fields (swiddens) large enough to meet household needs for grain.

Farmers intercropped a multitude of vegetables and herbs with the upland rice and maize. Once fallowed, fields could be opened by anyone in the village once trees were allowed 13 to 15 years to regenerate. Akha customary access practices involved fixed, enduring rules for areas of forest and flexible access to sites for shifting cultivation. Akha also raised livestock, hunted wild game and collected a plethora of wild fruits, herbs and medicines from surrounding fields and woods. In the decades since the 1950s, distinct state projects in China and Thailand have instituted property rights in land and forests to control local cultivation practices. In each country, the government tried to interrupt complex, mutable Akha land uses. The outcomes differed for trees and people in the two villages. The basic differences in policies and attitudes in China and Thailand are summarized in Table 3.5.

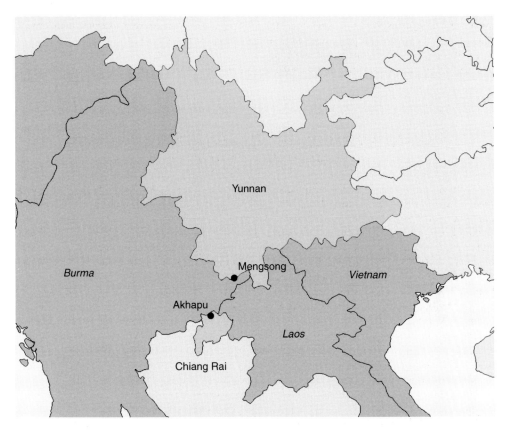

Fig. 3.3. Map location of the Akha villages of Mengsong, China, and Akhapu, Thailand.

Fig. 3.4. Akha village of Mengsong, China. Photo by Janet Sturgeon.

Fig. 3.5. Akha village of Akhapu, Thailand. Photo by Janet Sturgeon.

Fig. 3.6. Swidden and forests in Mengsong, China. Photo by Janet Sturgeon.

China

Following the 1949 revolution in China, Akha were made citizens of China. In the late 1950s, Mengsong Akha were organized into communes, as were farmers across China. Akha farmers continued to practise shifting cultivation, but state cadres forced them to open swidden fields on a larger scale than anyone had seen before. State planners viewed Akha as grain producers contributing to state goals. Akha intercropped vegetables with the grain, and in general their landscapes continued to be complex, mutable and marked by biodiversity.

Fig. 3.7. Swidden in Akhapu, Thailand. Photo by Janet Sturgeon.

Table 3.5. Differences in policies and attitudes.

China	Thailand
Akha as citizens	Akha as 'hill tribe' (not citizens)
Akha as grain producers	Akha as forest destroyers
Akha as property holders	Akha as interlopers without property rights
Akha as forest managers	State foresters as forest managers

In the early 1980s, Chinese policies shifted profoundly. Communes were dismantled and commune agricultural land was contracted to farmers under the household responsibility system. In Mengsong, farmers received both wet rice land and shifting-cultivation fields. Forests from communes were similarly allocated to villages and households. In Mengsong, state foresters designated as protected forests a 'scenic forest' right around the village, the watershed protection forest and the cemetery forest. Rules for forest use and protection were close to those from Akha elders, showing some communication about forest use, but state foresters now enforced them. At the same time, the village received a collective forest for house construction, and each household got a wooded plot for fuel wood.

Akha farmers appreciated the security of state-allocated property rights in land and trees. In the economic reform period, though, as polices

encouraged farmers to sell more goods in the market, Akha reworked access rights and land uses to suit emerging purposes. On low-elevation shifting-cultivation lands they opened terraces for wet rice. By switching from upland rice to wet rice, farmers freed up time to raise more vegetables for sale; to collect more wild fruits, mushrooms and medicinal herbs prized in lowland markets; and to engage in wage labour. Akha responded favourably to state exhortations to participate in markets, but they did so by altering land uses away from those set by the state. Farmers also began to raise more livestock for sale, and they burned large expanses of upland shifting-cultivation lands each year to produce grasses for grazing animals. Lands that the state had allocated to households had reverted to collective use. Many Akha allowed their water buffalo and cattle to graze in adjacent fields in Burma as well. Mengsong farmers were thus transgressing not only state-allocated property lines, but also an international boundary. The fluidity of their customary land uses had reasserted itself as Akha looked for new ways to contribute to growing markets.

In 1997, forestry station officials stated their disapproval of Akha burning fields each year for pasture, a use not designated by the state. Those fields would have to regenerate into trees and be reclassified as collective forest, but would still belong to the village. Villagers' remaking of the landscape went against state foresters' notions of settled land use under state regulation.

Overall, however, the land area in Mengsong was the same as in the 1950s. Farmers continued to cultivate wet rice, shifting-cultivation fields and vegetable plots. They had forest sites for both protection and use. And they raised large numbers of livestock and hunted wild game. Their land uses and livelihoods were complex and flexible, maintaining biodiversity and allowing them to shift labour around in response to infestations of pests, new markets and changing government policies.

Thailand

In the 1950s, the village of Akhapu served as a way station for caravans transporting opium from Burma to the Thai lowlands. Akha farmers sold livestock and grain to the caravan traders from their abundant harvests and large herds. At that time the Thai state was distant from their concerns. In the 1960s, as the Thai government began to log Thailand's forests to finance economic development, forests in the north of Thailand took on increasing importance in state plans. Ethnic-minority farmers who lived in the forest, such as Akha, were labelled 'hill tribes' and criminal users of state assets. Akhapu farmers continued to practise shifting cultivation, but they were incorporated into Thailand as 'not Thai' and not citizens of Thailand. As a result, they were not eligible for property rights in either land or trees.

From 1976 to 1982, Khun Sa, the most famous of the 'drug lords' in the Golden Triangle, occupied the northern Thai district where Akhapu is located. Khun Sa treated this as his domain, buying opium from farmers and producing heroin in local factories. In the early 1980s, under pressure from the US government to get rid of 'drug lords', Thai military troops routed Khun Sa, who retreated across the border into Burma.

To eliminate opium production in the north, in the 1980s the Thai government welcomed an array of international highland development projects. The projects had two main purposes: to end shifting cultivation and to replace opium with other cash crops. The projects introduced the intensive cultivation of cabbages and other vegetables on much reduced areas of land. In Akhapu, the cash crop was tea, which farmers planted throughout the understorey of their community forest. As tea became increasingly important to household incomes, farmers cut more trees to provide the right amount of sunlight for tea. By 1997, those areas of intensely managed tea looked like tea fields, not a forest.

In the 1990s, the Thai government celebrated the fiftieth year of the king's reign (1996) with a national reforestation campaign that reclaimed all shifting-cultivation fields in the north. In Akhapu, foresters paid farmers to plant tiny pine trees across the landscape, but the land was no longer theirs to use. With the loss of shifting-cultivation lands, farmers had nowhere to pasture livestock and sold off water buffalo and cattle in large numbers. Most farmers also lost their grain fields and scrambled to find poorly paid wage labour to pay for grain. The state took away large tracts of land that had previously alternated between fields and forest and produced the biodiversity that sustained Akha livelihoods. Without citizenship, Akha could not buy houses in town or work for major industries. Their 'hill tribe' status limited them to low-end jobs without security or benefits. They had rapidly lost their extensive, mutable landscapes with regenerating forests, and were struggling for tomorrow's meal instead.

Comparison of 1997 Outcomes in Mengsong and Akhapu

In Mengsong, land area was still extensive, with sites for wet rice and upland rice as well as forests for protection and use. Cultivation and collection practises were still complex and changing. Villagers' experienced broad-based gradual increases in income, with the highest-income household earning almost ten times more than the lowest-income household. As for forests, the species richness was 217 per 4.2 ha of sampled plots, with an average of 16.8 species per 78 m^2. Species dominance in protected forests was mainly old-growth species. These are indicators of forest age and abundant biodiversity. In Mengsong, Akha livelihoods and their forests were both in good condition.

In Akhapu, usable land had shrunk to less than a quarter of the former extensive landscape. Shifting cultivation was gone, with a few farmers planting wet rice and most people moving into wage labour to pay for grain. Their community forest had been planted in tea and was turning into tea fields with a few tall trees. Their shifting-cultivation fields, meanwhile, had been planted in pine trees that belonged to the forestry department. In Akhapu, most household incomes were declining. A few people did well – the highest-income household earned over 900 times more than the poorest household – but most people were falling through the cracks. In their forests, the species richness amounted to 87 species for 9.4 ha of sampled plots, with an average of 7.4 species per 78 m^2. Species dominance on protected sites showed mainly pioneer species. In comparison with China, Akha livelihoods and forests were in poor and declining condition.

The major differences between China and Thailand account for the distinct outcomes (summarized above). Akha in China were citizens and property holders and were viewed as grain producers and forest managers for the state. Akha in Thailand were 'hill tribe' farmers, without citizenship or property rights, seen as interlopers and forest destroyers. In Thailand, state foresters managed trees. Where Akha were included in the policy, given property rights and accommodated, to a certain degree, in policy implementation, Akha and their forests fared well. Where Akha were excluded from the policy, denied property rights and ignored in policy planning, Akha and their forests fared poorly.

Epilogue

A follow-up visit to Mengsong in 2002 revealed that the Chinese government had reclaimed all forest land in western China, including shifting-cultivation fields. The Mengsong forests were still in good condition, as in 1997, but household incomes had dropped by 25%. China was following Thailand in seeing shifting cultivators as forest destroyers – a view that leads to elimination of customary land uses and undermining of local livelihoods. Akhapu in Thailand has meanwhile been the site of armed conflicts between rebel armies from Burma, and was therefore off-limits for a return visit. Other researchers indicate that farmers there continue to plant tea. The take-home message is that many governments fail to recognize that shifting cultivators regenerate the forest and manage for biodiversity. A view of shifting cultivators as 'forest destroyers' obscures the value of accommodating them in ways that benefit not only the farmers but the forests around them as well.

Additional materials for this case can be found in Sturgeon, 1997, 2004, 2005.

Funding for this research was provided by the Tropical Resources Institute of the Yale School of Forestry and Environmental Studies, the East Asia Institute and the Agrarian Studies Program at Yale, the Committee on Scholarly Communication with China and Ford Foundation/Bangkok. The follow-up study was funded by Ford Foundation/ New York.

CASE 3.4. INDIAN FOREST: LAND IN TRUST (PHILIP H. RIGDON)

Introduction

Forests are a vital part of Indian communities due to the social, economic and cultural values the forests provide for tribal people. Reservation forest provides opportunities for economic development, employment and income, traditional hunting, fishing and food-gathering places and religious and cultural sanctuaries. Since time immemorial tribes have utilized and managed their forest for the resources they need.

Today, across the United States, tribes continue to use the forest resources in both modern and traditional ways. The uses of these forests are as diverse as the people that live on these lands. Indian forests range from the rainforest in Washington, the palms of Florida, the hardwoods of the Midwest and North-east and the juniper stands of the South-west to the mixed conifer of the interior mountain west. With the diversity of forest, individual tribes have different goals and objectives for their land. In some areas traditional foods and medicines are principal, while in others wildlife, fish and recreation are vital, and still others income and employment are the primary driving forces (Morishima, 1997).

Background of Native Forests

In the continental United States 193 Indian reservations in 33 states have 6.9 million ha or 17.1 million acres of forest land. Out of the 6.9 million ha, 3.8 million ha or 9.3 million acres are woodland (forested land with less than 5% crown cover of commercial timber species) and 3.2 million ha or 7.8 million acres are timberland. The timberlands contain a standing inventory of more than 44 billion board feet and support an annual allowable harvest of 810 million board feet (Morishima, 1997; Petruncio, 1998).

In the Pacific North-west, the annual allowable sale quantity from Indian forest is about 500 million board feet, almost half of the volume available from the Forest Service and Bureau of Land Management lands under the North-west Forest Plan (Morishima, 1997). From 1992 to 1996, 706 million board feet were harvested annually (Petruncio, 1998). Utilization of forest resources provides the backbone of economic activity for many tribes. In 1991, the Bureau of Indian Affairs (BIA) estimated that Indian forest and related programmes generated over $465 million and supported over 40,000 jobs for Indian communities and over 9000 jobs and $180 million contributed to non-Indian neighbour communities (IFMAT, 1993; Morishima, 1997; Peterson, 1998).

The commercial forestry on Indian reservations has immense potential for generating economic gain for tribes, but Native American values provide a unique context within which forestry must be conducted. On most tribal lands, the utilization of forest resources is balanced with important resource values such as religion and culture, water quality and wildlife and fish (Morishima, 1998). Meeting these objectives can be difficult and often impossible due to inadequate funding and the complex ownership pattern in Indian country. During the last 25 years the Intertribal Timber Council and the BIA have been advancing towards self-management and tribal government 'self-determination', but the funding needed to develop these goals is greatly inadequate (IFMAT, 1993). The direction

of Indian forestry has been unclear and the future entails redefining the US government's role and its trust responsibility in relation to tribal forestry. Understanding the historical policy and direction of Indian forest management is vital to understanding how tribes manage their lands today.

Historical Overview of Indian Forests

Indian reservations are federal lands held in trust for the benefit of Indian tribes. This trust responsibility is rooted in the Justice John Marshall 1830s court decisions involving the Cherokee Nation in Georgia. In *Cherokee Nation* v. *Georgia*, Marshall found that tribes existed as 'domestic dependent nations' within the United States. This created a ward–guardian relationship between tribes and the United States. With the treaties, establishment of reservations and statutory mandates by Congress, a trust responsibility relationship has been developed and continues today (Newell, 1998).

For much of the 19th and 20th century, policy toward native communities has sought to make them self-sufficient (Newell, 1998). Within this framework, the idea was to assimilate Indians by making them into farmers. The first policy of forest management on Indian land was established in 1873 when the Supreme Court ruled in *US* v. *Cook*. The ruling stated that Indians on the Tulalip Reservation in Washington State had no legal right to sell timber unless the clearing was for agriculture purposes; otherwise the logs belonged to the United States. The court viewed Indian rights to the reservation and the timber upon them as rights of occupancy only. This narrow view was based on rulings in *Johnson* v. *McIntosh* and *Cherokee Nation* v. *Georgia* that free title to tribal lands belongs to the United States (Schlosser, 1992; Newell, 1998; Peterson, 1998). While this established that timber on Indian lands could not be sold unless it was being cleared for agriculture, many Indian agents strongly disagreed with the outcome. Prior to this action many western tribes were already utilizing forest resources and many were already dependent on timber for employment and income. This case was the basis of policy on Indian forest and it took 15 more years before Congress would address this issue (Peterson, 1998).

The major policy of the federal government was to assimilate tribes and Indian people, so one approach was to move land out of communally held tribal land into land that is owned by individual people. In 1887, the General Allotment Act was passed by Congress, which gave individual Indian people the ownership of 80 acre parcels and made all remaining unused lands available for claim by non-Indians. This eventually caused millions of acres to leave Indian ownership (Morishima, 1998). Today, coordinated management on many reservations must contend with a chequerboard of ownership patterns, where free, tribal and individual allotment lands are distributed across a landscape. Nationally, approximately 10 million acres of trust lands are held in 80 to 160 acre parcels by allottees. Due to inheritance, some allotments have hundreds of owners, which further intensifies the problems of integrating management (Historical Research Associates, 1986; Morishima, 1998).

In 1889 Congress passed the 'Dead and Down Act', granting tribes the right to salvage dead timber for commercial purposes. Green timber could not be harvested unless it was being cleared for agriculture. This was the first time Congress or the federal government recognized the Indians' right to use their forest for commercial purposes (Schlosser, 1992; Peterson, 1998).

Two acts established the Division of Forestry in the Bureau of Indian Affairs. In 1909 the first act appropriated $100,000 for forestry on reservations. The second, in 1910, authorized the Secretary of Interior to approve timber harvesting on a sustained-yield basis (Motanic, 1998; Peterson, 1998). Even with the new approach, Congress refused to address the failing allotment policy or the possibility that reservation resources should be tribal rather than individual (Newell, 1998). By failing to address this issue, Congress missed an opportunity to reconsider how Indian communities can reach economic stability. The laws were minimal at the least, covering less than two pages and one short paragraph on tribal timber and another on sustained-yield management (Motanic, 1998; Newell, 1998).

In 1934, the passage of the reform-minded Indian Reorganization Act moved Indian policy in a new direction. The federal Indian policy shifted to a new emphasis on tribal political reconstruction (Newell, 1998). The act directed the 'Secretary of Interior to make rules and regulations for the operation and management of Indian forest units

on the principles of sustain-yield management' (Schlosser, 1992). The act also signified that tribes generally are the real owners of the land and resources. The act also gave tribal governments the power to stop unwanted activities (Schlosser, 1992). Furthermore, tribal constitutions and governments were developed, giving tribes their first opportunity to set policy and direction on their reservations. Many tribal governments began reacquiring lands they had lost during the 'allotment era'. The act also allowed all forestry activities, including clear-cutting, as a silvicultural tool (Schlosser, 1992; Newell, 1998). This was a new era for Indian communities and was called the 'Indian New Deal' (Newell, 1998).

The ultimate goal of the new deal was to develop tribes into independent self-governments. Within the extreme of this self-governance, various western congressmen moved towards a federal policy of termination during the 1950s. President Eisenhower wanted 'out of the Indian business' and the approach at that time was to eliminate the sovereignty status of tribes if they could economically and socially sustain themselves. This policy lasted until the mid-1950s, when nearly everyone involved recognized that this path was not working (Newell, 1998).

Following the failure of the termination policy, the executive branch embraced a policy of 'self-determination'. In 1975 Congress and Richard Nixon passed the Self-Determination Act. Under the act, federal agencies and Congress supported tribes assuming the responsibility of managing many of the programmes once staffed by the BIA (Newell, 1998). Within this new approach, tribes began developing tribal goals and addressing severe problems with federal trust responsibility, inadequate funding and the provision of services for Indian forest.

Throughout the history of the BIA's Division of Forestry, forest management was a forestry programme wrapped inside a social service agency that continued to develop new programmes aimed to help Indian communities. With much respect, the foresters and staff working in the BIA Division of Forestry had a difficult task during the first 100 years. They were developing an Indian forest industry during a time when the role of the federal government was constantly changing, which was further intensified by inadequate funding to meet the ultimate goal of sustainable forestry. Also during this time, tribes had developed a mistrust of the federal government due to poor management, little tribal involvement and, in some cases, outright corruption (*United States* v. *Mitchell I* and *United States* v. *Mitchell II*). Throughout Indian country, tribes were questioning if the federal government was meeting their trust responsibilities and, if the government was not meeting these responsibilities, whether the government was liable (Schlosser, 1992; Reynolds, 2001). Indian leaders throughout the country began gathering to discuss forestry policy and develop future policy that would combine tribal visions and federal trust responsibilities.

Intertribal Timber Council

As tribal governments moved towards 'self-determination', many tribes were questioning the direction and past management of their forest. Tribes were considering two options to address the Indian forest policy; the first was litigation, which would have cost both sides a considerable amount of time and money and the risk of not addressing the forestry and trust responsibility issues. The second option was the development of an organization that brought all the players to the table (Reynolds, 2001). It was a forum for tribal leaders and the BIA Forest Division to gather and develop strategies to meet tribal resource goals and management objectives. In 1976, Ken Smith of Warm Springs, Joe DeLaCruz of Quinalt, Gary Morishima of Quinalt, Bill Northover of Yakama and Ernie Clark of Colville brought BIA and tribal leaders together and established the Intertribal Timber Council (ITC) (Reynolds, 2001).

There were several important developments that came out of the organization; the first was the development of an annual symposium that covered a wide range of forestry issues across Indian land. The annual symposium has enabled tribal and BIA resource managers from across the country to get together, share information and discuss concerns about resource management on Indian land (Reynolds, 2001).

The other development was a new collective voice in Washington, DC. In the past, individual tribes were having difficult times addressing issues on funding, policy and trust responsibility but, collectively, leadership in Washington, DC, began paying attention. Throughout its history,

ITC has been influential in national Indian forest policy direction by advocating vital shortcomings of Indian resource management and the trust responsibility of the United States government. During the 1980s Congress appropriated more money to Indian (BIA and tribal) forestry programmes than at any time in history (Motanic, 1998, 2001; Peterson, 1998; Reynolds, 2001).

This influx of attention on Indian forest by Congress finally culminated with the passage of the National Indian Forest Resource Management Act in 1990. This was a great step in the right direction but it was also at a time when the federal government was beginning to downsize; as a result many of the funding requirements initially identified in the bill were stripped before its final passage (Newell, 1998). Although the act is a shift in the right direction, forestry programmes were and still are understaffed and inadequately funded (IFMAT, 1993). ITC continues to work on defining trust responsibilities and advocating the shortcomings of forest management on Indian lands. Its leadership position in Congress was secured by its involvement in a landmark investigation of forests and forestry on Indian lands called an Assessment of Indian Forest and Forest Management in the United States (Peterson, 1998). The ITC influence has been immense and vital; without the ITC, the National Indian Resource Management Act and its direction would have been minimal and funding problems would have been worse than they currently are.

Conclusion

Although the Indian Forest Management Assessment Team report identified many shortcomings within Indian forestry, the report also recognized the potential for Indian forest to serve as a model of sustainability for society as a whole. Tribal philosophies are grounded in traditions that reveal a fundamental respect for all the resources that share the earth. Due to the unique communal ownership, Indian lands must be used in ways that protect and enhance the resources for the generations of children yet unborn because they will bear the environmental and economic consequences of today.

The history of Indian forestry is important; it tells us where we have come from in dealing with these issues and displays some appreciation for the leaders and organizations that have worked so hard for Indian resources. It has also detailed the deficiencies and problems facing Indian forest and forestry today. The future should be a balance that will incorporate native culture and beliefs into resource utilization and economic development, a very difficult task. The outlook remains unclear with the dynamic federal–tribal trust relationship, but a model of sustainable forestry is possible if Indian communities continue to work together to solve these very difficult issues.

CASE 3.5. FOREST MANAGEMENT AND INDIGENOUS PEOPLES IN WESTERN CANADA (JOHN L. INNES)

Canada is a large country with an enormous forest area (over 400 million ha). At the same time, it is a country where remarkable tensions have developed between the original occupiers of the land and the various immigrants who now occupy much of the country. In the late 1800s, a series of treaties were signed with indigenous peoples, and, more recently, significant areas of land have been ceded to indigenous control. These include the territory of Nunavut and the Nass Valley (the subject of the Nisga'a agreement). In British Columbia, with the exception of Treaty No. 8 in the north-east of the province, only a small number of local treaties were signed in the mid-1800s, and the province is now in a long and tortuous treaty negotiation process with the First Nations that have occupied the land 'since time immemorial'.

Background to the Current Negotiations

The Royal Proclamation by King George III in October 1763 required the consent of 'Indian' peoples before their land was occupied and gave the Crown the sole authority to negotiate such land settlements. In addition, it specified that 'Indian' peoples should not be disturbed in their use and enjoyment of the land. This proclamation has never been repealed and was recognized in the Constitution Act of 1867, which was updated in 1982 to further affirm and protect aboriginal and treaty rights.

James Douglas, as Chief Factor of the Hudson's Bay Company and then as Governor of the Crown Colony of Vancouver Island, arranged 14 treaties to buy 358 square miles on Vancouver Island. The occupiers were paid in blankets and promised rights to hunt on unsettled lands and to carry on fisheries 'as formerly'. In 1861, as Governor of Vancouver Island colony, Douglas instructed his staff to ensure that 'the extent of the Indian Reserves . . . be defined as they may severally be pointed out by the Natives themselves'. The apparently sympathetic Douglas was succeeded by Lands Commissioner Joseph Trutch. Trutch in 1870 prohibited the pre-emption of Crown land by aboriginal people and denied the existence of aboriginal rights or the need for treaties.

After almost 100 years of repeated attempts to establish title, the Supreme Court of Canada finally determined in the 1973 Calder case that the Nisga'a had held aboriginal title before settlers came. No decision was made as to whether title continued to exist. However, negotiations on a modern treaty with the Nisga'a were started the following year. The Constitution Act (1982) finally recognized and affirmed aboriginal and treaty rights, both those that exist and those that may be acquired. This was contested by the British Columbia (BC) government, which only recognized the existence of aboriginal rights in 1993, following the Delgamuukw Decision by the British Columbia Court of Appeal.

The Nisga'a Treaty Forestry Provisions

The Nisga'a lands comprise 1992 km^2 in the lower Nass River area, 45,000 ha of which is productive forest land. The Nisga'a Treaty transferred control of this resource from the Province of British Columbia to the Nisga'a government. Existing forest licensees were given a 5-year transition period to continue harvesting, and the Nisga'a agreed to maintain the wood-fibre supply to local mills. The Nisga'a were given the right to establish rules and standards that meet or exceed provincial standards to govern forest practices on Nisga'a lands following the transition period. Importantly, the Treaty also dealt with non-timber forest products; the Nisga'a have the right to control the harvest of pine mushrooms and some other botanical forest products on Nisga'a lands.

Other Treaties

The Province of British Columbia has entered into treaty negotiations with a substantial number of other First Nations, although no treaty has been signed. Six agreements in principle (AIP) (as of September 2005) have been signed, and these provide an indication of the likely forestry provisions. Individual First Nations can make laws regarding the management of forests on their lands (except for timber marks and scaling). The responsibility for forest health and for fire protection and suppression is under negotiation with the Sliammon and Maa-Nulth First Nations. Under the Lheidli T'enneh AIP, the Lheidli T'enneh Government (LTG) will be responsible for the management and control of forest health problems on their lands. When a forest health issue threatens Crown lands, the LTG must use reasonable efforts to reach an agreement on an appropriate cooperative response to minimize the impact on Crown land. Forest practices and standards must meet or exceed those established under provincial legislation.

As the treaty-making process is ongoing, the government of British Columbia has developed a process called 'interim measures' (IM). The parties negotiate agreements before or during the treaty negotiations when an interest is being affected that could undermine the process. They include capacity-building initiatives, economic opportunities, governance-related initiatives and measures to protect land and resources. An example is provided by the Cowichan–Hul'qumi'num TRM and Cowichan IM, signed on 7 March 2001. This protects 1700 ha of Crown land near Duncan and Cowichan Lake within TimberWest's Tree Farm Licence 46. It was held for 2 years, with renewal clauses, and would be part of the treaty settlement. Support was provided for the Cowichan tribes to become involved in forestry. This constituted direct financial support and an invitation to submit a proposal for a community forest pilot agreement. On 4 February 2004, the government announced that it had signed a $13.6 million revenue-sharing deal with the Cowichan tribes for the next 6 years, and an additional $600,000 over 3 years will come from the Treaty Negotiations Office for forest-sector training. The money came from a $95 million fund set up to pay off First Nations in return for their agreement not to block logging on Crown land. The deals are independent of the Treaty talks.

Cultural Aspects

The 1992 National Forest Strategy recognized the importance of aboriginal rights in relation to forest management, and recognized the role of aboriginal peoples in forests, traditional use, knowledge and ways of life. The duty to consult arose from the Haida and Taku rulings by the BC Court of Appeal. Government and third parties must properly consult with and accommodate the interests of First Nations, pre-treaty, before proceeding with development on their traditional territories. The decisions were derived from the 'trust-like' relationship between First Nations and the Crown. The duty exists when a First Nation asserts title through entering the treaty process and continues until after a treaty is signed or aboriginal rights and title are defined through the courts. The extent of the duty depends on the strength of the First Nation's connection to the land.

There are some fundamental differences between forest management as viewed by many First Nations and forest management as viewed by the government of British Columbia and the logging industry. Curiously, the First Nations concept of forestry is generally much more closely attuned with modern forestry than that of the government, and the first licensee to obtain certification under the Forest Stewardship Council was a joint venture with a First Nation. The company, Iisaak Natural Resources Ltd, which operates in the Clayoquot Sound area on the west coast of Vancouver Island, is based on the premise of *Hishuk-ish ts'awalk* (everything is connected). Their forestry practices (which have yet to be shown to be economically viable) emphasize smaller cut blocks and a much greater concern about the maintenance of ecosystem functions than many industrial logging operations.

In 2005, the province of BC started allocating a portion of the annual allowable cut to First Nations, and this trend is likely to continue. The nature of the forestry that will emerge is not yet evident, but it is likely to be very different from the industrial logging that has dominated British Columbia to date.

4 Ecology and Conservation of Forests

JON M. HONEA AND KRISTIINA A. VOGT

Introduction: the Interconnectivities of Forest Ecosystems

When we think of forests, we typically think of the trees and other plants that give a forest its basic characteristics of height, shape and colour and even its many sounds and smells. In addition, the word forest brings to mind the animals that live there. A great diversity of animals are found living in forests: large animals, such as bears, deer, wolves and elephants, as well as the countless smaller ones, including rodents, primates, snakes, birds of every colour and size and a multitude of insects. These animals fill us with wonder, irritation and fear. Forests also contain streams, seeps, swamps, lakes and mighty rivers that supply vital moisture, refuge and food for forest organisms. The many occupants and features of forests are interconnected in myriad ways, some obvious, like predator–prey relationships, and others almost invisible, such as dung beetles' role in fertilizing soils by rolling animal dung into their underground dwellings.

The concept of interconnectivity must be kept in mind when we think of another important aspect of forests: their role as the provider of products and services critical to our survival and happiness. All of these are interconnected since conservation of animals and microbes living in forests and the acquisition of environmental services that they provide can be impacted by human activities (see Case 4.1 – Mycorrhizal Symbioses in Forest Ecosystems: the Ties that Bind and Case 4.2 – Small Mammals and their Relationships to Forests in the Pacific North-west United States). Since the goal of sustainable forest management is to provide habitat, services and products all from the same forest, understanding the interconnectivity of structures and functions in them becomes crucial.

Our use of forests has a great potential to alter the other uses or functions of forests. This means we must be good forest stewards. Failure to manage them wisely has repeatedly resulted in social instability. This instability results when important resources are no longer available because they were either misused or used irresponsibly (see Chapter 1, this volume). Beyond the practical motivation of self-interest, many people want to protect forests and their inhabitants based on either cultural or ethical motivations. Management that aims to sustain healthy forests and the provision of forest services into the future is based on forest ecology, the science that studies the many parts of forests and the ways in which they interact (see Chapter 7, this volume – 'Continuing Challenge for Sustainability: Linking the Social and Natural Sciences and Codifying Indicators').

Forest ecology, like all branches of ecology, approaches its subjects in terms of their structure and function. Structure refers to what is there, both living and non-living, how much of each structure there is, and how they are spatially distributed. For example, forest structure in far-northern latitudes is characterized by large expanses of small-diameter trees of medium height, often dominated by one or two species of conifers, which have a competitive edge over other trees in such cold, harsh conditions (Persson, 1980). Function refers to what the parts of an ecosystem do and how they interact with one another, with various functions influencing other functions and ultimately determining the overall structure of the system. Energy capture and use, nutrient cycling and species interactions are all ecosystem functions and are described in more detail below.

Ecologists study a forest as an ecosystem, which includes every organism and structure found in that forest. An ecosystem includes both the living and non-living parts. Depending on one's questions, an ecosystem can vary in size from smaller than the canopy of a single tree to a city park or the riparian forest of a wilderness river to the entire world. The varying spatial scales that can be included in an ecosystem make it clear that ecosystems are not closed systems; that is, they are inevitably influenced to some degree by things going on outside the defined boundaries of the observer.

In the past, some ecologists viewed ecosystems as a kind of super-organism with the parts acting like the organs and tissues that make up a conventional organism (Golley, 1993). The 'ecosystem as super-organism' metaphor is persuasive because the parts of an ecosystem are so tightly linked that changes in one aspect of the system often affect all parts in some way. Some ecosystems seem to maintain a dynamic equilibrium of repeating change that is evocative of the metabolism of a living organism.

Today, we recognize that the metaphor of a 'super-organism' is not a valid term for ecosystems. The 'super-organism' metaphor assumes that the organism controls the changes and direction of the ecosystem as a whole when perturbed. We recognize that this is not correct for several reasons:

- This concept does not adequately incorporate the chance introduction of a new organism into an ecosystem, such as an invasive species that dominates a site without having been part of the super-organism.
- It does not consider that an ecosystem can be affected by many activities that occur outside the boundary of the ecosystem, such as agricultural fields adjacent to a forest, increasing the disturbance impacts on forests (Franklin and Forman, 1987; Saunders *et al.*, 1991; Case 4.3 – Puerto Rico and Hawaii: the Dilemma of Coqui Frog Conservation or Eradication in Wet Tropical Forests).
- It does not include the legacies or imprints of disturbances that might have occurred over the past 1000 years or even in the last 10 years in one part of the ecosystem (Beard *et al.*, 2005; Case 2.2; Case 2.4; Case 4.3 – Puerto Rico and Hawaii: the Dilemma of Coqui Frog Conservation or Eradication in Wet Tropical Forests).

The metaphor of a super-organism also does not adequately depict the complexity of responses of organisms adapting or moving to a new area when environmental conditions change. In the past, ecosystems were thought to respond to global temperature increases by simply moving to track their preferred temperature regimes. However, current research indicates that individual species respond in different ways to environmental change: one species may be affected by a change in the timing of spring temperature increase, another by changes in summer rainfall, a third by a decrease in the numbers of its competitors, and so on. Investigations of how communities responded to past climate change have shown that the species react by moving individualistically, each tracking conditions favourable to itself, so that the ecosystem does not move as a whole unit (e.g. Pielou, 1991).

New ecosystems form across the landscape as new species enter an ecosystem and old species leave and otherwise change their distributions. Of course, some species disappear altogether if there are no favourable locations within a distance that they or their propagules (such as seeds) can travel. However, while ecosystems do not appear to be comprised of integrated parts forming a living entity in the way that we understand organisms to, ecologists agree that the parts of an ecosystem are indeed linked in ways that result in a widespread ripple effect when any one part is altered. For example, Case 4.4 describes the unexpected influence of forest clearing on malaria outbreaks (Case 4.4 – Malaria and Land Modifications in the Kenyan Highlands).

We depend on many different aspects of forest ecosystems, from wood, wildlife and livestock forage to clean water and diverse recreational experiences. Each may influence the others in many complex ways as well as some that are more direct. This means that forest ecology, the study of the structure and function of forests, must be at the foundation of any plan to alter a forest by using its resources.

One Tree is Not Just Like the Next Tree in an Ecosystem: Terminology, Taxonomy and Regeneration Trade-offs

The terminology and taxonomy of trees

Let's return to the subject of our first image when thinking of forests, that of trees. Trees come in countless shapes, sizes and even colours. One way of breaking down this great variety so that it is more manageable is to group trees taxonomically (the classification of organisms based on how they are related to one another). The major division is between gymnosperms (which means 'naked seed', because the egg and seed is exposed on a scale, usually in a cone) and angiosperms (which means 'enclosed seed', because their eggs are enclosed in an ovary, which develops into a fruit after the eggs are fertilized). Some existing ferns, a more primitive plant than the previous groups, grow large enough to be considered trees, but, due to their current rarity, they are not covered here.

Gymnosperms include the conifers, ginkgos and cycads. Trees from these divisions of plants dominated the Earth's landscapes during the drier climate of the Mesozoic era. This era lasted for about 200 million years and ended about 65 million years ago with the mass extinction that included the loss of the dinosaurs. The ginkgo or maidenhair tree (*Ginkgo biloba*) is native to eastern China and is the last remaining species of the ginkgos. Cycads are palm-like plants found in the tropics and sub-tropics and are the most primitive of the gymnosperms.

Conifers make up the most diverse group of gymnosperms. Although they are less common today than they were 100 million years ago, conifers are nevertheless the dominant tree type in many large areas such as western North America and most of the northern parts of Asia, Europe and North America. They are also the most import-ant tree type for the production of lumber, paper and resins. Other notable points about gymnosperms are that they include some of the world's largest organisms (e.g. California redwood, *Sequoia sempervirens*) as well as what may be the oldest organisms, the bristlecone pines (*Pinus longaeva*), some found to be over 4500 years old!

The fruit of angiosperms occur in an incredible range of varieties, often asso-ciated with their means of dispersal. If an angiosperm reproductive structure has seeds (or nuts), it's technically a 'fruit' to a botanist, the scientist who specializes in plants. The fruit, nuts and many of the vegetables that we commonly eat – such as cherries, walnuts (including the shell), oranges, olives and avocados – come from angiosperms that developed a strategy to include tasty flesh (some enhanced by human-directed selective breeding) with their fruit to entice animals to carry away their offspring, perhaps even depositing it with a fertilizing heap of excrement. The ovaries of other angiosperm species develop wings to carry away their seeds on the wind – think of maple (*Acer* spp.) 'helicopters'. Still others, such as the lethal bird-catcher tree (*Pisonia* spp.) of New Zealand, develop sticky surfaces on its seeds so that they are carried away by any bird that gets too close (Burger, 2005). And, no matter what you may have heard, coconuts did not develop their characteristic hollow 'nut' to encourage dispersal by swallows, even large African swallows.

Angiosperms have the greatest number of species among the groups of plants alive today. This seems to be related to their very efficient means of getting sperm or pollen to the egg. Fertilization is often done with the aid of a pollinator moving from flower to flower, rather than haphazardly via wind, as is the case with conifers. Although birds can be important pollinators in the tropics and subtropics and small mammals can be important in some desert areas, the most important pollinators throughout the world are the insects, especially bees, butterflies and flies.

Pollinators are typically encouraged to carry pollen by small amounts of nectar in flowers where the pollen and/or eggs are located. As the pollinator is collecting the nectar for food, pollen attaches to its body at the perfect location to become detached at the next flower where it is best able to fertilize an egg. Co-evolution, or the simul-taneous change in two species over time as each influences the other, is a common theme in angiosperm–pollinator relations. As a result of co-evolution, a specific type of pollinator can only pollinate the flowers of some plants. Flowers that have evolved to be available to only one pollinator increase the chances that their pollen will be delivered only to other flowers of the same species. The downside to this is if either plant or pollinator declines or dies out, the other inevitably follows.

Two other ways of categorizing trees can be confusing, but are mentioned here because they are still widely used. One is softwood versus hardwood and the other is evergreen versus deciduous. Both sets of terms came into common usage in Europe and were relatively coherent synonyms for gymnosperm and angiosperm; however, the number of exceptions to the rules increased when applied to other regions of the world. These exceptions have reduced the value of the terms. Softwood is used as a synonym for conifers and hardwood for angiosperms. This is because most conifers have soft wood and most angiosperms have hard wood, due in large part to the greater diversity of cell types that make up the water-conducting tissues of angiosperms. Unfortunately for this taxonomic method, some softwoods have harder wood than hardwoods, as is the case with longleaf pine (*Pinus palustris*) and yew (*Taxus* spp.). Another example is balsa (*Ochroma lagopus*), which is technically a 'hardwood' but has wood that is softer than most softwoods.

Evergreen is a term used to refer to plants that retain leaves all year round. Evergreen plants usually do lose their leaves, but do so individually and grow new ones throughout a year. On the other hand, deciduous means 'fall off' and refers to plants that lose their leaves all at once, usually before winter in temperate regions and at the beginning of the dry season in tropical and subtropical regions. Evergreen is often used to refer to conifers in general and deciduous to refer to angiosperms. One must be careful with such generalizations, however, as there are a few deciduous conifers (e.g. larch (*Larix* spp.) and the dawn redwood (*Metasequoia glyptostroboides*)) and many evergreen angiosperms (e.g. trees in the genera *Eucalyptus* and *Rhododendron*).

The trade-offs of being a gymnosperm or an angiosperm

The biological structure and function of gymnosperms and angiosperms, as well as all other living organisms, can be considered in terms of trade-offs. This is an instructive idea to keep in mind when studying ecological and most other complex systems. Given a finite quantity of resources, an organism must choose the best resource allocation strategy for a risky and changing environment. Of course, organisms do not actually make conscious choices about how to spend their resources; rather, a population of organisms of the same species changes over many generations in ways that permit them to better survive and reproduce in their environment.

Such 'descent with modification' works because each species is made up of individuals that are not exactly alike. Traits in individuals that allow them to be successful in particular environmental conditions and produce more offspring are more likely to be passed down with each generation. Within the limited capacity of each species to change, the genome of a species will evolve over time to track environmental change. Those species that cannot adapt fast enough to change or move to a more favourable environment become extinct.

Over time, this process has led to some tree species that produce many, small seeds and others that produce a few large seeds. The production of many small seeds is characteristic of plants that colonize temporary habitats such as clearings made by some disturbances. Often clearings grow back to resemble the surrounding forest in a relatively short time, so plants that favour clearings generally have seeds that can be readily carried on the wind and they should produce lots of these seeds to increase the odds that more will find another clearing. Frequently plants able to colonize clearings are considered weedy because they grow so quickly and dominate a site. This is a characteristic of an invasive species (see Chapter 5, this volume).

In contrast, conditions in the interior areas of many forests are stable, so trees that are successful growing in forests tend to produce fewer larger seeds. The larger seeds often contain more food that can be used by the seedling as it germinates and tries to establish itself with little sunlight from which to synthesize its own energy (Howe and Richter, 1982). The large seeds do not need to be carried away on the wind. In fact, it may be dangerous for them to do so because they risk entering environmental conditions in which their survival may not be successful. There are countless other examples of trade-offs in resource allocation, including annuals versus perennials, delayed versus immediate germination, hibernation versus migration, generalists versus specialists, and so on.

Key Processes that Interconnect Organisms in Forest Ecosystems

Plants capturing carbon from the atmosphere

Of the many valued resources produced by forests, most are biological. Even those that are not – such as mining, reservoir construction or water withdrawal – have clear and potentially large influences on the biological resources that we value. The ultimate driver of the production of biological resources in forests and most other biological systems is photosynthesis, the capture of energy from sunlight and its conversion to chemical energy, which can be used by organisms to fuel growth and reproduction (see Chapter 6 for the link between photosynthesis and plant carbon sequestration).

Photosynthesis occurs in plants. Although sunlight is multispectral (i.e. contains all colours or wavelengths of light), most plants are green because they use mostly violet, blue, orange and red light and reflect green and green-yellow light. In photosynthesis, light energy is absorbed by pigment molecules (mostly chlorophyll) on special membranes in plant cells located mostly in the leaves. Absorption of energy from sunlight causes a pigment molecule to become energetically excited. This unstable state stimulates the pigment molecule to release a negatively charged particle, called an electron. The loss of an electron places the pigment molecule in a different type of unstable state and, like a vacuum, the pigment pulls an electron from a nearby molecule of water thus stabilizing itself and making it ready to absorb more energy from sunlight. After this happens twice and a water molecule has lost two electrons, the now-unstable water molecule splits to release hydrogen and oxygen (water is composed of two atoms of hydrogen and one of oxygen). As the process continues, hydrogen ions ('ions' are charged particles and these hydrogen ions are positively charged after losing those negatively charged electrons) build up and oxygen ions will pair with one another to form stable molecules of oxygen gas (O_2), which we and all other aerobic organisms depend on to survive.

As ions of hydrogen build up on one side of the membrane where the pigments are located, the hydrogen ions are forced through a channel out of the membrane. Like water pushing a turbine to produce electricity for our homes and industry, the flow of hydrogen ions through the channel provides energy, which is captured in a molecule called adenosine triphosphate (ATP), an important carrier of energy in all organisms. Recall that before the pigment molecule robbed water of an electron, it lost one of its own after absorbing energy from light. Electrons moving away from the pigment molecules result in another flow of charged particles, this one of negatively charged electrons. This second flow produces energy that is captured in another important molecular energy carrier called nicotinamide adenine dinucleotide phosphate (NADPH). ATP and NADPH carry energy to a part of the plant cell where they provide the energy to join molecules of CO_2 gas together to synthesize three- to four-carbon molecules called carbohydrates (technically any compound containing carbon, hydrogen and oxygen made metabolically by an organism), which are the building blocks of more complex carbon molecules such as amino acids (which make up proteins), nucleic acids (which make up DNA and RNA) and fatty acids (which make up the lipids that act as protective barriers around cells as well as stores of energy).

These carbon molecules are used to build the bodies of all organisms and to drive their metabolic processes. Plants often convert the initial three- to four-carbon molecules into sucrose (a simple sugar), which is easily transported to parts of the plant that need carbohydrates, or plants join the initial carbohydrates together into long chains of starch, which is a stable storage product for later use. The various larger carbohydrates produced by energy resulting from photosynthesis are in turn the source of energy for animals that consume plants (or other organisms, living or dead). Organisms split large carbohydrate molecules in a process termed respiration to release the energy bound in them to fuel movement and metabolism and to produce the smaller molecules used to synthesize the particular carbohydrates they need for growth and reproduction.

Minerals and water interconnect different organisms in an ecosystem

Nutrient and water cycling are important processes linking organisms in an ecosystem. Nutrient cycling is the process in which elements necessary for life, such as nitrogen and phosphorus, are acquired from atmospheric or geological reservoirs and transferred from organism to organism through consumption, waste production, death and decomposition. At any point, nutrients may return to their abiotic reservoirs or continue cycling through other organisms.

Nitrogen

Other than carbon, hydrogen and oxygen, mentioned above, the most abundant element in living organisms is nitrogen. It is used in such key molecules as amino acids and nucleic acids. The largest reservoir of nitrogen on Earth is dinitrogen gas (molecules of two nitrogen atoms bound together – N_2) in the atmosphere but it is not available for use by most organisms in this form.

Some bacteria and other specialized organisms are able to 'fix' atmospheric N_2, in other words to capture it and convert it to an organic form that can be used by other organisms unable to use N_2 directly. Some plants have evolved into a complex relationship with bacteria able to fix N_2. Such plants typically house nitrogen-fixing bacteria in reservoirs in their roots. In addition to protecting the bacteria, plants provide the bacteria with sugars produced by photosynthesis in exchange for the organic nitrogen fixed by the nitrogen-fixing bacteria.

Trees or shrubs with nitrogen-fixing bacteria are adapted to a broad range of environments and are often capable of growing in dry, nutrient-poor areas. They have been planted in many regions of the world to try to stop desertification or to produce animal fodder. They are being planted to stop desertification in areas where no other plants will grow. Nitrogen-fixing trees are also common in the tropics and can be found growing scattered throughout the forest (Beard *et al.*, 2005). In forest agriculture, they are used as shade plants for lower-growing plants such as coffee, thus mimicking how coffee grows naturally in a tropical forest.

Nitrogen-fixing trees or shrubs are a good food source for domesticated animals because their ability to fix atmospheric nitrogen means that the plant tissues have higher protein content than other plants that are only able to acquire nitrogen from soils, where it is in short supply. Since forest-dwelling people commonly collect leaves from forests to feed their animals, planting nitrogen-fixing trees or shrubs adjacent to villages is effective at reducing the pressures to collect leaves from the forests. The higher nutrient quality of nitrogen-fixing trees or shrubs is one of the reasons that international organizations have been encouraging the planting of these species in the tropics. Since tropical forests are being over-harvested to provide fodder for domesticated animals and the supply of plant foliage is decreasing, nitrogen-fixing plants can reduce poverty rates and reduce unsustainable forest collecting. This practice should allow forests to be restored to more healthy conditions. It should also reduce the over-harvesting that is common today.

Since nitrogen-fixing plants have very fast growth rates, they are able to produce more fodder than if leaves were collected from a natural forest. These fast growth rates also mean that nitrogen-fixing trees or shrubs can be a good source of fuel wood. However, high growth rates also make nitrogen-fixing plants very effective as invasive species. The classic example is the introduction of a nitrogen-fixing plant to control soil erosion in Hawaii. This plant grew so well that it spread widely and altered the soil nitrogen levels significantly. As a result native species are being crowded out of their natural habitat (Vitousek and Walker, 1989). Such replacement of native species can cascade through an ecosystem to affect other organisms that depend on the native plants for habitat or food and are not able to use the new plants in the same way (Carpenter and Cappuccino, 2005).

Other nutrients

Phosphorus is a vital component of nucleic acids and phospholipids and it is the phosphorus bond in ATP that stores and releases energy used for biological work. Most phosphorus is in the Earth's crust and is liberated when it comes into contact with water. Plants and many microbes are able to assimilate dissolved phosphorus and bring it into the food web. Other important nutrients include sulphur, iron, potassium, magnesium, calcium and all the other ingredients you find on the back of your multivitamin bottle.

Animals, including humans, unable to use these chemicals in their most abundant form, rely on plants or microbes to take up and attach them to carbohydrates before they can use the nutrients. Too little of these nutrients can lead to characteristic symptoms, such as yellowing in the leaves of plants experiencing nitrogen deficiencies. For mammals eating foods deficient in vitamin C, common symptoms are fatigue, sore joints and swollen gums. This latter syndrome in humans is called scurvy and was once common on long sea voyages lacking fresh fruit and vegetables.

An interesting interconnectivity has been found between small mammals and fungi that produce mushrooms or sporocarps in forests (see Case 4.1 – Mycorrhizal Symbioses in Forest Ecosystems: the Ties that Bind). Some mammals, like the red-backed vole, are almost totally dependent on eating mushrooms as their only food source. The fungi in turn need the mammals to disperse their spores. So the fungi produce an aroma in their fruiting body, called a sporocarp, to attract these mammals to eat them. Sporocarps are also high in sodium, which is not required by fungi but provides mammals with a mineral that is essential to maintain their nervous system. In this way, voles acquire an essential nutrient and the fungus has its spores dispersed. Changes to the forest ecosystem that affect voles (e.g. increase in its predators, loss of habitat, etc.) can affect the fungi. Such changes would in turn affect organisms that interact with the fungi, including trees that they have a mycorrhizal relationship with.

On the other hand, the presence of too much of an element can also be unhealthy. This is the reason why soils with excesses of essential salts (e.g. potassium chloride, calcium chloride) only have a few plants that are able to grow in the environment they provide. Plants growing in saline habitats are restricted to a few specialists with adaptations to resist the negative effects of salt toxicity (see Chapter 5, for more detail).

The water cycle

Another crucial cycle influencing all organisms is the water cycle. Water has many essential roles, including as a physical force for transporting material such as wood, sediment and organisms. It is also a key component of probably every internal metabolic process and is known as the universal solvent (e.g. the leaching of phosphorus from rock).

The cycle of water through the environment is driven by the sun and gravity, with organisms acting as temporary reservoirs. The largest reservoir, containing about 97.5% of the water of Earth, is the ocean, but terrestrial plants and animals cannot consume this water because of its salt content. Heat from the sun causes water to evaporate from the surface of the ocean and other bodies of water and, as it evaporates, water vapour is separated from the salts. When atmospheric water vapour cools, it condenses to form clouds and eventually precipitates as rain, snow, hail, sleet and anything in between.

Precipitation that isn't directly intercepted by plants infiltrates into soils if they are sufficiently porous and unsaturated. The remainder of the water that does not filter through the soil moves over the surface of soils as runoff. Surface runoff moves more rapidly than soil water, so has a greater capacity for erosion. Liquid water in soils enters plant roots, following osmotic gradients (i.e. from higher to lower concentration). Plants maintain water flow from the roots, up stems and into leaves by opening pores on leaf surfaces called stomates: as water evaporates out of the plant through its stomates – a process called transpiration – a vacuum is created and water rises up the plant as in a siphon.

Besides participating in crucial metabolic reactions, water inside a plant maintains its turgor pressure and thus its rigidity. This is why plants wilt when they do not get enough water.

The water cycle is intimately intertwined with nutrient cycles. As water evaporates from a surface, the nutrients in the water body become more concentrated (think of the ocean or any salt lake). As water vapour condenses to liquid water, it again becomes a solvent for dissolved minerals. Even in the atmosphere, water droplets accumulate dissolved minerals from particulates carried on the winds and from gases such as sulphur dioxide or nitrogen oxides from industrial exhaust, which become strong acids and fall as acid rain (see Chapter 5, this volume, on disturbance). Precipitation can be a major source of nutrients in some forests.

Intra- and interspecific interactions

Photosynthesis, respiration and the cycling of nutrients and water tie organisms to one another and to their physical environment in a rich web of interaction. Organisms influence one another directly through interactions within and between species, intraspecific and interspecific interactions, respectively. Important intra- and interspecific interactions include competition, predation, parasitism and mutualism. The following paragraphs will explain these terms in greater detail.

Competition occurs when more than one organism requires a resource that is in short supply. Such resources can include anything from food and water to habitat and mating opportunities. An example would be plants competing for sunlight in a forest opening recently created by the fall of a large tree or by a browsing herd of elephants. Plants that first establish in an open area or survive the disturbance have an advantage over latecomers. Those plants that grow faster or taller have an advantage over slow growers or shorter plants.

Interestingly, plants that get outcompeted by others in full sunlight may be the best competitors in more shaded areas, especially those allocating more energy to the production of defences. They have traded fast growth for protection. Plants growing in the shade receive less energy from the sun so cannot quickly grow tall enough to escape herbivores. This risk has driven natural selection in dense forests to favour plants that allocate more resources to defences such as spines, bad taste and poisonous chemicals, such as tannins, nicotine and caffeine, another example of a trade-off in important resources.

Predation occurs when one organism consumes another living organism. An obvious example would be one animal capturing and consuming another. This is called carnivory or meat-eating. Herbivory, when organisms consume plants for energy and nutrients, can be considered a form of predation; however, herbivory also resembles parasitism when the whole plant is not killed by the herbivore (also called grazing, or browsing in some cases). Some organisms eat a variety of animals and plants and these are called omnivores; humans and most bears are examples of omnivores.

Detritivores are not predators in the strict sense, because they consume organisms or parts of them that are no longer alive. Detritivores play a key role in nutrient cycling. Detritivores – including scavenging mammals, reptiles and birds as well as insects, fungi and bacteria and other microbes – are the organisms that consume others after they die. They also include organisms that consume parts of organisms as their cells and tissues die and fall away. As detritivores consume and digest organic matter, they break it down, releasing nutrients back into soils and water, where they are available to plants and microbes. Plants and microbes have the ability to take up and utilize the nutrients by incorporating them into carbohydrates such as amino acids and nucleic acids, which are essential for metabolism, growth and reproduction. In this way, they are cycled back into the living components of ecosystems.

Parasitism occurs when an organism acquires some of its resources directly from another organism without killing it, but in the process reduces the fitness (ability to survive and reproduce) of the second organism. Some parasites, such as tapeworms and nematodes, live within their hosts and others, such as leeches and fleas, live externally. The reduction in host fitness can occur because the parasite consumes host tissues or even includes cases when the parasite steals host resources, such as with social parasites of ants, bees and termites. The most diverse group of *Formica* ants in the northern hemisphere is the *fusca* group. These ants have social parasites that enslave their workers, either by capturing and bringing them to the parasite's colony or by completely taking over the *fusca* group colony, as when a queen of the *Formica rufa* ant group displaces a *fusca* group queen and slowly replaces the previous queen's brood with her own. A parasitoid is similar to a parasite except that the host is killed as the parasitoid develops.

Mutualism occurs when the interactions between two species improve the fitness of both. The relationship between some plants and nitrogen-fixing bacteria mentioned above is a mutualism. Another excellent example of a mutualism is mycorrhizae, a fungal association with the roots of plants (see Case 4.1 – Mycorrhizal Symbioses in Forest Ecosystems: the Ties that Bind). Probably the most famous example of a mutualism would be lichens, a complex of fungi and blue-green algae. The fungi are capable of taking up nutrients from rocks and they share these nutrients with the blue-green algae, an organism that can photosynthesize. As you might guess, the algae in turn share the sugars they produce through photosynthesis. This combination of capabilities and cooperation allows the lichen complex to survive on bare mineral surfaces in environments too harsh for many other organisms. For this reason they are often early colonists following severe disturbances, such as glaciation, landslides and volcanic eruptions.

In some cases the herbivore–plant relationship resembles a mutualistic one. In this case, although the individual plant may have its fitness reduced or even be killed, the herbivore is often responsible for spreading the range of its prey as it passes seeds through its digestive system (i.e. potentially transporting plant propagules over great distances).

In addition to the direct interactions described above, there are endless combinations of indirect interactions. As an example, when large populations of pinhead-sized bark beetles emerge as adults from their host trees in the spring and summer and go on the hunt for new trees, they use pheromones as chemical signals to coordinate mass attacks on susceptible trees. They fatally wound the selected trees as they burrow under the bark to mate and lay their eggs. Soon after, the next generation of beetle larvae hatch and begin to extend their parents' burrows just under the bark surface as they consume the sugar-rich vascular tissues called phloem that occur there. In doing so, the beetle larvae sever the vital flow of sugars throughout the tree, finishing the job their parents started. Although this interaction is hardly favourable to the tree, it benefits the many animals that make their homes in dead trees (particularly those that eat beetles) and, as the trees eventually fall and decompose into the soil, many organisms there are enriched by the recycled nutrients.

However, during periods of widespread drought, bark beetles can wipe out entire forests, as many water-stressed trees become weak and susceptible to their attack. In such cases, beetles can damage forest ecosystems, particularly as they also create a large fuel base of dead wood and thereby increase the chance of a catastrophic fire (Whitfield, 2003b). Ecosystems are densely tangled webs of such complex inter-relationships, which are dynamic over space and time as species composition changes cyclically with seasons as well as progressively, as unique contingencies inevitably occur at various scales.

Environmental Change and Succession

Why we should care about climate change

Environmental conditions change over time, either as organisms modify them or as physical processes change. Examples of changing physical processes would be abiotic disturbance events and climate regimes, covered in Chapter 5, this volume. Climate regimes vary over a large range of scales, including higher-frequency phenomena, such as El Niño/La Niña – a 3–7-year cycle of warm to cool ocean currents, which has a great influence on weather in regions surrounding the Pacific Ocean – as well as lower-frequency phenomena, such as the ice ages. The current ice age has been going on for about 3 million years, with a series of global cold periods lasting about 40,000 to 100,000 years each, interspersed by briefer (tens of thousands of years) warm periods termed interglacials; fortunately for us, the current interglacial has lasted about 15,000 years.

The new global warming trend, which began about 50 years ago and is correlated with human production of greenhouse gases, such as CO_2, appears to be a large-scale phenomenon that will affect the entire planet (see additional detail in Chapters 5 and 6, this volume). The predicted increase in temperature of 1.4°C to 5.8°C between 1990 and 2100 (IPCC, 2001) seems small until one learns that:

- not only is the average a result of variation over the surface of the Earth, with some places being cooler and some warmer, but
- the continental ice sheets that occurred during the height of the last global cold period resulted from an average decrease of ~5°C.

What sort of world would an increase in that range produce? The complexity of the world climate system and its many feedbacks make it impossible to be certain at this point; however, the importance of climate for ecological systems that humans rely on means that this issue certainly demands attention and care.

Climate change can be expressed in many ways. For example, a warming trend could result from warmer evening temperatures or warmer winter temperatures, with temperatures at other times remaining in the previous range. Such changes can have a variety of consequences for organisms, in part depending on which environmental cues are important to each. Animals that enter a form of stasis in the winter, such as hibernating bats or bears, will experience an increased metabolic rate with warmer temperatures during the winter months. As a result, they will consume their fat stores earlier than usual and be forced to emerge early to seek food, perhaps before accustomed food sources are available. Or perhaps they will be required to increase the amount of food they must eat before they go into hibernation. In either case, the animals will have to modify their behaviour in order to survive.

Changes in temperature are usually accompanied by changes in precipitation. Due to the vital importance of water to all organisms – as metabolic substrate, carrier of food and nutrients, habitat medium and agent of environmental change (and maintenance) through disturbance processes such as storms, floods and landslides – changes in the amount and timing of its availability will have profound consequences for ecosystem structure and function.

Physical processes are not the only things that cause environmental conditions to change. Organisms are also a major agent of change in ecosystems. As they grow, move and make physical changes to their habitat, organisms shape ecosystem structure. Ecosystem function is affected as organisms influence nutrient cycling and the water cycle and participate in such interactions as competition, predation, parasitism and mutualism, described above. As organisms change their environment, each must respond to the changes made by it and others. Over time, the species present and the numbers of each may change in response to this environmental change. The change in species composition in an area is called succession.

Natural ecosystem dynamics: succession in forests

The initial stages of forest succession occur after some disturbance creates a gap in a forest. The gap can be small, such as that created as a tree falls, or it can be large, such as that created by a glacier or volcanic eruption. Primary succession begins after severe disturbances that actually or at least functionally remove the soil layer. This occurs after such events as glaciation, volcanic eruptions, landslides and even some intense fires that burn away the organic carbon in the soil. Some of the first colonists in such harsh environments are lichens, which produce their own energy and assimilate nutrients directly from rock or mineral soils. As lichens grow and die, organic material accumulates and may be supplemented by debris and propagules carried on the wind, in water or by passing animals.

Early food webs are based on lichens and associated organic material, so detritivores as well as their predators and parasites may be important components. As more organic material accumulates, sun-loving (also called shade-intolerant) plants can become established, adding more organic material as they grow, die and decompose. As more and more plants become established, there is intense competition for sunlight and, over time, the plant community grows taller and taller. After much time has passed, centuries in some cases, the forest returns. In an established forest, trees and other plants that compete well in the shade dominate the understorey, as one might expect, but eventually they also dominate the overstorey as well, because the shade-tolerant trees are best able to survive years in the understorey and grow into the upper canopy.

In contrast to primary succession, secondary succession begins after a less severe disturbance that opens a gap in the forest without destroying the soil layer. Secondary succession retains elements of the previous ecosystem, called legacies, which can include an intact soil layer, coarse woody debris, some plant species and microbes.

These gaps can be created by low-intensity fires, many types of animal activities, windstorms and anything else that results in fallen trees. If some of the original trees remain after a disturbance, then the survivors growing out to capture the newly available sunlight may simply fill the gap.

The size of the gap also has a big influence on which plants grow to fill it. If it is sufficiently small, then trees on the perimeter may reach out to fill in the gap. With larger gaps, new colonists become more and more important. In such cases, the timing of gap creation influences the availability of seeds entering the opening. If a species of tree doesn't have seeds available at that time, it cannot take advantage of the newly opened space. Large seeds, if available, may contribute colonists to small gaps and along the perimeter of large ones, but, with larger gaps, smaller seeds that can be carried greater distances become more important. Usually, seeds already present in the soil, known as the seed bank, also play an important role in recolonization during secondary succession. In this way, seed dormancy can be a successful strategy. Some seeds can lie dormant in soils for decades or longer, waiting for sunlight to warm them or fire or some other trigger to signal that it is time to germinate.

In traditional succession studies, the concepts of climax forest and later dynamic equilibrium were central to how succession was understood to occur (Golley, 1993). A climax forest would be the final stage of succession, in which a dominant shade-tolerant tree species constantly replaces itself as old members are replaced by the same species growing up through the understorey. When early botanists looked around and found no extensive climax forests, they believed that they were observing the result of human intervention. In the absence of human disturbances, they reasoned, climax forests would occur everywhere forests could grow. Over time, empirical studies began to reveal the prevalence of natural disturbances that alter the course of succession. However, even after it became clear that natural disturbances of some sort occur everywhere as continual agents of change in forests and other ecosystems, the metaphor of nature in balance was retained in the idea of dynamic equilibrium.

The concept of dynamic equilibrium describes a cycle of stages constantly repeated as regular disturbances that are characteristic for each region reset the clock of succession. While forest change may resemble a cycle of recurring stages over scales of seasons to years to decades, this is not the case at larger scales of centuries and longer. These larger scales are still only a few generations to most trees. The unique responses of each species to climate change and the changes in the cast of species caused by evolution work together with the occurrence of surprising contingencies on all scales to rearrange ecosystem structure and function. Ecological succession is impelled to move continuously into new directions as the species remaining at each point in time interact with one another and their environment as they attempt to acquire the resources they need to survive and reproduce.

Describing natural ecosystems as being in dynamic disequilibrium at grand time scales may appear to give licence to efforts to exploit forest resources without paying attention to potential consequences. That would be a great mistake because, at shorter time scales more meaningful to individual humans and even human societies, natural forest ecosystems change very slowly. The quality of the lives of individual humans depends substantially on the relationship of their society to the natural systems that produce their resources. Rapid change in those ecosystems requires rapid change in the human societies dependent on them and this often leads to upheaval in the lives of individuals in the societies affected (see Chapters 1 and 2, this volume).

Biodiversity: the Glue that Holds Ecosystems Together

What is biodiversity?

One of the most striking expressions of ecosystem structure is biodiversity. Biodiversity typically includes the number of species in an area, as well as the relative abundances of each. For example, an area that has 100 species with 90% of the organisms all the same species is less diverse than an area with 100 species in which all occur in equal numbers. Given the incredible complexity of many natural ecosystems, often only one or some small number of groups is included in an assessment of biodiversity – perhaps only the plants or birds or insects. In addition to taxonomic biodiversity, other important aspects of biological variation include:

- genetic diversity (i.e. the number of alleles for genes in a population, species or some wider taxonomic grouping), and
- ecological or functional diversity (i.e. focusing on apparent redundancies in species interactions, nutrient cycling and energy flow due to multiple species performing similar roles in an ecosystem).

One of the earliest observed patterns of biodiversity was the direct correlation between the number of species and the size of the area examined, now known as the species–area relation. Such a pattern might be expected because larger areas are likely to contain a greater variety of habitats, which should support a greater diversity of species. The species–area relation generally holds, however, even in ever-expanding areas of seemingly similar habitat. Perhaps the largest-scale pattern that has been observed is the gradient of increasing diversity from the poles to the tropics, commonly called the latitudinal gradient of biodiversity.

Scientists have long wondered what factors cause the observed patterns in biodiversity. As early as the 1920s a German biologist named August Thienemann argued that the species–area relation and the latitudinal gradient are the result of spatial heterogeneity and the length of time that a locality's conditions are relatively constant. Therefore, a combination of these factors would allow time for processes of natural selection to increase the number of species capable of filling the available niches (i.e. ecological roles or positions) in any ecosystem.

One of the earliest hypotheses to explain the latitudinal gradient of diversity was presented by Alfred Russel Wallace in 1878. As an aside, it was Wallace's independent proposal of a theory of evolution by the process of natural selection that prompted Charles Darwin to publish his own more developed theory in 1858. On the topic of the latitudinal gradient, Wallace proposed that the increasingly variable and extreme climate with distance from the tropics would increase the likelihood of species extinctions, thus causing the gradient of fewer species near the poles. Klopfer and MacArthur (1960) found that birds they regarded as having more specialized habits were more common in the tropics and birds with more general habits became more common farther from the tropics. They reasoned that the greater climatic stability of the tropics encouraged niche specialization and that, because the resulting species had smaller niches, more would occur per unit area.

In a richly detailed argument in opposition to Klopfer and MacArthur, Connell and Orias (1964) contended that there is a direct positive relation between productivity and biodiversity. They argued that there was not enough evidence to support the hypothesis that known biodiversity gradients are caused by disturbance or harsh environmental conditions. They further stated that Klopfer and MacArthur's findings are questionable due to 'the absence of an operational definition of niche' and to the apparently unspecialized bill structure in some of the tropical birds that were claimed to be specialized. They argued that, in stable environments, more biological energy is allocated to production than in less stable environments, where more energy must be used for coping mechanisms, such as heat production or movement. As production increases, larger populations can be supported. With larger populations, there is greater genetic diversity. In order for intraspecific diversity to translate to interspecific diversity, partial or complete isolation of small groups must occur. This is possible in high-energy, stable environments because primary productivity would be accordingly high. High productivity would decrease the need to move very much to acquire resources and thus increase isolation and increase rates of speciation. They argued that these conditions should support the greatest biodiversity and that biodiversity would be limited at its highest end by the danger of overspecialization, which would tend to lead to extinction.

The species–area relationship is an important component of the 'island biogeography theory' (Preston, 1962; MacArthur and Wilson, 1963). This theory seeks to explain the patterns of diversity found on islands based on their size and distance from the nearest mainland, both influencing the probability of colonists from the mainland making it to the island. The size of the island also influences extinction probability because larger sizes have greater resource availability. The principles of island biogeography theory have been extensively used to design protected areas or reserves, and to assess the impacts of landscape fragmentation on species abundance and survival (Vogt *et al.*, 1997).

However, in the 1980s it became increasingly clear that the species–area relationship was too simple a description of the relationship between species and their habitat space. It did not consider the fact that islands or forest patches of different area sizes vary in habitat complexity even when they appear to look the same. This is partly due to the fact that disturbances are highly variable in space and time, which, in combination with differing pre-disturbance conditions, leads to varying changes following a given type of disturbance (Beard *et al.*, 2005). Despite the issues of oversimplification, this theory is still commonly used because managers have to evaluate the impacts of land-use activities on species survival and the existing data available to conduct such assessments are quite poor.

There were many other contributions to biodiversity theory during this active mid-20th-century period, but those discussed above introduce the main ideas held today. Most current theories of biodiversity agree that productivity is a major controlling factor; however, most describe the relationship somewhat differently from the above and acknowledge that disturbance and heterogeneity are important as well. One influential synthesis of factors influencing biodiversity is David Tilman's (1982) theory of resource competition in a heterogeneous environment. His model makes these predictions:

- biodiversity should be highest in relatively resource-poor habitats and should decline rapidly as resources decline further and decline slowly as resources increase;
- high-biodiversity communities should have many co-dominant species, with fewer co-dominants as resources increase because a small number of species will be able to exploit the large resource base and exclude others; and
- at a given resource level, increasing spatial heterogeneity increases biodiversity, with the strongest effects in resource-poor habitats.

Even today, there is no consensus among scientists on the ideas originally published by Tilman in 1982 and their general applicability to all ecosystems globally. This means that new theories are still being developed and most of these are probably not going to be general theories broadly applicable to all ecosystems.

Why is biodiversity relevant?

Beyond the observed patterns of biodiversity and proposed explanations for those patterns, why is biodiversity relevant? The most important reason, known as the insurance hypothesis, is that increased functional diversity (i.e. more species performing a similar ecological role) should increase the rate of an ecosystem's recovery from disturbance. This is because, as stated above, species tend to respond to environmental change in different ways. So, although one species performing an important ecological function – such as photosynthesis or producing dead-wood habitat by killing trees – may be reduced in number or effectiveness by some disturbance, the system is more likely to continue functioning as it had before the disturbance if there are multiple species performing that role.

Reductions in biodiversity may also increase the susceptibility of an ecosystem to invasion by exotic species, which can cause extensive disruptions in ecological function (see Chapter 5, this volume). For many people, the protection of biodiversity is an important ethical issue as well. The ethical motivation may arise from a feeling of responsibility for the effects of our actions on other living things as deserving such regard for their own sake, for use or appreciation by future generations of humans, or for both.

Challenges in conserving biodiversity

Appreciation of the importance of biodiversity has led to attempts to conserve it as we recognize that it is being lost. Early attempts at species conservation focused on the endangered species themselves, by avoiding killing, capturing or harming them ('take' in the language of the United States Endangered Species Act). However, many endangered species are in decline not because people are killing them directly, but because their habitat is being changed in ways that reduce their survival. In some cases, reduction in habitat has resulted in loss or decline in many species. For this reason, more successful efforts at endangered species conservation emphasize preserving the habitat critical to the species in decline. Focus on habitat and ecosystem processes that maintain it is a more efficient approach than protecting one species at a time.

In practice, large well-recognized species – cynically called 'charismatic megafauna' – receive the most public attention and are therefore the most likely to receive protection. Because of the political, economic and bureaucratic complications of explicitly protecting multiple species or focusing on habitat, one approach to this challenge has been to select one species in particular to protect in the hope of such efforts acting as an umbrella to protect all the others. In this case, either indicator or keystone species are good candidates for protection. An indicator species is one whose presence indicates a healthy habitat that includes approximately the full suite of biodiversity present historically as well as the ecosystem function and range of disturbance processes necessary to support it. An example of an indicator species would be the northern spotted owl (*Strix occidentalis caurina*) in the old-growth Douglas-fir forests of the Pacific North-west United States.

A keystone species is one that performs a vital role in an ecosystem that has great influence on many other species. An example of a keystone species would be elephants in many varied African landscapes. Some of the key roles that elephants play are seed dispersal of some large trees with thick shells around nuts that only elephants can crack (the tree embryo survives the intestinal journey), gap creation and maintenance by destructively browsing trees, and nutrient cycling, as most of their dung is returned to the soil as barely digested, still nutrient-rich plant matter. Salmon are another example of a keystone species in the riparian forests where their spawning streams occur. See Case 4.5 for more on this remarkable fish (Case 4.5 – Salmon: Fish of the Forest). Another example of a keystone species would be the brown lemmings found in the tundra, since a diversity of animals (including moose) seem to eat brown lemmings as a food source and their grazing of the vegetation increases its growth rates in cycles related to their population densities. While a focus on indicator and keystone species can be a useful solution to biodiversity challenges, resource managers and voters should not lose sight of the larger goal of ecosystem conservation as the only viable means of sustaining the ecosystem functions that provide the diverse range of products and services we value. It is all the parts surviving together as a dynamically integrated whole that maintains the healthy system and sustains the production of goods and services that we value.

Another important challenge in conserving biodiversity is the fact that some species that play important roles in their native habitats become threats to ecosystem health when transported into new habitats. The classic example here is a frog that has an important role in accelerating the recovery rates of tropical forests in Puerto Rico but is displacing native frogs in other tropical forests on the island of Hawaii (Case 4.3 – Puerto Rico and Hawaii: the Dilemma of Coqui Frog Conservation or Eradication in Wet Tropical Forests). The coqui has characteristics that allow it to increase its population density following a hurricane, which has made it a very successful species in Puerto Rico. Even though Hawaii has hurricanes, the assemblage of species differs and Hawaii's native species do not respond as aggressively after a hurricane.

CASE 4.1. MYCORRHIZAL SYMBIOSES IN FOREST ECOSYSTEMS: THE TIES THAT BIND (RANDY MOLINA)

The vast majority of land plants enter into a highly evolved root symbiosis with specialized soil fungi. This mutualistic symbiosis is termed mycorrhiza – literally fungus–root (see Fig. 4.1). The fungi take up nutrients, such as phosphorus and nitrogen, from the soil and transfer a significant amount to roots. In return, the fungi receive their main energy source from the plant in the form of simple sugars from photosynthate. The primary benefit to plants comes from the ability of the microscopic fungal hyphae to explore ten to 100 times or even greater volumes of soil for nutrients compared with roots. The fungi also produce a suite of enzymes to attack both mineral and organic material in the soil, thus making nutrients available for uptake.

Most forest plants, and certainly all forest trees, depend on functioning mycorrhizal symbioses for survival and growth. This dependency became evident early in the 20th century when foresters tried to introduce exotic pines into regions such as Australia, New Zealand, Puerto Rico and Africa, where compatible pine mycorrhizal fungi did not occur. These early plantations typically failed until mycorrhizal fungi from the native host habitat were also introduced. Applied research on mycorrhizae of forest trees from 1950 to 1980 focused on how to take advantage of the mycorrhizal symbiosis through artificial fungus inoculations to improve tree seedling performance in nurseries and after outplanting; some research on that technology continues today. Overall, artificial inoculations with mycorrhizal fungi produced positive results in some situations, but not in all studies (see reviews by Perry *et al.*, 1987, and Castellano, 1996).

By the early 1980s, emphasis in mycorrhizal research in forestry shifted from the reductionist approach of harnessing the benefits of mycorrhizal fungi for reforestation programmes to a holistic approach of understanding how mycorrhizal symbioses and the fungi in particular influence ecosystem processes and community dynamics. New knowledge gained from this approach would enable managers to sustain these important soil organisms and processes and thus better meet emerging ecosystem management goals. The wealth of new knowledge gained on the ecology and function of mycorrhizal symbioses over the last 20 years has vastly altered our understanding and appreciation

Fig. 4.1.   Ectomycorrhizal roots of ponderosa pine colonized by *Rhizopogon idahoensis*. Photo credit to Jim Trappe.

of how they affect and govern critical ecosystem processes and overall resilience. Two examples below illustrate the value of this new awareness.

Forest Food Webs

Ecologists often view species interdependencies and interactions through food-web dynamics. Given our proclivity to study large, charismatic fauna and flora, small and often cryptic species such as fungi are typically overlooked in this regard. In forest ecosystems, however, fungi profoundly affect food-web dynamics by influencing several trophic levels.

In the temperate forests of western North America, for example, hundreds of species of ectomycorrhizal fungi form hypogeal (below-ground) fruiting bodies or sporocarps, commonly called truffles. These fleshy structures are equivalent to mushrooms in their reproductive function, but have lost the ability (through evolutionary adaptation) to develop a stalk, lift themselves out of the soil and disperse their spores to the air. Instead, as truffles mature below ground, they become aromatic and easily detected by diverse small mammals (squirrels, chipmunks, voles, etc.) that forage on the forest floor. Small mammals excavate and consume the truffles, and later defecate and thereby disperse the spores throughout their home range (see Fig. 4.2). Fungal spores pass through the digestive system unharmed and, once back in the soil, may form new mycorrhizae with host tree roots.

Many small mammals depend on this food source through much of the year; large mammals, such as deer, elk and bear, also seek out these fungal treats. But the importance of this food source does not stop with mammals. Small mammals are the main prey for many other forest animals. The classic example in the Pacific North-west is the threatened northern spotted owl. It consumes great numbers of these small mammals, particularly the northern flying squirrel. Much of the diet of northern flying squirrels consists of truffles. In summary, then, trees produce photosynthate and fuel the mycorrhizal fungus life cycle, including truffle reproduction; mammals eat the truffles, and predators consume the small mammals – all are connected and interdependent within these food-web dynamics. When managers are trying to maintain habitat for the owl, they need to think beyond a simplistic approach that focuses on roost and nest

Fig. 4.2. The northern flying squirrel eating a truffle. Photo credit to Jim Grace.

sites and factor in habitat needs of the mammals and truffle fungi. This exemplifies a holistic ecosystem management approach (see papers by Carey *et al.*, 1999, and Luoma *et al.*, 2003, for more details on mammal mycophagy and management implications).

The food-web story does not end, however, at these upper trophic levels in forest ecosystems. If one looks at food webs below ground, a similar fungal dependency occurs, but this time for the smallest of animals, soil microarthropods. These tiny insects are well known as the final shredders of soil organic matter, providing the substrate for final mineralization by soil microbes. Species diversity of soil microarthropods is staggering, numbering in the tens of thousands, and about 80% of those species are fungivores (Moldenke, 1999). Because much of the fungal hyphae in forest soils belong to mycorrhizal fungi, the importance of this fungal biomass and nutrient content (particularly the sugars they are receiving from the host trees) becomes evident in this pervasive soil food-web linkage. The mycorrhizal linkage that shuttles the carbon from the plants to below-ground processes is thus a critical element in creating and sustaining healthy soils.

Important Inter-plant Connections in Space and Time

Mycorrhizal associations are made complex due to the thousands of possible combinations of plants and fungi that can enter into the symbioses. For example, over 5000 fungal species form ectomycorrhizae with forest plants and express different levels of host specificity (Molina *et al.*, 1992). Individual plants can form mycorrhizae with dozens of fungal species at any one time. Some fungi specialize in forming mycorrhizae with few host plants (for example, with a single plant genus like *Pseudotsuga* or *Pinus*) while others form mycorrhizae with many plant species from different families or orders. The important outcome of these potential mycorrhizal associations is that plants of the same or different species often share common mycorrhizal fungi *in situ* and this sharing influences plant inter-actions and community dynamics.

In a groundbreaking field study, Simard *et al.* (1997) demonstrated that carbon (labelled photo-synthate) can be translocated from one plant to a neighbouring plant via their shared mycorrhizal fungal connections, and the amount of carbon increases when the recipient plant is shaded. These and similar results of inter-plant sharing of resources via common mycelial networks remain controversial as to the magnitude and importance of the transfer (see review by Simard and Durall, 2004), but the possibilities have challenged current paradigms about competition in plant communities.

Beyond providing the potential of plants to directly share resources via mycorrhizal linkages, common mycelial networks play an important role affecting forest resiliency in space and time.

Following disturbance, early seral plants can maintain mycorrhizal fungus legacies in the soil by providing immediate energy (photosynthate) below ground. Later seral plants can then take advantage of active fungal colonies that are already well supported. In some forests of western North America, for example, *Arctostaphlos* (manzanita) species are common understorey shrubs or ground covers and they form mycorrhizae with many of the same fungi as the overstorey trees (Molina and Trappe, 1982). Following disturbance, many manzanita species recover and grow rapidly, often from resistant burls or buried roots, and can quickly support the fungal community. As such, these mycorrhizal associations function as biological legacies to sustain both soil processes and the soil microbial communities needed by later seral tree species. Linkages of seral forest communities as mediated through mycorrhizal symbioses thus contribute significantly to ecosystem recovery and sustainability (see reviews by Perry *et al.*, 1989, and Molina *et al.*, 1992).

Concluding Remarks

Mycorrhizal symbioses affect forest ecosystems in many additional ways, particularly in nutrient cycling, retention and soil aggregation, and thus influence overall health and productivity. With the advent of new molecular tools (see Horton and Bruns, 2001), we are now able to probe into the identities of the individual fungi that occur on single roots, better understand the composition and seral changes of the fungal community, determine the size of fungal individuals in soil and tease apart their population dynamics. Most importantly, we are now beginning to unravel the complex functional diversity of these critical soil organisms and learn more about how they provide the ties that bind individual plants into plant communities. For further readings on the ecology of mycorrhizae consult the books by Smith and Read (1997), van der Heijden and Sanders (2002) and Dighton *et al.* (2005).

CASE 4.2. SMALL MAMMALS AND THEIR RELATIONSHIP TO FORESTS IN THE PACIFIC NORTH-WEST UNITED STATES (STEPHEN D. WEST)

Old-growth forests in the Pacific North-west United States were considered biological deserts in the 1950s because of the lack of large mammals living in them. This view considered large mammals and

game animals as synonymous (i.e. mainly a good
food source for humans) with biological diversity.
Thus the lack of large mammals meant that these
old-growth forests were not good at providing food
for human survival in this region. When large
mammals are found in Washington, they are more
commonly found in young forest landscapes. A
recent assessment in the state of Washington, United
States, showed that large mammals comprised
about 20% of the total mammal species found in
this state (Fig. 4.3). Approximately 80% of the
mammal species in Washington are small.

Focusing on large mammals with high food
value as the desirable species to inhabit these
forests failed to recognize the importance and
diversity of small mammals found here. In this
region of the world, where most people are of
European descent, small mammals are not considered a good food source unless you are desperate
for something to eat. This contrasts with other
global societies who have a culture of eating small
mammals.

In Washington, 80% of the mammal species
found in these forests are small and include species such as rodents, shrews, moles and bats
(Fig. 4.4). Of the small mammals, the rodents are
well represented in Washington's forests, with
48% of the state's terrestrial mammals being
rodents (51 species in total). In fact, rodents
are more numerous than any other species of
mammals at global scales. There are 16 species of
bats in Washington and 12 of these are forest
dwelling species.

The survival of small mammals in the Pacific
North-west United States is closely linked to the
forests and their uses (O'Connell *et al.*, 2000). Any
land-use activity that reduces the presence of
large live trees, coarse, large, woody debris or the
shrub layers within the forest will have negative

repercussions on the viability of small mammals
that depend upon these forest structures.

To maintain the full complement of small
mammal species in this region, stand-level forest
management alone is not adequate. Retaining viable
habitat for small mammals requires maintaining
the structural complexity and habitat diversity that
forests provide at the landscape level. For instance,
some small mammals forage in one part of the landscape but then den or bed down in another part.
An excellent example of this is bats, which use
successional stages at the ends of the successional
gradient by foraging in the unobstructed space
over early stages and roosting in huge trees and
snags in late stages.

The structural complexity and habitat diversity
that small mammals need from forests will be discussed in the next section. This will be followed by
a brief summary of different forest management
activities and how they impact small mammals.
Also how forest managers can improve the retention
of small mammals will be explored.

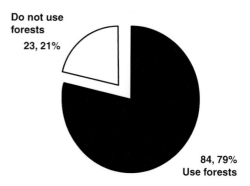

Fig. 4.4. The number and per cent of
mammal species that use and do not use
forests in Washington.

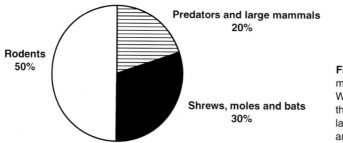

Fig. 4.3. The per cent of
mammals that are found in
Washington in the following
three categories: predators and
large mammals; shrews, moles
and bats; and rodents.

What do Small Mammals Need from Forests?

Almost four-fifths of the mammal species in the Pacific North-west use forests during part of their life cycle (Fig. 4.4). Mammals use every stratum of the forest, such as in the soil, on the ground, in the shrubs, in the trees and some in the air. The different strata that are found in forests are one key to the high diversity of small mammals found there.

Many of the small mammals are insectivores, such as shrews, shrew-moles and moles. The Townsend mole and coast mole feed exclusively on soil invertebrates and not the roots of our garden plants as thought by many people.

Rodents are found in a variety of habitats and eat a variety of foods. Most of the rodents eat plant material, especially fruits and seeds. Some small mammals (e.g. ground squirrels and gophers) are most abundant along forest edges and in open-canopy forests and are considered 'generalist' feeders. These mammals feed on the above- and below-ground parts of plants. Many of the voles or meadow mice are most numerous in the pre-canopy stages, where they construct runways and burrow systems. They feed on leafy vegetation and are known to girdle many young trees, killing them and altering forest succession.

Some of the mammals are much more specialized in their feeding habits. For example, the flying squirrel is a fungus specialist and eats mushrooms found in the forests (see Case 4.1 – Mycorrhizal Symbioses in Forest Ecosystems: the Ties that Bind). The fungus forms symbiotic associations with the tree root systems and thereby interconnects plants and animals in these forests. The mushrooms eaten by the flying squirrels are found mostly in certain forest developmental stages, especially older forests. This is because more mushrooms are produced in later developmental stages of a forest.

Small mammals also need good habitat or roosting areas, which forests are able to provide. Logs or large woody debris and standing snags are very suitable habitats for small mammals as well as for many other animals. Logs provide small mammals the following:

- protection from the physical environment and from other predators;

- a good source of food because many other animals live in the wood, especially insects;
- good lookout points for viewing who or what else might be in the vicinity; and
- a convenient travel route to move through the forest.

The larger the log, the greater the number of habitats available to small mammals. This means that, to maintain small mammal populations in forests will require retaining a sufficient number of large decaying pieces of wood. The larger logs also provide habitats that will be available for a longer time period because they decompose slowly and can exist for a couple of centuries. In the Pacific North-west United States, many of the relics of stumps and logs from the harvest of old-growth forests are still providing habitat. Since these large stumps and logs are not being replaced under current management practises, it is believed that eventually the quality and quantity of the habitat for small mammals will decrease.

The link between small mammals and forest trees (live or dead) is exemplified by the bats. Bats can use large living trees as day roosts. The exfoliating bark of old large trees provides excellent roosting locations for bats. Because young conifers have relatively smooth bark, they do not provide good roosting locations for bats. It is a challenge to provide roosting areas for bats since they prefer southern exposures on trees that are located on forest edges. Forest edges and southern exposures are prone to disturbance by wind throw and fire. Large snags work as well for bat day roosts since they provide thermal choices for bats and the cavities in these snags provide space for maternity colonies.

Forest Management Activities that Threaten the Diversity and Survival of Small Mammals

Minimize the simplification of forests to maintain small mammals

The most detrimental long-term forest management activities with regard to maintaining small mammal populations result from the simplification and elimination of habitat elements needed by small mammals. Any simplification that eliminates large living trees, large woody debris and vegetation strata will decrease the species richness of small mammals dependent on these structures. Forest structure is simpler when forest stands are in the earlier successional stages of growth (typically younger than 50 years of age for Douglas-fir forests). Once forests are in the 'understorey reinitiation' and 'stem exclusion' stages of stand development, many of the understorey species become more prevalent and thus provide more vegetation strata (Oliver and Larson, 1995).

The worst forest habitats for small mammals are what are called 'dog-hair' stands. Dog-hair stands are very dense forest stands of small-diameter trees, typically found in 30–40-year-old forest stages. These types of stands have poorly developed understorey and ground-level vegetation and will therefore provide poor habitat for mammals. When forests are too dense, they need to be thinned to increase the production of the lower strata (e.g. sub-canopies, shrubs or ground vegetation).

If managers want to increase the local diversity of small mammals, there will be a need to enhance the critical habitat elements that are needed by them. Some trees will need to be retained when cutting forests so that there are legacies of trees and woody debris that carry over into the next forest stand that will grow on the site. Systematic efforts will need to occur to either retain or create large snags and downed logs in these newly growing forests. When forest patches are being created during harvesting, it is also important that forest patch sizes exist that are 8–10 ha (or 20–25 acres) or larger in size. Forest patches in these sizes will provide a core of interior forest conditions needed by some of the mammals.

Maintain all stages of forest succession

Ideally forests should be managed as a landscape so that there is a mix of early and late successional forest stages present at any given time. Bats tend to use the early successional areas and water bodies for foraging, while they roost in forests where there are suitable trees and snags. For a bat, high-quality habitat provides them food, water and roosting areas that are located close to one another in the landscape. So land-use activities that increase the distance between food, water and roosting habitat end up being detrimental to bats.

The changes in species of small mammals along with the changes in successional stages of the forest are shown in Fig. 4.5. There is a species shift in small mammals when forest stands are around 15–25 years old. The timing varies with each

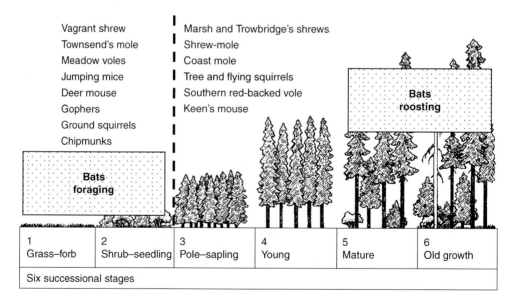

Fig. 4.5. Changes in the dominant species of small mammals with forest succession in the Pacific North-west United States (modified from Thomas, 1979).

forest but this shift generally coincides with the closing of the forest canopy. At that time, small mammals of fields are replaced by those that live predominantly in closed-canopy forests. Another group of small mammals are habitat generalists and tend to be present across successional stages.

Urban development threatens small-mammal populations

The growth of urban/suburban development and agriculture is resulting in the conversion of forests to other uses. This is becoming one of the biggest threats to retaining small mammals in our landscapes. Regional declines in small mammal populations are being attributed to these land conversions. Many of the forest patches remaining in the wild-land–urban interface are not suitable for small mammals because they do not contain a sufficient structural diversity (large trees, large woody debris and vegetation strata).

CASE 4.3. PUERTO RICO AND HAWAII: THE DILEMMA OF COQUI FROG CONSERVATION OR ERADICATION IN WET TROPICAL FORESTS (KAREN H. BEARD)

Introducing the Species of Conservation Interest and an Invader

As the sun sets behind El Yunque peak, the national forest in north-eastern Puerto Rico, and the day turns slowly to dusk, the hillsides gradually come alive with tropical sounds. These tropical sounds are mostly from the frogs, endemic to Puerto Rico, of the genus *Eleutherodactylus*. These small frogs (often less than 3 cm in length) fill nights in the Puerto Rican countryside with mating calls that have given the frogs their common name, coqui (pronounced koo-KEE). There are 16 frogs in the genus *Eleutherodactylus* in Puerto Rico, but the most widespread and abundant is *E. coqui*, also referred to as the coqui. These frogs are perhaps the most conspicuous animals that reside in the wet forests of Puerto Rico.

Coqui frogs have been in the consciousness of island dwellers for thousands of years, as evidenced by Taino Indians' petroglyphs and pictographs depicting coquis. Children in Puerto Rico grow up learning of these frogs, not only because of their ubiquity around the island and conspicuous mating calls, but because the song of the coqui is the focus of Puerto Rican folk tales. One of these tales concludes that, if coqui frogs were ever to leave Puerto Rico, they would no longer sing. Furthermore, because Puerto Rico has no native ground mammals or other such charismatic fauna, this small frog has become Puerto Rico's unofficial mascot. Many tourists and others who visit Puerto

Rico have heard the coqui's call and remember the frog fondly. However, there is more to the coqui than myths and souvenirs.

Why the Coqui is so Important for Puerto Rico

The coqui plays an important ecological role in Puerto Rican forests. It has one of the highest densities (roughly 20,000 individuals per ha on average) ever recorded for a terrestrial frog and has the greatest biomass of any vertebrate species in the Puerto Rican forests (Stewart and Woolbright, 1996). In addition, it represents a unique functional role in the nocturnal forest: it is the dominant insectivore. I conducted research on the coqui in Puerto Rico and found that their high consumption of arthropods (roughly 120,000/ha/night) reduces arthropod populations and converts nutrients into more available forms for plants and microbes. I also found that coquis increase the cycling rates of limiting nutrients, which results in greater leaf litter decomposition rates and plant growth rates (Beard *et al.*, 2003).

Coqui effects are most evident following the frequent hurricanes that disturb Puerto Rican forests. Hurricanes create their habitat (fallen dead plant matter or necromass), which can increase coqui densities up to sixfold (Woolbright, 1991). Following a hurricane, for the ecosystem to recover and for net primary productivity to re-establish, plants must regain the foliage lost. Heavy grazing of burgeoning arthropods would slow this recovery; however, it appears that post-hurricane booms in coqui populations partially control these arthropods. Greater coqui densities would also increase nutrient availability; this is important because nutrients are limited following hurricanes, and thus coquis may assist plants in regaining photosynthetic capacity and microbes in decomposing increased necromass or dead plant materials. Therefore, coquis appear to play an important role in the recovery of Puerto Rican forests following a disturbance.

Coqui as an Invader and Anti-conservation in Hawaii

Yet, according to the International Union for the Conservation of Nature (IUCN), amphibians are the most threatened vertebrates in the world and Puerto Rico is no exception. In 1986, Rafael Joglar noticed that coqui populations of some species had declined. Then, over a period of 25 years, three of the island's endemic frog species had become extinct, and several more species were found to be endangered (Joglar, 1998). Even *E. coqui* appeared to be declining at high elevations. Because *E. coqui* is not federally listed as threatened or endangered, there has been little conservation effort focused directly on this species. However, there are other threatened and endangered species, including other *Eleutherodactylus* species, that inhabit similar forest types, and habitat protection for these species benefits *E. coqui*. Furthermore, there has been a fair amount of research conducted to understand the cause of amphibian declines in Puerto Rico.

Based on this background, it might come as a surprise that *E. coqui* was not greeted with open arms when it was accidentally introduced in a shipment of nursery plants to Hawaii in the late 1980s. Because the coqui has direct development (no tadpole phase), it was able to spread quickly, especially on the islands of Hawaii and Maui, where there are now hundreds of populations. There are several reasons that the coqui invasion is of economic concern to the state of Hawaii. It threatens multi-million-dollar floriculture and nursery industries due to quarantine restrictions and de-infestation measures now required on plants

before they can be exported. It also threatens private property value and tourism because of its loud mating calls (80–90 decibels), which exceed levels set to minimize interference with the enjoyment of life (70 decibels, Hawaii Department of Health). The very same call that endears the coqui to Puerto Ricans has made it one of the most loathed species in Hawaii.

Challenges to Eradicating Coqui in Hawaii

In addition, the invasion of the coqui into Hawaii may be of ecological concern. There are thought to be three potential ecological consequences of the invasion: (i) the coqui may reduce endemic arthropods; (ii) the coqui may compete with endemic birds, the majority of which are insectivorous; and (iii) the coqui may contribute to endemic bird declines by bolstering populations of bird predators (e.g. rats and mongoose) (Kraus *et al.*, 1999). The potential to reduce endemic arthropods is the most likely to occur and could be devastating because arthropods comprise the large majority of Hawaii's endemic fauna. Even though coquis play an important ecological role in Puerto Rican forests, it is important to realize that coquis did not evolve in Hawaii's unique communities and ecosystems, and their presence could have serious negative consequences.

Thus, the coqui became the focus of a massive eradication attempt in Hawaii. In fact, in April 2004, the Mayor of Hilo declared the coqui situation a state of emergency in order to secure more federal funds to eradicate the frog. Numerous coqui control measures have been evaluated in Hawaii, including hand-capturing, habitat modification, biological control and chemical control; however, chemical control has been the most successful at reducing frog populations. More than 50 chemicals have been evaluated for their effectiveness as frog toxicants. Although several have proved effective (e.g. caffeine), citric acid and hydrated lime are the only products presently approved and labelled for use on coqui frogs. The situation surrounding the approval of the use of caffeine to control coquis illustrates the complexities of the coqui problem in Hawaii.

In 2001, the US Environmental Protection Agency granted a 1-year emergency registration to use caffeine for coqui control in response to an application prepared and submitted by the Hawaii Department of Agriculture; however, restrictions imposed by the latter agency made compliance impossible. In addition, there was a public outcry against the use of caffeine because there was an unfounded belief that it might harm humans. The use of caffeine also caused protest by people who did not believe that frogs should be killed for ethical reasons. Finally, there was advocacy for the coqui by individuals who believed that the mating calls enhanced property aesthetics, while others believed that the frog might control non-native pests, such as mosquitoes; data show that coquis do not consume mosquitoes (Beard and Pitt, 2005). These beliefs led to deliberate coqui introductions in conspicuous areas, such as the entrance of Hawaii Volcanoes National Park.

Thus, for several multifaceted reasons, coqui eradication has been difficult. The just described restrictions, protests and deliberate introductions slowed eradication efforts while the frog continued to spread. In addition, eradication was made difficult by a general lack of belief that a small frog could constitute a real threat, ecologically, economically or socially, to Hawaii. Finally, at a time when amphibian populations are declining globally, the idea of eradicating an amphibian seemed to some people to be inconsistent with conservation efforts. At this point, it is unlikely that the coqui frog will ever be eradicated from the Hawaiian Islands. In this case, social constraints, imposed by the response to eradication, played a greater role in the final outcome than the potential negative economic or ecological consequences of this introduced frog.

CASE 4.4. MALARIA AND LAND MODIFICATIONS IN THE KENYAN HIGHLANDS (NOBORU MINAKAWA)

Malaria and Forests

The highland areas (above ~1500 m in elevation) of western Kenya are characterized by undulating topography, with a series of small valleys and gently sloping hills. Most inhabitants still live in traditional huts, which are constructed with mud walls on the flat hilltops and slopes. Although European settlers established large tea plantations in the highlands during the colonial era, local inhabitants still depend on small-scale farms, mainly producing maize. The highland areas were once covered with the vast tropical rainforest that stretched across the girth of Africa from the Congo Basin. The recent rapid population growth in the highlands (Kenya's population has almost doubled since 1980) drastically reduced the forest, which had already been fragmented by cultivation and past climate changes. Now the remnants of the forest are seen as only two distinct forest patches, Kakamega Forest and Malava Forest, and tiny fragments of riparian forests in valleys. These remaining forest patches still maintain a variety of unique animal and plant species that are related to those in West Africa.

Since the late 1980s, malaria outbreaks have occurred frequently in the East African highlands, where malaria was previously rare. It is believed that the cool highland climate has lowered vector (*Anopheles* spp.) densities and slowed parasite (*Plasmodium falciparum*) development. Despite the long history of control efforts, malaria remains a major threat to human health, causing over 1 million deaths in sub-Saharan Africa every year (over 75% of them are children). Thus the threat is mounting, as evidenced by the recent malaria outbreaks in the highlands. Unlike their counterparts in malaria-endemic regions (lowland areas), the residents of highland areas generally lack immunity to *P. falciparum* and are particularly vulnerable to malaria infection. Several hypotheses have been proposed to explain the increased malaria transmission in the highlands, including land modifications, global climate changes, increased drug resistance, cessation of malaria control activities, demographic changes and poverty (Mouchet *et al.*, 1998).

The Ecology of Malaria

Anopheles gambiae is the major malaria-vector mosquito in the highland areas. Although this species is relatively better adapted to the cool highland climate than its sibling species, its densities still remain much lower in the highland areas than in the lower areas. Therefore, it has been suggested that global warming has increased vector densities high enough to lead to a malaria outbreak in the highlands (Stone, 1995). However, recent studies have suggested that the malaria outbreaks in the highlands are not explained by only global warming (Hay *et al.*, 2002; Zhou *et al.*, 2004a). Although the effects of global climate changes on highland malaria remain debatable, in general, the number of malaria cases is positively correlated with amount of rainfall. In fact, the worst malaria outbreak in the highlands followed the unexpected long rain caused by the 1997–98 El Niño (Lindblade *et al.*, 1999). Numerous aquatic habitats for *A. gambiae* appear after heavy rains, and the vector populations quickly develop.

Larvae of *A. gambiae* are known to occur mostly in sunlit temporary water pools in the adjacent lowland area, but not in wetlands that are characterized by tall aquatic plants such as papyrus (*Cyperus papyrus*). In the highlands, recent studies found that *A. gambiae* larvae seldom occur in aquatic habitats within dense forests or within wetlands; however, they mostly occur in sunlit temporary pools in farmlands and pastures that have been converted from wetlands and riparian forests in valley bottoms (Minakawa *et al.*, 2005). The temporary pools in the cultivated areas are mainly standing waters in ditches, animal footprints and human-made holes. Larval development times are faster and survival rates are greater in the cultivated areas (Minakawa *et al.*, unpublished; Munga *et al.*, unpublished). These findings suggest that such land modifications create a suitable environment for *A. gambiae* by increasing the amount of sunlight over the aquatic habitats. Increased sunlight increases water temperature and possibly food sources (algae), which enhance vector production in the highland areas. Such small sunlit temporary pools have not only warm water but also fewer predators of *A. gambiae*.

Most stagnant aquatic habitats are formed in valley bottoms by means of surface runoff from uphill and from springs and groundwater seepage. Flooded water from streams also forms several standing water areas in valley bottoms. These aquatic habitats are rather small and short-lived in

the land where trees have been cleared, because water evaporates faster with increased solar radiation in the open area. A wetland is also fragmented to smaller pools in the process of cultivation. Farmers create ditches to draw water from a wetland and often plant introduced eucalyptus (*Eucalyptus* spp.) in the wetland. Eucalyptus is a fast-growing tree species, and it quickly drains a large amount of water from the soil, which accelerates the drying up of the wetland.

Consequently densities of malaria vectors become higher in houses near the aquatic habitats in the valley bottom where forests and wetlands have been cultivated (Lindblade *et al.*, 2000; Zhou *et al.*, 2004b). Increased vector densities increase malaria transmission near the valley bottom, and malaria parasite prevalence and densities in the local human population increase with a decrease of distance from the valley bottom (Munyekenye *et al.*, 2005). It has been suspected that the increased local ambient temperature caused by such land modifications also enhances survivorship of adult mosquitoes and development of parasites in vectors. Thus, there is enough evidence to support the notion that the land modifications have contributed to recent malaria outbreaks in the highland areas.

Searching for a Solution

Based on these findings, reforestation may be the way to reduce vector densities by increasing shade over aquatic habitats. However, it will be very difficult to regenerate a dense forest in the cultivated areas in the highlands of western Kenya. Regeneration of a forest with indigenous trees takes decades. The fast-growing eucalyptus may not produce a crown dense enough to reduce solar radiation significantly. Moreover, farmers frequently cut eucalyptus trees for firewood. More than anything, poverty will not allow local inhabitants to give up the lands that have already been cultivated. Thus, it is becoming clear that poverty, land modifications and malaria are closely related in the highland areas. While we are still searching for an effective solution, the tiny forest fragments are turning into smoke little by little, and mosquitoes are ready for another malaria outbreak in the highlands.

CASE 4.5. SALMON: FISH OF THE FOREST (JON M. HONEA)

Introduction to a Keystone Species

The behaviour of migratory organisms allows them to benefit from environments where key resources such as food, mating opportunities and safety from predators are in rich supply for relatively short periods of time. Often migratory species not only use different habitats, but actually play key roles in ecosystem processes in the various habitats they use. Migrants may increase the productivity of some habitats they use by transferring energy and nutrients across ecosystem 'boundaries'. Salmon are an excellent example of this phenomenon.

Before extensive human urbanization as recent as 300 years ago, salmon spawned in enormous numbers in forest streams along coasts and even 100s of kilometres into the interior of Europe, North America and East Asia. During that time, it was not uncommon to find as many as four spawners per square metre of stream. Whole societies, particularly along the northern coasts of the Pacific Ocean, were organized around the seasonal abundance of this rich and highly predictable resource. The ubiquity of salmon in early Europe is indicated by the very low value of salmon noted in historical records. Everywhere the relationship is documented, salmon are so abundant initially that they are used as crop fertilizer. Over the centuries in Europe and eventually in much of North America and Asia as well, a wide range of changes occurred that reduced the numbers of salmon in streams and rivers, thus driving up the price of this resource, which is both highly nutritious and easily acquired where it is available. By the 19th century, salmon had become so rare in Europe that only the richest could afford it (Montgomery, 2003).

Salmon require cold, clear, fresh water for their eggs to develop buried in nests called redds, which the female adult digs in the gravel and cobble of her spawning stream or lake. More than 98% of the time, the adults follow unique scents that they learned as juveniles to return to the same stream where they had developed as eggs several years before (Quinn and Fresh, 1984). Such fidelity over time allows salmon to become adapted to the particular conditions that occur in their spawning streams. The temperature regime influences how long the eggs take to develop. Those that hatch too early or too late do not survive as fry (the first life stage above the gravel) due to dangerous flow conditions, lack of prey, occurrence of predators or other factors that vary seasonally. As a result, the most successful adults spawn all within the same time frame (usually a few weeks' span), giving the next generation of fry the greatest chance to emerge into conditions that they can survive and thrive in.

Temperature optimization must be balanced with many other factors, including access by adults to the spawning stream. In many coastal streams of the Pacific North-west of North America, spawners must await the first rains of autumn before the streams are deep enough to ascend. In other regions, winter ice-up is a key factor influencing the timing of spawning. These factors and others interact with incubation temperature to determine the timing of fry emergence and the number produced.

One remarkable characteristic of salmon is that the great majority die a few days after spawning (see Fig. 4.6). In this way, huge numbers of dead salmon, and a great many live ones before and soon after spawning, become food for nearly every other organism living in or around spawning streams. The aquatic scientist Jeff Cederholm and his colleagues summarized 88 different species of wildlife regularly feeding on live or dead salmon (Cederholm et al., 2000).

Salmon consumption can be important to aquatic insects, as is clear from findings of increased density and growth rates of insects feeding on salmon (Chaloner and Wipfli, 2002; Minakawa et al., 2002). More aquatic insects should mean more food for salmon fry, which feed on them. Nitrogen from salmon has even been found in the leaves of riparian (stream-side) vegetation (Bilby et al., 1996). In many cases the nutrients dissolved in water after decomposition may move through subsurface flow pathways (e.g. O'Keefe and Edwards, 2003). However, in areas where human disturbance has not reduced or eliminated their populations, bears and other predators and scavengers are major agents moving salmon nutrients from streams into the riparian forest (Gende et al., 2004). This evidence supports the argument that salmon are a 'keystone species' by supplying vital nutrients, historically in massive quantities, to organisms in and around forests (Willson and

Fig. 4.6. Salmon dying and returning nutrients into the stream. Photo by Jon Honea.

Halupka, 1995). To complicate the issue, some researchers have pointed out that, in some lake systems where sockeye are the main salmon species, salmon may actually export nutrients from lakes when the nutrients of one group of spawners are consumed directly and indirectly by the juveniles of a previous spawning year and then carried away upon their seaward migration (Moore and Schindler, 2004).

One important question that arises upon the discovery that salmon-derived nutrients are taken up even by riparian vegetation is 'What difference does it make?' In other words, do salmon provide a significant source of fertilizer that increases the growth rate of riparian vegetation and perhaps even influences forest structure? Jim Helfield, completing his doctorate with Professor Bob Naiman, found that spruce trees growing near salmon streams not only contained salmon nitrogen, but also grew over 250% faster than at sites without salmon. As a result, spruce on a salmon diet would grow to a diameter of 50 cm in only 86 years rather than the 307 years required for spruce with no salmon subsidy (Helfield and Naiman, 2001). This has important implications not only

for forest succession and structure, but also for forest–stream interactions due to the greater influence of large downed trees on water flow relative to smaller trees. Because pools created by large trees are important rearing habitat for juveniles of some salmon species, healthy salmon runs may contribute to self-sustainability by supporting greater forest productivity.

Although most current research into the effects of salmon on ecosystem processes focuses on nutrient enrichment, we are beginning to understand that salmon are also agents of disturbance to stream ecosystems. The average depth of most salmon redds is 20–30 cm, depending on the size of the female and average diameter of the gravel (Steen and Quinn, 1999). Not only does redd excavation change the contours of the bed surface and even the width of the stream in some cases (Montgomery *et al.*, 1996), but, by digging into the stream-bed so deeply, salmon can devastate the aquatic insect community (Minakawa and Gara, 2003; Honea, 2005). In fact, one researcher found that juvenile salmon moved upstream following adults into spawning areas, apparently to feed on aquatic insects as they were kicked up into the current during redd excavation (Joe Ebersole, US Environmental Protection Agency (EPA), personal communication).

Although aquatic insect populations can be severely reduced initially, over time insects capitalize on the availability of dead salmon to fuel a rapid recovery. Some aquatic insects, such as blackflies, probably benefit from the entire process because redd excavation cleans the surfaces of cobbles on the streambed and thus provides more of the type of surface they require to attach themselves securely so they can extend their heads into the current and filter out organic matter drifting downstream (Honea, 2005).

Loss of Wild Salmon

The reasons for decline in salmon numbers are the same from Portugal to Poland to Norway, from Connecticut to New Brunswick, from California to British Columbia and from Korea to Siberia. Over-harvesting of the fish was probably the first human activity to substantially decrease the number of salmon. Improved preservation techniques, from curing in smoke or brine to canning to refrigeration, have allowed local catches to be sold in distant markets, increasing the number of potential consumers. As human populations increased in an area, the salmon populations decreased, causing the price there to increase. Higher prices increased exploitation in outlying areas as the fishing industry sought to take advantage of the market. The technology for actually catching the fish has also advanced to the point where nearly all of the fish in a river can be harvested. Throughout the salmon's range, to limit the impact, harvesting has become heavily regulated. For example, once widely used techniques (e.g. waterwheels and salmon traps that caught and scooped up salmon with destructive efficiency) have been eliminated.

Dams built across rivers for hydropower and to create reservoirs for crop irrigation, transportation and flood control are another major factor in the reduction of salmon population numbers. Dams impede or entirely block access to upstream migration of adults and downstream migration of juveniles. In addition, they transform swift cool rivers into warmer slow-moving lakes, which bury spawning gravels and increase the risk to predation on juveniles by increasing downstream transit time and visibility to predators such as the northern pike minnow and non-native walleye, catfish and small-mouth bass.

The third most important reason for the massive reduction in salmon numbers was the destruction of spawning and rearing habitat as forests were cleared for timber, to produce more agricultural land and to make way for expanding urbanization. When trees are removed, there is more erosion of upland soils into streams. This is especially bad where the soils are compacted due to roads, landings and log-skidding trails, which reduce the rate of infiltration of water and therefore increase surface runoff. Increased sedimentation in streams reduces salmon egg survivorship as clogged gravels slow the delivery of cool oxygenated water and removal of metabolic wastes.

The loss of trees also decreases transpiration, which typically means that more water enters streams, thus increasing the frequency of high-flow events, which can be detrimental to stream-dwelling organisms. Many streams in forested areas served as conduits for moving timber to market. Splash dams, now illegal, were constructed by large piles of logs, which accumulated a large body of water behind them. When the log piles were dynamited, the resulting flood of water carried the timber downstream, but in the process scoured out spawning gravel, in some cases down to bedrock, a disturbance that requires decades or centuries to recover from. Finally, the loss of stream-side trees results in a decline in the frequency of important pool habitat as older in-stream wood is washed away in floods and no new recruits are available.

Can Wild Salmon be Recovered?

By reducing the number of adults through overharvesting, obstructing their migrations to and from the ocean and reducing the area and quality of the spawning and rearing habitat that remained, the reduced availability of a highly valued resource was inescapable. Each of the factors negatively affecting salmon provides substantial positive contributions to important aspects of human societies, such as nutritious food, cheap and relatively clean electricity, irrigation for valued crops and wood products for a wide diversity of needs, as well as the livelihoods of those employed in producing these goods. This has made change difficult and slow.

The value of wild salmon, however, has spurred us to closely examine the actions that contribute to their decline. In some cases, we have found that change is possible without substantial loss of benefits from the other factors. For example irrigation can be improved with techniques that prevent loss to evaporation as well as with screens to prevent salmon from being diverted into fields. The impact of logging can be reduced by reducing impaction through improved roads and landing practises, by placing buffer zones around streams and by switching from clear-cuts to techniques that retain some living and dead trees. In a few cases, the value of salmon has been found to be greater than that of particular dams and they are being removed to give access to historical habitat.

We have demonstrated that we can change when we want to, but so far it has not been enough. If we do not recognize the connections between forests and water and salmon and fail to acknowledge our effects on all of them, we will be unable to restore salmon and will continue to lose those that remain. If, on the other hand, we make space for salmon, many scientists believe that recovery is possible. They are resilient and highly productive fish. If we are responsible and concerned cohabitants, this rich cultural icon, symbolizing triumph over adversity, will continue to be represented by real fish struggling upstream.

5 Human and Natural Disturbances Impacting Forests

Robert L. Edmonds, Kristiina A. Vogt and Toral Patel-Weynand

Introduction

Forests are subjected to natural and human disturbances that affect the landscape (Pickett and White, 1985; Sprugel, 1991; Edmonds et al., 2005). Natural disturbances, such as wildfire, maintain forest health and affect their normal structure and function, as well as impacting wildlife and species diversity. On the other hand, they can cause economic hardships through loss of timber and diminished tourist income in affected areas.

Human disturbances can also have devastating effects on forest health and the persistence of forests in our landscapes. In forests, human disturbances can also feed back to decrease human health and the health of forest-dwelling animals. In a survey conducted in 1993, people in industrialized and developing countries in responding to the survey overwhelmingly stated that environmental problems affected their health (Dunlap et al., 1993). This link of human activities impacting human health was especially prevalent in the developing countries (47–89% of the people held this view), which have major pollution problems; this contrasted with industrialized countries, where a lower proportion of the population (14–73%) saw a link between the environment and human health. When respondents were asked how environmental problems affected their health 10 years earlier, people did not link human disturbances and environmental problems as strongly to human health. For example, 27% of the public in Great Britain felt that there was a link between environmental problems and human health in 1983 and this had increased to 53% in 1993 (Dunlap et al., 1993). In India, 40% of the people linked environmental problems to human health problems in 1983 but this had increased to 74% in 1993. Globally, human perceptions that the condition of the environment affects their health is quite recent and is rapidly increasing (Dunlap et al., 1993). For this reason, it is important to include a discussion of human health in a chapter that covers disturbances since they are closely linked to one another (see below – 'Disturbances in Forests and Human Health'). There is a scientific basis for this perception held by many people around the world.

Human disturbances are caused by changes to normal fire cycles, forest harvesting, air pollution, global warming and the moving of plants, insects or pathogens around the world. The most devastating human disturbances are the result of changing land uses (e.g. conversion of a forest into agricultural fields) or because humans modify the chemical environment of the atmosphere or the soils (e.g. over-fertilization of agricultural fields, combustion of fossil fuels, which increases greenhouse gas emissions into the atmosphere, application of herbicides to kill vegetation, which is applied indiscriminately or during a war).

War and social conflicts tend to increase the magnitude of the impacts of natural and human disturbances. Both the intensity and the magnitude of natural and human disturbances are affected by land-use changes that over-harvest forests (e.g. exposing bare ground to higher rates of erosion) or increase the application of chemicals that can be toxic to humans and other forest-dwelling animals (e.g. defoliants used in Vietnam to eliminate forest cover so that people could not hide in forests). Wars and social conflict have disastrous impacts on the health of both humans and forests, particularly in tropical countries that are less industrialized. Since so many people living in the tropics are dependent on resource extraction for survival, these changes in the environment further exacerbate the incidence of disease in humans and other forest-dwelling animals.

It is important to understand human and natural disturbances since they determine how forests can be managed to be sustainable, how effective conservation efforts will be and the level of resource extraction that is possible from forests. Humans are currently changing disturbance cycles at an unprecedented rate so it behoves us to understand the impact of the decisions that we are making in forests.

Recognition that Natural Disturbances Need to Occur

Until the early 1970s, natural disturbances (i.e. fires, windstorms, insect and pathogen outbreaks, volcanic eruptions, drought, flooding, avalanches, etc.) were generally thought of as freak, unpredictable events in ecosystems. However, in the mid-1970s, numerous groups of ecologists working in several different areas simultaneously pointed out that natural disturbances were both common and predictable (Pickett and White, 1985). In addition, few, if any, sites where natural disturbances were common ever progressed all the way through the 'normal' course of succession (i.e. change in the composition of dominant plant or animal species through time on a given site) because of these recurrent disturbances.

As previously mentioned, natural disturbances maintain forest health and are an important part of maintaining forests and particular species of trees in our landscapes. For example, lodgepole pine forests are kept healthy by infrequent catastrophic fires that occur in the Rocky Mountains of the United States (including in the Yellowstone National Park) (Agee, 1993). In other parts of the United States, frequent surface fires maintain the mixed conifer forests in the Sierra Nevada Mountains in California and the ponderosa pine forests in Arizona and on the east slopes of the Cascade Mountains in the states of Washington and Oregon (Agee, 1998). Hurricanes frequently disturb forests in the south-east and eastern United States, in the western and eastern parts of Mexico and on the Caribbean islands (Beard *et al.*, 2005).

In many ecosystems, natural disturbances are recurring and generally predictable events. While one cannot predict that any given forest will burn in any given year, one can predict that a forest on a naturally dry site (because of either the soil properties or the climate, or both) will probably burn within the next 50 years. Furthermore, forests found growing on wetter sites are less likely to burn within that same time period, although occasionally forests growing in the riparian areas in dry areas may burn before the forests growing in upslope areas.

Disturbance is a very natural component of most ecosystems, and ecologists now talk about the natural disturbance regime for an area, i.e. the pattern of natural disturbance that one would see in the area if it were not affected by humans. Many plants living in frequently disturbed ecosystems have developed adaptations that help them thrive in the presence of repeated disturbances (e.g. serotinous cones in jack and lodgepole pines, where the reproductive structures of each tree species need fire to open their seed cones for seeds to be able to disperse so that these species can regenerate). While disturbance type and frequency commonly control vegetation characteristics, the reverse is also true: vegetation characteristics have a strong effect on disturbance type and frequency (see Edmonds *et al.*, 2005).

Disturbance is thus a normal component of most natural ecosystems and exerts a powerful control on vegetation characteristics. Many plants and animals are well adapted to particular kinds of disturbances and often cannot survive without them.

Forest Health

Disturbance and forest health are intertwined. Forest health is a term that is increasingly being used in forest management. It is a concept that was adapted from human health, but forest health is not as easily defined. Forest health is not measured by the number of dead trees present. Dead trees play an important ecological role in forests, particularly for wildlife, so some dead stumps or snags are desired.

Definitions of forest health range from utilitarian to ecological. From a utilitarian perspective a healthy forest is one where biotic and abiotic influences on forests (e.g. pests, pollution, thinning, fertilization or harvesting) do not threaten management objectives now and in the future. It is a condition of forest ecosystems that sustains their complexity of organisms and habitats while still being capable of providing for human needs and environmental services. From an ecological perspective, a healthy forest is one that has a fully functioning community of plants and animals. In addition, it is resilient to disturbances, both natural and human-generated. Maintenance of biodiversity is an important aspect of forest health.

A forest health problem is suspected if trees appear stressed or sickly under normal weather conditions (including periodic droughts), there are high mortality rates of trees and there is a lack of regeneration of tree species typical for that forest. Other symptoms of health problems can include:

- higher or lower tree density than the historical range for the ecosystem involved;
- higher proportions of particular tree species, understorey plants, fish or wildlife than historically reported or some of these species are no longer present;

- low numbers of land or aquatic indicator species that have been shown to reflect the health of the ecosystem (e.g. salmon in the rivers of the western continental United States, Canada, Alaska, Norway and the east coast of Russia);
- one or more plant or animal species excessively impacting or dominating the ecosystem so that the natural changes in plant or animal species will no longer occur (see Case 5.1 – Kudzu); and
- excessive soil erosion.

The majority of forest health issues involve species loss, invasive species, insect or disease epidemics, excessive intense wildfires, air pollution, water quality or quantity problems, impacted wildlife populations, nutrient imbalances and soil or watershed damage. These are the same factors affected during a war, when people use forests as sites of refuge as happened in Rwanda (see below – 'Wars').

Forest Declines

Tree mortality occurs naturally in forests (Ciesla and Donaubauer, 1994). In old-growth forests of the Pacific North-west United States as well as in the American tropics, approximately 1–2% of the large canopy trees die naturally each year (Franklin *et al.*, 1987; Beard *et al.*, 2005). Forest declines can be defined when tree mortality is higher than the natural rates of mortality for any given ecosystem. Humans have altered these natural rates of tree death. Higher rates of tree mortality have resulted from human-generated air pollutants and the introduction of exotic pathogens and insects. Some forest harvesting practises and site-burning procedures have also contributed to higher rates of tree mortality when not used properly. In most cases, scientists have been able to provide an explanation for excessive tree death. There have been many situations, however, where episodes of tree mortality over wide areas have been difficult to explain (Manion and LaChance, 1992). These have typically been called 'declines' and the cause or causes of this mortality are unknown.

In other cases, scientists attributed forest decline to a cause (e.g. acid precipitation) when the facts did not exist to show that this was the primary factor responsible for this decline. In Case 5.2 (Acid Rain, Air Pollution and Forest Decline), John Innes provides an excellent example of how scientists around the world can accept a causal link to forest decline that had not been verified using a systematic scientific approach. This resulted in acid precipitation being implicated as the causal factor for forest decline when it was not the causal factor but contributed to decline symptoms. Identifying the causal factor in forest decline is important since attempts to manage or mitigate decline will not work if secondary factors that exacerbate decline symptoms are in fact being treated.

Forest declines have occurred on all continents and even on the Pacific islands (Edmonds *et al.*, 2005). Forest declines can occur in native species as well as in exotic plantation species, such as *Eucalyptus* spp. in Brazil. Symptoms of decline may be subtle, but are persistent, progressive and usually occur over longer time periods, ranging from 3 to 30 years. Periods of decline may be followed by periods of recovery, which can be temporary or complete. Forest declines usually involve only one tree species. Thus the term species decline might be a better term than forest decline (Skelly, 1989).

Diebacks and declines were noted as early as the 1700s in Europe (Manion and LaChance, 1992). In the 20th century, a number of diebacks of ash, birch and maple occurred in hardwood forests in North America. Oak and maple declines have also been recorded, as well as maple blight in these same forests. Declines have also taken place in conifer forests, such as Alaska yellow cedar in south-east Alaska. No principal pathogen capable of causing diseases in healthy trees was identified in the majority of these cases. Rather, they seemed to be caused by a variety of factors. Changing climate or unusual weather conditions have commonly been blamed for these declines.

Since we don't fully understand forest declines a number of theories have evolved (Manion and LaChance, 1992). Each theory defines a complex set of factors that created the decline. At least five theories or concepts have emerged:

- In the 'environmental stress and secondary organisms' theory (Houston, 1992), healthy trees are affected by environmental stress and secondary organisms attack the weakened trees. As the disease progresses, tissues are killed or their functioning impaired, and the trees may ultimately die.
- A three-step chain reaction of causal factors is involved in the 'predisposing, inciting and contributing factors' theory (Manion, 1991). Predisposing factors include tree age and genetic potential, soil conditions, planting species outside their natural ranges and air pollution. Inciting factors include defoliating insects, drought, air pollution and human activities such as site manipulations or the application of road salt. Contributing factors, which usually cause mortality, include *Armillaria* root rot, canker fungi, wood- and bark-boring insects, nematodes and perhaps viruses and phytoplasmas.
- The 'climate change or climate perturbation' theory (Auclair *et al.*, 1992) is focused largely on climate change and deviation from the natural variation in climates as being the dominant mechanism of forest dieback. There is strong evidence that climate change is linked to the progressive dieback and mortality of yellow birch, white ash and red spruce in north-eastern North America and silver fir and Norway spruce in Europe.
- The 'air pollution' theory is largely linked to the large-scale European forest decline that occurred, particularly in Germany and Switzerland, in the 1970s (Schutt and Cowling, 1985). This decline was blamed on air pollution and acid precipitation. Subsequently, similar claims were made to explain forest declines in the 1980s in the eastern United States, particularly in high-elevation forests. During the same period there were 'unprecedented' declines in high-elevation spruce–fir and sugar maple forests in the eastern United States, which were specifically blamed on acid precipitation. However, most of the mortality in the spruce–fir forests turned out to be the result of an introduced insect pest, i.e. balsam woolly adelgid. This is not to say, however, that air pollution and acid precipitation have not affected forest ecosystems. These trees may have been weakened by either one of these two conditions and became more susceptible to disease or insects.
- The 'ecological or natural dieback' theory (Mueller-Dombois, 1992) is based on four principal factors: simplification of forest structure, edaphically (influenced

by the soils) extreme sites, periodically recurring perturbations and biotic agents. Some plant species accumulate heavy metals, such as aluminium, in high concentrations in their tissues (e.g. tea has been recorded to have 1% aluminium in its leaves). By accumulating aluminium in their tissues in higher concentrations than normal, they alter nutrient cycling in forest ecosystems and increase the biological cycling of aluminium, which is toxic to other plants and microbes. This ability of some plants to modify the chemistry of the soil has been recognized in agriculture and led to the practice of crop rotation. When soil chemistry is sufficiently modified by an aluminium-accumulating plant, susceptible species may experience dieback because they are poorly adapted to the changed chemistry of the soil environments. This is similar to naturally occurring serpentine soils, which are high in manganese. Some plants are tolerant and can grow in soils high in manganese, while most other plants find this element toxic even in small concentrations. Periodically recurring perturbations, such as drought, may temporarily stress a forest stand. During long drought events, biotic factors (e.g. insects and fungal pathogens) typically provide the *coup de grâce* and hasten dieback and death of plants. Global change could accelerate the rate at which plant death occurs.

Forest Disturbance Agents

Natural disturbance agents

The most common natural disturbance agents that affect forest health are either abiotic (wildfire, wind, extremes of temperature and moisture and volcanic eruptions) or biotic (pathogens, insects and other animals – birds, mammals and marsupials) variables.

Wildfire

Wildfire is a natural process that is influenced by weather, topography and fuels (Agee, 1993). Hot, dry, windy weather, steep terrain and high quantities of fuel all favour the occurrence of wildfires. Wildfires are the classical disturbance agent in forests throughout the world (Pyne, 1995). Wildfire also interacts with and increases the prevalence of other disturbance agents, such as insects, disease and wind. It is not independent of the ecosystem in which it occurs and is influenced by the amount, arrangement and structure of fuels or dead forest materials present in any given site. Its effects on vegetation depend on how plants are adapted to fire, such as having thick bark, serotinous cones, stem sprouting and lignotubers (see Edmonds *et al.*, 2005). How wildfires, insect outbreaks and global warming are interacting together towards decreasing the health of boreal forests in Alaska is discussed in Case 5.3 – Wildfire in the Boreal Forests of Alaska.

Each forest ecosystem has a fire regime that it is adapted to withstand that is based on the severity of the fire. There are three broad classes of fire severity (low, mixed or moderate, and high) (Agee, 1993). A low-severity fire regime is one in which fires occur frequently (< 20 years) and are of low intensity. Forests dominated by ponderosa pine and longleaf pine are adapted to low-intensity fire regimes. Mixed or moderate regime fires are of intermediate frequency (25–100 years) and range from low to high intensity. Forests dominated by coastal Douglas fir, such as in Washington and Oregon, are typical examples of a forest adapted to moderate- to high-intensity fires. High-severity regime fires occur infrequently (> 100 years) and are usually of high intensity or very hot fires. In high-intensity regime fires, many trees are often killed so that these fires are called 'stand replacement fires'. These fires completely destroy large areas of forests where no survivors are left and regeneration has to occur from areas not burned; hence the name 'stand replacement fires'. Forests dominated by jack pine, lodgepole pine and subalpine fir are adapted to these high-intensity fire regimes.

Plants have developed life strategies to combat fire and are termed avoiders, endurers, invaders, resisters and evaders (Rowe, 1983). Intense hot fires tend to favour endurers, evaders and invaders, while low-intensity fires favour plants that are resisters and may need fire to persist in these landscapes (e.g. lodgepole pine is unable to compete with other tree species growing on the same site and needs fires to kill this competing vegetation). Fire exclusion (not allowing fires to occur) favours avoiders or species that are not adapted to survive any intensity of fire.

Wind

Wind can blow down individual trees or large areas of forests in the landscape (Kimmins, 1997). Different names and characteristics are given to windstorms depending on where one is in the world (e.g. hurricanes, cyclones, tornadoes, typhoons). Hurricanes occur in North and South America, cyclones occur in eastern Australia and typhoons in South-east Asia. Tornadoes damage swathes of forests in the United States every year.

Hurricanes frequently level large areas of forests in the south-east and eastern parts of the United States, in the Caribbean islands and in Central America. The 1938 hurricane that occurred in the New England states in the United States is often mentioned because its impact (e.g. dead trees) can still be seen in these forests (Foster, 1988). Hurricane Hugo in 1989 destroyed more than 1.6 million ha of trees in South Carolina (Gresham *et al.*, 1991).

Windstorms occur in the coastal United States Pacific North-west. A windstorm that is frequently cited is the '1921 blow' on the Olympic Peninsula in the state of Washington (Boyce, 1929). Windstorms have also been recorded in many other parts of the world and even in locations not considered to be susceptible to windstorms (e.g. Brazilian Amazon).

Extremes of temperature and moisture

Very high or very low temperatures can cause plant injuries (Sinclair *et al.*, 1987). High-temperature injuries include heat defoliation, sugar exudation and pole blight, which are exhibited as resin flow in pole-sized western white pine. Sun scald (similar to the sunburn that humans have when exposed to too much radiation from the sun) is also common, particularly on thinned and pruned trees. Cedar flagging occurs after a hot, dry summer and affects western red cedar (*Thuja plicata*). Cedar flagging is a condition resulting from these climatic conditions where older foliage turns red/orange in later summer or autumn and is eventually shed (Sinclair *et al.*, 1987). Western red cedar typically recovers from 'cedar flagging'.

Low-temperature injuries fall into two categories; frost injuries and winter damage (Tainter and Baker, 1996). Both early and late frosts can cause extensive damage. Early frosts in autumn cause injury because tree tissues are not hardened off and buds may not be set. Young trees, particularly seedlings in open areas or in frost pockets, are more susceptible than older trees. Late frosts occurring after bud burst cause extensive damage because new tissues are very susceptible. Vertical frost cracks can occur along the stem of susceptible species due to differential swelling and shrinkage of tissues. The mechanism of freeze damage is not completely understood (Larcher, 1995).

Winter damage is generally not as much of a problem as is frost damage for species growing in their natural range because native tree species naturally harden off and set their buds before these cold events occur. But winter damage can occur. Conifer winter damage can be extensive in areas that typically have mild winter climates such as the Puget Sound area in the state of Washington in the United States. Many exotic species from warmer climates have been planted in the Puget Sound area and may die when winter frosts do occur because they are not adapted to these temperatures.

On the slopes of broad mountain valleys in cold northerly environments (e.g. the Rocky Mountains in the United States, the Alps in Europe), winter damage known as 'red belt' sometimes occurs in conifers (Edmonds *et al.*, 2005). Frozen soils in combination with warming air temperatures create a situation where trees start to grow, triggered by the warmer air temperatures. To grow, trees need to take up water from the soil to support foliage function and to keep foliage temperatures from overheating. Since the soils are still frozen, trees are unable to acquire soil water and move this water to the foliage. Foliage is cooled by water evaporating from the surface of the foliage. When trees cannot obtain soil water, they subsequently dehydrate, the foliage gets too hot and is killed, resulting in the foliage turning red in colour. Occasionally very cold winter conditions will affect trees over extremely wide areas. For example, temperature extremes during the winter of 1950/51 resulted in visible bands of red-coloured trees in the south-eastern United States (Tainter and Baker, 1996). A similar situation occurred in western Oregon and Washington in 1955.

Many tree species are adapted for handling heavy snow loads, e.g. high-elevation or high-latitude spruces and firs, which have a pyramidal shape that reduces snow accumulation. During heavy snow years, damage of the canopy of trees is common. At low elevations, snow can cause stem and branch breakage, particularly in low-elevation species like Douglas fir.

Ice build-up on branches can also cause breakage. In the New England area in the United States ice build-up is common and results in extensive damage to the canopies of hardwood and coniferous trees. Tree damage from hail is common in some areas, particularly where thunderstorms are common.

Trees struck by lightning may be shattered or large branches may break away from the main stem of a tree. Burn scars are common on lightning-struck trees. Bark beetles are often attracted to conifers struck by lightning, and many isolated southern pine beetle attacks are related to lightning strikes.

Extremely dry conditions commonly cause drought distress symptoms in trees, including wilting, loss of foliage, dead branches and even tree death. In areas where rainfall is unreliable, drought stress is common. Drought stress is commonly found in the eastern slopes of the Cascade Mountain Range in the states of Oregon and Washington and in the Sierra Nevada Mountains in California in the United States. Fire suppression in these regions has resulted in increased tree density, which has exacerbated drought symptoms during times of low rainfall. Drought stress also occurs in other areas of the world, especially in Australia.

Excess moisture or flooding also causes tree stress, the appearance of wilting symptoms on a tree and even death (Larcher, 1995). Soils may become anaerobic (absence of free oxygen) so that plant roots become oxygen-stressed. Few tree species are well adapted to withstand long periods of flooding. Swamp cypress in the southern United States is an exception of a species thriving in wet environments. Hardwoods generally grow better in wetter soils than conifers. Wet soils also favour the development of root diseases caused by *Phytophthora* spp., which kill or weaken plants.

Volcanic eruptions

Volcanic eruptions are relatively rare events, but can strongly influence forests in their vicinity. Active volcanoes have erupted many times in the past and will continue to do so in the Pacific Rim of Fire in South and Central America, the western United States (including Alaska) and Canada, Siberian Russia, Japan, South-east Asia and New Zealand. Volcanoes produce tephra or ash, lava, pyroclastic flows, mud flows and gases, such as SO_2, that are toxic to plants. Soils around volcanoes can be superheated, which can make them unfavourable for plant growth because soil physical and chemical characteristics have changed (e.g. soils become impervious to water).

Pyroclastic flows of superheated air moving down the side of a volcano can blow down trees or kill them but where they remain standing up. This happened during the eruption of Washington State's Mt. St Helens in 1980. During this eruption, tephra (up to 30 cm deep) was deposited on a large area of forest to the north-east of the mountain and winds carried it high enough to circumnavigate the globe.

Mudflows are particularly devastating to riparian forests and are capable of travelling many 10s of kilometres from the source. Lava has less of an impact on forests because relatively few volcanoes have viscous lava that can move long distances. Good examples of volcanoes with viscous lava are found on the big island of Hawaii.

Pathogens

Plant pathogens include fungi, fungus-like organisms, bacteria, viruses, phytoplasmas, parasitic plants, nematodes and protozoans (Agrios, 1997). Fungi tend to be the dominant cause of diseases in forest ecosystems and strongly influence ecological processes, such as tree mortality, decomposition rates of plant materials, nutrient cycling and forest succession or the natural transition in the composition of vegetative communities with time (Edmonds *et al.*, 2005). Fungi have beneficial roles as well as causing diseases. The diversity of fungi found in natural ecosystems is high. It has been estimated that there are about 1 million species of fungi on earth, but only about 100,000 have been named (Alexopolous *et al.*, 1996). Most fungi are saprophytic or form beneficial relationships with trees (such as mycorrhizal fungi associated with the fine roots of plants, which increase the uptake of nutrients needed by a plant to grow; see Case 4.1 – Mycorrhizal Symbioses in Forest Ecosystems: the Ties that Bind).

About 10% of known fungal species can cause plant diseases and a smaller number have been documented to cause forest diseases.

Pathogens, insects and fire are strongly linked in forest ecosystems (Schowalter and Filip, 1993). Many insects disperse fungal spores that cause disease (Alexopolous *et al.*, 1996; Tainter and Baker, 1996). Other defoliating insects may weaken trees to the point where they are susceptible to root-rot fungi, like *Armillaria* spp. Root disease fungi, such as *Armillaria* spp. (Shaw and Kile, 1991) and *Heterobasidion annosum* (Woodward *et al.*, 1998), have a great impact throughout the world. *Phytophthora* spp. (fungal-like organisms) are important pathogens (Agrios, 1997). Pathogens can attack all the different parts of a plant, including foliage, twigs, branches, cones, stems and roots (Hansen and Lewis, 1997). Interactions between pathogens, insects and fire are strongest in areas where fire frequency and moisture stress are high (e.g. inland conifer ecosystems in western North America) (see Case 5.3 – Wildfire in the Boreal Forests of Alaska). Pathogen- and insect-killed trees increase the amount of dead biomass in forests, which increases the risk that higher-intensity forest fires will occur.

Insects

Insects are more numerous and have a greater number of species than any other form of life on earth (Pedigo, 1999). Thus it is not surprising that they are well represented in the species that live in forests. Similarly to fungi, insects have beneficial as well as detrimental impacts on forest health. Insects affect every plant part, like pathogens, and also play important ecological roles with regard to tree mortality, decomposition and nutrient cycling and tree succession. There are 13 orders of insects. Insects that cause forest problems are found in the following orders: Coleoptera (beetles and weevils), Lepidoptera (moths and butterflies), Diptera (flies), Hymenoptera (sawflies, wasps, ants and bees), Dictyoptera suborder Isoptera (termites) and Hemiptera suborder Homoptera (aphids and aldelgids) (Pedigo, 1999).

In forests, the insects that cause the majority of the tree mortality and defoliation are the beetles and moths. Beetles embrace the largest insect order, with perhaps 500,000 species. Beetles associated with trees are very diverse and specialized with respect to the part of the tree attacked (e.g. predators, root feeders, foliage feeders and stem feeders in bark and wood). During bark beetle outbreaks, trees have been killed across thousands of square kilometres of forest land in western North America.

Conifers are particularly susceptible to bark beetles in the genera *Dendroctonus*, *Ips* and *Scolytus* (Furniss and Carolin, 1977). Five important *Dendroctonus* genera attack conifers in North America: mountain pine beetle (*D. ponderosae*), Douglas-fir beetle (*D. pseudotsugae*), spruce beetle (*D. rufipennis*), southern pine beetle (*D. frontalis*) and western pine beetle (*D. brevicomis*). Engraver beetles belonging to the genera *Ips* and *Scolytus* can also cause tree mortality. Forest management practices such as fire suppression have generally increased the populations of these insects in their native environments.

Animals

Animals are valued for their diversity in forests, but are also important disturbance agents. They influence plant production and species succession by the consumption of one species' seedlings or seeds while leaving another species to grow. Many birds, such as nuthatches and grosbeaks in North American forests, consume seeds as do mammals such as shrews and ground squirrels (Black, 1994). Voles feed on grass roots and forbs during the growing season, but move to rhizomes, bark and cached food during the winter. Deer and elk have a considerable impact by feeding on seedlings and young trees. Some tree species are particularly vulnerable to deer browse, such as western red cedar. Tree squirrels feed on foliage buds, elongating shoots and cones, as well as phloem and xylem in the spring. They typically strip short pieces of bark. The presence of squirrels is indicated by discarded tree parts littering the ground under tree crowns.

Porcupines are found throughout North American forests and feed on the bark and sapwood of all ages of conifers. Black bears cause similar damage to conifers in the spring by stripping bark from trunks of vigorously growing trees. Snowshoe hares occur widely and when populations are high cause considerable seedling mortality by eating the tops of seedlings.

Ever since humans began growing trees in plantations, animal damage has been recognized as a problem for forest regeneration. Seeds, seedlings and older trees are subject to damage by animals. Birds and mammals are responsible for seed destruction, cone severing, browsing, clipping, bud injury, seed pulling, tree cutting and bark damage. Similar examples of the disturbance effects of birds and mammals occur throughout the world. For example, damage to vegetation due to cockatoos and marsupials has been recorded in Australia.

Free-range livestock, including both sheep and cattle, have a considerable impact on forest lands in the western United States. Grazing on public lands is common. Cattle cause damage by trampling, browsing and rubbing against sapling-sized trees. They can also cause soil compaction and site degradation, particularly in sensitive riparian zones. On the other hand, both cattle and sheep can be used to control competing vegetation when their presence is properly monitored.

Human disturbance agents

Human-caused disturbances in forests result from air pollution, acid rain, climate change, road salt applications, forest management (including forest harvesting and fire suppression), invasive plants, insects, pathogens, land-use changes and chemical applications that occur during a war. Many of these disturbances are found to co-occur in the same forest and decreases in forest health are usually a combination of multiple disturbance agents functioning in the same forest.

Air pollution

Air pollution has caused injuries to trees ever since the industrial revolution, which goes back to the 1800s (Treshow and Anderson, 1989). The main types of air pollutants are gases, particulates and acid precipitation. Air pollutants can be classified as primary or secondary. Primary pollutants are those that, when released from their sources, are already toxic to plants (e.g. sulphur dioxide (SO_2) and particulates). Secondary pollutants are those that develop as a result of pollutants reacting with one another, such as are found with ozone (O_3) and with acid precipitation. Sulphur dioxide and O_3 are the major gases influencing trees, but nitrogen oxides (NO_x), peroxyacetyl nitrate (PAN), hydrogen fluoride (HF), silicon tetrafluoride (SiF_4), ammonia (NH_3), chlorine (Cl_2) and freon can also cause forest health problems.

Sulphur dioxide occurs naturally in the atmosphere as a result of volcanic activity, but industrial sources, such as copper smelters and coal-burning power generators, produce significant levels of SO_2 on a regular basis. Recently scrubbers have been installed at most plants to reduce SO_2 emissions at the source and smoke stacks have been raised to reduce pollutant concentrations near the ground. Sulphur dioxide causes interveinal necrosis and chlorosis of broadleaved trees, tip necrosis and banding of conifers (see Edmonds *et al.*, 2005). In the early stages, conifers may shed their needles prematurely; however, high concentrations of sulphur dioxide will cause tree death.

Particulates do not generally cause serious problems to vegetation and most effects are localized (Legge and Kruper, 1986). Particulates are usually derived from industrial activities such as cement and coal-burning power plants. When vegetation grows close to roads, their leaves are commonly coated with particulates that can contain toxic chemicals, damaging plant leaves. Particulates have also been shown to have positive effects when they add nutrients (e.g. calcium (Ca)) back into forests as part of atmospheric deposition. In forests where acid precipitation has caused soil acidification and the leaching of nutrients from the soil, these atmospheric inputs can compensate for the air pollution-caused soil losses of nutrients. This has been recorded in the New England states in the United States, where cement plants and road building occurring in the Midwestern states have put supplemental Ca back into the forests (Hedin *et al.*, 1994; Likens *et al.*, 1998).

The major secondary pollutants are ozone, acid precipitation and excess nitrogen. Ozone is a major air pollutant gas throughout the world. Near the ground surface, ozone is a pollutant gas and is largely a result of photochemical oxidation of nitrogen oxides from the exhaust of internal combustion engines in the presence of hydrocarbons. This seems contradictory since ozone is an important natural component of the upper atmosphere but is a pollutant at the ground level. At the ground level, concentrations of O_3 are particularly high in areas where solar radiation and vehicular traffic are high and where topography confines the movement of pollutants (e.g. southern California). Other areas with high O_3 concentrations occur in the southern and eastern United States, Europe, Mexico, Central and South America and Asia.

High O_3 concentrations cause flecking or stippling of the upper surface on broadleaved plants, while tip necrosis and chlorotic mottling with banding are common on conifers. Older needles and leaves are more susceptible to damage, leading to premature defoliation. Typically visible damage to vegetation is thought to occur at ozone concentrations greater than 80 parts per billion by volume (ppbv). Not all tree species are equally affected, however, and susceptibility varies greatly. Ponderosa pine is extremely susceptible to ozone, as well as the white pines.

In southern California, mortality of ponderosa pine due to ozone initially averaged about 8% per year (Miller and McBride, 1999). This mortality rate has been reduced due to stronger emission regulations and the survival of individual resistant trees. Species succession and composition have been affected in southern California, so there are now a higher proportion of oaks and fewer ponderosa pine trees growing in the landscape. Ecosystem functioning has also been affected by this shift in species dominance. For example, oak leaves decompose at a slower rate and therefore release fewer nutrients to other plants growing in the same environment. Forest-floor decomposition and nutrient cycling rates have also decreased, which means that only species more efficient at acquiring nutrients can grow in these environments.

Acid precipitation is formed by interaction of sulphur dioxide and nitrous oxides in the atmosphere with water vapour, forming sulphuric and nitric acids, respectively. Normal rainfall pH is 5.3. Acid precipitation occurs in rain, snow and fog, where pH can be 4 or less. Fog pH can be as low as 2.5. Areas downwind of industrial and power-generation sources (e.g. eastern United States) are likely locations where acid precipitation damage will be found. Automobiles are largely the source of nitrous oxides. Acid precipitation has been implicated in forest decline in Europe and the eastern United States. Acid precipitation causes soil acidification, which increases the levels of available aluminium, which is toxic to plant roots. Aquatic ecosystems are particularly susceptible to acid precipitation, resulting in fish mortality.

Excess nitrogen (N) from both dry and wet deposition is now considered an air pollution problem affecting the functioning of forest ecosystems in eastern North America and Europe (Aber *et al.*, 1998). Trees in areas that are saturated with N have reduced net primary production because of the death of fine roots and mycorrhizae needed by plants to take up nutrients from the soil. Excess N in the soil is transformed to a nitrate form, which complexes with nutrient cations, such as Ca, and is easily washed out of the rooting zone. This results in a loss of critical nutrients needed by plants to grow.

Global change

Trees are very responsive to weather conditions and the current distributions of tree species are largely determined by climate. In recent years there has been great concern that human activities are causing global change, particularly global warming due to human activities that are increasing the emissions of greenhouse gases into the atmosphere (e.g. CO_2) (see Chapter 6, this volume). Several human activities have been implicated in increasing atmospheric CO_2 concentrations: fossil-fuel combustion in the process of generating electricity and when combusting transportation fuels, forest burning and increased decomposition rates of organic materials when forest cutting increases the temperature of the soils, which increases microbial activity.

Carbon dioxide is a greenhouse gas that traps long-wave radiation in the atmosphere. Atmospheric CO_2 has increased from about 275 parts per million by volume (ppmv) in the 1850s to over 350 ppmv today. Other greenhouse gases, such as nitrous oxide, chlorofluorocarbons, methane and tropospheric ozone, are also increasing. Chlorofluorocarbons also reduce stratospheric ozone, allowing more radiation to reach the earth's surface, contributing to global warming. Global temperatures have been predicted to rise 1.5 to 3.5°C, with the greatest increases expected at higher latitudes.

Rainfall patterns will also change as a result of global warming. Some computer models predict a decrease in summer moisture over middle and high latitudes. Such a change may cause this region to become drier and have longer growing seasons.

Global change has occurred in the past, but the rate of change caused by recent human activities is thought to be more rapid than natural. This rate of change has great implications for trees. Trees adapt to change slowly because of their long generation time. An individual tree can survive from less than 100 years to over 800 years. Many of the diebacks discussed earlier might be related to climate change. Diseasecausing organisms and insects, because of their short life cycles, are able to adapt more quickly to climate change than trees. This quicker adaptation by disease organisms might pose a more serious threat to forest health than global change alone. Plantation forestry might become more important in the future if we are forced to switch to species better adapted to the new climates created by these changes.

Salt injury

There are a number of sources of toxic salt injuries, including highway de-icing salts, ocean salt spray, flooding with ocean salt water, soil salinity, irrigation with saline water and cooling-tower drift. Plant injury due to salt comes from both direct spray on the foliage and uptake through the roots. The major sources of salt are from the ocean and the application of salt to roads during the winter months in parts of North America and Europe. Sodium chloride is the major salt causing these problems.

Trees growing close to the ocean are usually adapted to higher levels of salt exposure, such as shore pine (*Pinus contorta* var. *contorta*) in western North America. Windstorms, and especially hurricanes along the Atlantic and Gulf coasts of the United States, have been recorded to carry salt water as much as 80 km inland. During hurricanes, storm surges can trap salt water that cannot drain back to the ocean. This is generally more damaging than salt spray. Salt burn occurs as a result of salt uptake by trees and mortality may result. Generally, salts are leached beyond the rooting zone relatively rapidly so the damage is not long-lasting. Species vary in their sensitivity to salts. For example, loblolly pine is more sensitive to salts than slash pine and other conifers in the south-east United States.

In northern latitudes of North America, especially in the eastern mid-continent regions, sodium chloride and sometimes calcium chloride are used on roads to prevent ice formation on road surfaces. Trees growing adjacent to highways are often covered with salt spray during the winter months, especially conifers. During the spring foliage may turn brown, especially in the mid-portion of the crown. During the winter months, the lower crowns are usually protected from the salt because they are covered by accumulating snow. Buds are not usually killed by salts so that the new foliage that eventually emerges is healthy, once the older needles drop off. Even though trees are not killed by the salt, the vigour of these trees is lower and mortality may result in the long run.

High salt concentrations influence plants due to osmotic retention of water and to specific ion effects on the protoplasm. As salt concentrations increase, water becomes less available to the plant and high concentrations change the normal ratio of K (potassium) and Ca (calcium) to Na (sodium). This influences proteins and membranes. Photosynthesis is impaired by both stomatal closure and salt effects on chloroplasts. Dwarfism and necrosis of the roots, buds, leaf margins and shoot tips result, while growth rates and biomass production are impaired.

Just as tree species have varying tolerances to ocean salt, they also have varying tolerances to road salt. Conifers tend to be more damaged than deciduous hardwoods.

Since conifers retain their foliage in the winter, their foliage and branches are more efficient at intercepting airborne particles and spray. Eastern white and red pines are severely damaged, while other pines are more tolerant. Eastern hemlock is also severely damaged while eastern red cedar and spruce seem to be tolerant. Hardwoods also vary in their susceptibility to salt. Red and white oak and birch species seem to be tolerant while red and sugar maples are susceptible. Apparently tolerant species are able to withstand above-normal concentrations of chloride in their foliage. In the western inland of North America and Australia, salt damage is more common in irrigated agricultural areas, where high water evaporation brings salts to the top of the soil, which can then be dispersed by wind.

Introduced plants

Non-native or introduced plants are agents of forest disturbance throughout the world and go largely unchecked and unmonitored. They can be trees, shrubs, vines, grasses, ferns and forbs (see Case 5.1 – Kudzu). Some were introduced accidentally, but most were introduced as ornamentals or for livestock forage. They arrived without their natural predators of insects and diseases that tend to keep them in check. Now they are increasing across the landscape with little opposition, beyond the control and reclamation measures applied by landowners and managers on individual landholdings. They grow under and beside forest canopies and occupy small forest openings.

Exotic or introduced plants increasingly reduce tree productivity of native species. They hinder forest management activities such as when the understorey is replaced by thorny shrubs (e.g. *Barberis barberis* has replaced the natural understorey in spruce forests in the state of Maine in the United States). Invasive species also degrade wildlife habitat and habitats needed by biodiversity of conservation interest.

In the Pacific North-west United States, approximately 500,000 acres of national forests and grasslands are affected by non-native plants, which mostly originated from Asia and Europe. About 95 invasive species have been reported in this region alone. The major invasive species are: Scotch broom, gorse, English ivy, Scotch thistle, Japanese knotweed, Himalayan blackberry and tansy ragwort. At least 33 species are rapidly invading the 13 southern states in the United States; these include the vines kudzu (see Case 5.1 – Kudzu) and English ivy and the grasses giant reed and several species of bamboo. The same story can be told in other regions of North America and throughout the world.

Introduced insects

Many insects have been introduced into North America from Europe and Asia since the mid-1800s. Insects have also been introduced from the northern to the southern hemisphere, such as the wood wasp (*Sirex noctilio*). The major insects introduced to North America are the gypsy moth (*Lymantria dispar* – both European and Asian forms), larch sawfly, larch casebearer, beech scale, spruce aphid, balsam woolly adelgid, lesser European elm bark beetle, European pine shoot moth, pine sawfly, European spruce sawfly, hemlock woolly adelgid, European pine sawfly, winter moth, eucalyptus long-horned borer, common European pine shoot beetle and Asian long-horned beetle.

Many introduced insects are defoliators, which eat the foliage of plants for food. Examples of insect defoliators include the gypsy moth and the larch sawfly. Two introduced adelgid species with piercing, sucking mouthparts are also causing problems: hemlock woolly adelgid on foliage in the eastern United States and Canada and balsam woolly adelgid on stems and twigs of true firs in North America. Their effects are occurring at the same time as many native defoliators (e.g. spruce budworm, Douglas-fir tussock moth, hemlock looper, Pandora moth and western pine butterfly) are also reducing the health of forests. Recently, greater concerns are being raised about the impact of introduced beetles (e.g. wood borers, like the Asian long-horned beetle), which are killing trees in several parts of the United States. Ambrosia beetles are also of concern, since they cause wood products to deteriorate.

Introduced pathogens

Similarly to insects, a series of pathogens have been introduced to North America from Europe and Asia over the last 100 years. Forest management practices have increased fungal problems, but the greatest impact has come from the introduction of fungal pathogens into ecosystems where they are not native (Tainter and Baker, 1996). Examples of fungal pathogens are: *Chryphonectria parasitica* (the cause of chestnut blight), *Cronartium ribicola* (the cause of white pine blister rust), *Ophiostoma ulmi* (the cause of Dutch elm disease) and *Phytophthora* spp. (*P. cinnamomi* and *P. lateralis* both cause a number of root diseases and more recently *P. ramorum* has been causing sudden oak death in California) (Rizzo and Garbelotto, 2003). *Phytophthora* is also involved in jarrah dieback in Western Australia and little leaf disease in the United States. Fungi causing larch canker and dogwood anthracnose have also been introduced.

Forest management activities

Forest management activities can also act as disturbance agents. These include forest harvesting, thinning, pruning and site preparation, the use of fertilizers to increase plant growth rates and the use of herbicides to control unwanted plant growth considered to be a weed or competing with the growth of a commercial species (e.g. hardwood trees compete with loblolly pine growth in the south-eastern states in the United States).

Forest harvesting, particularly clear-cut harvesting, which consists of removing all trees in large land areas, is a strong disturbance agent. Clear-cutting was a common practice to harvest trees, but it was found to be an unacceptable practice by the public in the Pacific North-west United States and is not commonly practised today. Clear-cuts had the following effects in forests:

- changed microclimates so that temperatures were hotter in the cleared areas and along forest boundaries;
- changed forest structure adjacent to the clear-cut areas since these boundaries are more susceptible to disturbances such as wind and are knocked down more frequently;
- increased decomposition and nutrient cycling rates so that nutrients were lost from the soil rooting zone when vegetative regrowth was slow (vegetation captures nutrients from being lost from the biologically active zone of an ecosystem);

- changed hydrologic cycles as more water flowed out into streams and rivers;
- increased erosion rates because less vegetation was present to retain soil;
- changed biodiversity, with clear-cut areas having higher biodiversity of plant species but species that were considered to be weedy species.

Soil compaction also occurs as a result of harvesting activities. The degree of soil compaction from using heavy machinery varies, depending on soil type. The use of rubber-tyred skidders reduced the level of soil compaction in forests. Partial harvesting also reduces the impact on soils and the placement of green branches on the forest floor during harvesting prevents root damage from tracked machinery. Often, tree harvesting occurs during the winter months in those regions with snow accumulations.

During thinning operations, wounding of stems and roots by machinery or logs being pulled through the forest has been of great concern. Trees are particularly susceptible to damage during the spring when the sap begins to flow and the bark is loose. Young trees and thin-barked species are also very susceptible to being damaged or killed during thinning operations. Large open wounds can easily be infected by decay fungi. These wounds have occurred during thinning operations or when cutting trees in tropical forests when the desired tree is growing in a forest where most of the remaining trees have no commercial potential.

Herbicides are used in forestry to control competing vegetation during stand establishment, especially when growing conifers. Herbicides were also used in Vietnam to kill vegetation (see below – 'Disturbances in Forests and Human Health'). Herbicides can fall into two groups: those used for broadleaved vegetation (e.g. red alder, big-leaf maple) and those used for grasses and herbs. Triclopyr, 2, 4-D and dicamba are in the first group and act as plant hormones that disrupt normal growth forms. In sublethal doses they cause aberrant growth forms, but conifers may outgrow the problem within 1 or 2 years. Some other symptoms can include cupped leaves, curled leaf margins and swollen buds. These herbicides are sprayed aerially or from the roadside and can cause injury to non-target plants, because they can drift away from the target. The amount of drift depends on release height, droplet size and wind velocity.

For herbaceous vegetation, soil-active residual herbicides (e.g. atrazine, simazine, hexazinone, sulphometuron) and foliar-active glyphosate are effective. Simazine is a much used pre-emergence herbicide, which may cause problems after application to nursery soils because it persists in the soil for long time periods.

Conifer seedlings are sensitive to many herbicides. This sensitivity depends on the plant species. Severe injury can result when shoot growth is in progress. Foliar-active herbicides should not be applied over planted seedlings in the first growing season. Soil-active herbicides can be safely broadcast over dormant conifers in the autumn.

Wars

The impacts of human activities in forests during a war are expressed in multiple ways and many of these impacts occur beyond the boundaries of the forests themselves. To understand the severity of these impacts, one must first consider the constraints imposed by the climate, geology and ecology on people and animals surviving in forests when no war is happening (see Chapter 2, this volume). These constraints have to be included since they provide a history indicating the resilience of a forest to human and natural disturbances. They also provide clues to the intensity with which a forest can be used to extract resources.

In many of the tropical forest regions, disturbances that have not been a natural part of these ecosystems will greatly reduce the ability of people to survive as natural resource extractors. Since most people are already surviving at subsistence levels (e.g. collecting wood for heating or cooking) and growing food crops using trees to replenish soil nutrients (e.g. slash–burn shifting cultivation) (see Chapter 3, this volume), there are few alternative options available for them to pursue.

The impact of human disturbances in forests during a war can be positive and/or negative, depending on the circumstances. Therefore, understanding how these disturbances affect forest sustainability requires focusing on many other factors that initially may appear irrelevant.

The high number of variables present during the Rwandan civil war reflects the multitude of factors that have to be considered (Sato *et al.*, 2000) when refugees flee into forests during war years. The surrounding forest landscapes were devastated by the large influx of people, even though international organizations had provided tents and installed water pipes and public toilets in the refugee camps. During this civil war, refugees increased: deforestation rates; the removal of dead trees from the forests (e.g. refugees dug up stumps of trees for firewood even though they were supplied with fuel wood); soil erosion rates and the frequency of landslides; the frequency of large-scale forest fires (e.g. armed conflicts resulted in at least three large-scale forest fires in 1996–1997); poaching and reduction in animal populations (e.g. many refugees took shelter in national parks and hunted protected species for food; in the eastern part of one national park, half of the elephants were poached). In addition, plant species suited to the Congolese environment and coffee varieties were lost during the fighting that occurred during the Rwandan civil war (Sato *et al.*, 2000).

Many of the impacts of human activities during a war are a result of too many people attempting to survive as resource extractors in forests. The civil war in Rwanda is a good example of this. The large influx of people into the forests was devastating for them and for the conservation efforts that had been occurring here prior to the war. The multitude of impacts of human activities in forests during wars is further covered in a subsequent section – 'Influence of wars'. The influence of changing forest land uses on human health is covered below – 'Disturbances in Forests and Human Health'.

Influence of Disturbance Agents on Forests

Natural and human disturbance agents influence forest productivity, decomposition and nutrient cycling, biodiversity, forest structure and species succession. Human activities such as forest management, fire suppression, the introduction of invasive plants, air pollution, global change, insects and pathogens and wars have changed the dynamics of natural disturbances.

In some cases human activities have changed the course of forest succession, changed species composition and landscape biodiversity and even made some ecosystems unsustainable. Many individual tree species have been negatively affected and their populations reduced, such as American chestnut, white pine, white bark pine and eastern hemlock.

Forest health has generally been negatively affected by human activities. Furthermore, human health and the incidence of disease outbreaks have been affected by how forests were disturbed.

Influence of fire and wind: a United States example

In the inland western United States, trees across wide areas of the landscape are dying faster than they are growing or being replaced. Because of this, tree mortality conditions exist that almost guarantee the occurrence of large and severe wildfires (Agee, 1993; Edmonds *et al.*, 2005). In these ecosystems not only are the trees at risk but aquatic resources, wildlife and other forest amenities are also affected. Managers of public and private forests are being challenged to take rapid preventive action to restore these forests to conditions more similar to their historical range of variability or at least to a socially desired condition.

How did we arrive at this situation? Patterns of fire, insect, disease and wind disturbance have all been altered by forest management practices, particularly fire suppression. Plant species distributions, forest structure and composition, succession, biodiversity and landscape pattern have all been changed, and the incidences of diseases and insects and the risk of catastrophic fire have increased.

The forests at greatest risk are composed of unsustainable combinations of tree species, densities and structures that are susceptible to the drought and fire regimes that occur in the region. However, the health problems in the forests dominated by ponderosa pine are different from those in forests dominated by lodgepole and white pine.

Ponderosa pine forests that occur in the driest areas are particularly susceptible to forest health problems. Prior to the European settlement of North America, many of these forests consisted of large, widely spaced trees visited by low-intensity wildfire every 5 to 15 years. Burning by Native Americans may also have contributed to the appearance of these stands. With wildfire control, vegetation developed vigorously under the mature trees. On the driest sites, an understorey of ponderosa pine developed. On slightly wetter and cooler sites, shade-tolerant Douglas fir developed. The understorey then provided a 'ladder' for devastating crown fires. The problem was worse where selective logging quickly shifted the species mix from ponderosa and other long-needled pines towards firs. Unless these forests are restored to a species mixture more resembling pre-European conditions, these forests will be increasingly subjected to disease and insect attack and catastrophic fires.

There are vast areas of lodgepole pine forests in the inland west United States, which provide a different scenario. Lodgepole pine has evolved with outbreaks of mountain pine beetle and large-scale, intense, stand-replacing fires. When stands are young, the beetle poses little threat, but as trees age the inner bark thickens and they become more suitable hosts. As tree growth rates slow, the beetle population explodes and extensive tree mortality follows. Dead trees become fuel and sooner or later lightning or humans ignite the forest and it burns in spectacular fashion; witness the Yellowstone National Park fires of 1988. The period between such fires typically varies from < 100 to 300 years, depending on the site.

Lodgepole pine in the inland west United States has adapted to intense fire and a high proportion of its cones are serotinous (i.e. its resin-closed cones are opened by the intense heat from fires). These forests recover quickly from fire and within 5 years there is a thick carpet of lodgepole pine seedlings and many of the standing dead trees have fallen down. In the case of the Yellowstone National Park fires of 1988, this is a healthy forest, given the park's land management direction to allow natural processes to operate. Managing a lodgepole pine forest is difficult since the public does not appreciate that some tree mortality is needed in these forests if they are going to continue to exist in the landscape. The public is not used to thinking of either a large-scale bark beetle infestation or a large-scale intense fire as natural disturbances that should be allowed to happen, particularly if there is no risk to life or property. Thus risk assessment and management are important components of forest health and ecosystem management.

Mixed conifer forests with a strong western white pine component were common at mid-elevations in the inland west United States. Western larch, grand fir and Douglas fir were also present. Periodic low-intensity fires also burned through these forests naturally, maintaining the white pine. However, as the white pine aged to old growth it became more susceptible to insects, particularly to mountain pine beetle, which killed trees, leading eventually to occasional high-intensity stand-replacing fires (Agee, 1993). Root diseases also tended to thin stands and maintain white pine by selectively killing the more susceptible Douglas fir and true firs. With European settlement and selective cutting of white pine, fire suppression and the introduction of white pine blister rust in the early 1900s, the health of these forests dropped dramatically (Campbell and Liegel, 1996). Douglas fir is now the main species in this zone, while western white pine is poorly represented. Root diseases and the Douglas fir beetle are the major agents reducing productivity in the Douglas fir and true fir stands. Restoring these ecosystems is also going to be difficult, and perhaps it can never be accomplished because of the presence of the blister rust.

The influence of introduced plants, pathogens and insects on forest health

Introduced plants greatly impact forests by reducing their productivity and biodiversity. In North America, regenerating and restoring forest landscapes is difficult because of the widespread dominance of several invasive plant species (e.g. kudzu, Scotch broom, gorse, Himalayan blackberry) (see Case 5.1 – Kudzu). Riparian areas are particularly susceptible to invasive plants like Japanese and giant knotweed and tamarisk.

Introduced pathogens and insects have taken an even larger toll on the health of forests throughout the world. A handful of diseases and insects have caused tremendous damage in North America. White pine blister rust, chestnut blight, Dutch elm disease, gypsy moth and, recently, dogwood anthracnose have caused a large amount of mortality and changed tree species composition. *P. cinnamomi*, which may have originated in South-east Asia or north-east Australia, has now spread throughout Australia and the world. It has influenced the regional distribution of *Eucalyptus* spp.

in Australia. For example, susceptible species that would be killed by *P. cinnamomi* occur on drier ridges, where the pathogen is less active, while resistant species occur in moist depressions and swales, where the pathogen is more active.

Plantations of radiata pine in the southern hemisphere are also very susceptible to introduced pathogens from the radiata pine's own native ranges. For example, needle pathogens from North America cause much damage in New Zealand, Chile and Ecuador. *Sirex noctilio*, a wood wasp, is of considerable concern in radiata pine in Australia. The pine wood nematode, which has caused serious damage to native pines in Japan, was introduced from the United States and its spread is of concern elsewhere.

Serious quarantine and inspection measures were implemented by the United States and many other countries after the impacts of these introduced pests were noted, and they have been relatively successful. The gypsy moth was introduced to Massachusetts from Europe in 1869 before inspection measures were implemented. The gypsy moth produces a silky string and was originally brought to Massachusetts because someone wanted to start a silk business in the United States. It has been slowly expanding its range to include the entire north-eastern United States, portions of the south, the mid-west and even the west. It feeds on more than 300 shrubs and trees, but prefers oaks, and has caused serious defoliation in oak forests in the north-western states of the United States, although it has yet to invade the most susceptible forests.

There is also great concern that the Asian gypsy moth, which was recently discovered in both eastern and western North America, could cause even greater problems because it attacks conifers as well as hardwoods (the European gypsy moth rarely feeds on conifers). Unlike the European gypsy moth, the female Asian gypsy moth can fly, which greatly enhances its ability to spread. With increased world trade of lumber and logs, we see the possibility of increased pest and pathogen movement. There is a very large concern that pests and pathogens from the southern hemisphere or Siberia imported on logs could impact North American forests.

The influence of air pollution on forest health

As well as changing the forest through management, humans have inadvertently changed it through air pollution. Gases, such as sulphur dioxide and ozone, acid rain and excess nitrogen are the major air pollutants of concern. In the early days of industrialization, large areas of forest were killed or greatly impacted in North America, Europe and even Australia (Edmonds *et al.*, 2005). Most of this poor forest health was the result of the burning of high-sulphur coal and ore-smelting activities, which produced toxic sulphur dioxide gas.

Emission controls on new power plants and smelting operations have greatly reduced concentrations of SO_2. However, it is still adversely impacting forests in Eastern Europe. Ground-level ozone is now thought to be one of the major air pollutants stressing forests throughout the world. Ozone causes major forest health problems where there are large concentrations of automobiles and high radiation intensities in the growing season, such as southern California, but it is a worldwide problem.

Effects of acid precipitation

A massive research programme into the effects of acid precipitation on forest ecosystems in the United States and Europe was mounted in the 1980s. The original hypotheses proposed for how acid precipitation affected forests were that increased acidity in rain and fog would:

1. Decrease soil pH and the availability of nutrient ions;
2. Increase soil Al concentrations, resulting in the killing of fine roots and mycorrhizae, resulting in forest decline; and
3. Leach cations from foliage and soils, which would decrease plant growth rates and increase their susceptibility to *Armillaria* root disease.

It has, however, proved to be extremely difficult to verify all these hypotheses in the field (see Case 5.2 – Acid Rain, Air Pollution and Forest Decline). The observed forest decline problems in Europe and North America are due to multiple causes and not to one agent (e.g. acid precipitation, other air pollutants such as ozone and excess N, insects such as the balsam woolly adelgid, weather, forest age and global change).

The main conclusions from acid rain research were that:

1. Aquatic systems were being adversely affected in areas where streams and lakes had low acid-neutralizing capacity;
2. Fish populations were lowered or even decimated, particularly in the north-eastern United States, eastern Canada and Scandinavia, where the bedrock is low in bases such as Ca and Mg;
3. Sulphur and nitrogen deposition caused adverse impacts on certain highly sensitive ecosystems, especially high-elevation spruce–fir ecosystems; and
4. Gradual leaching of soil nutrients, especially Ca, Mg and K, from sustained inputs of acid rain could eventually affect the nutrition and growth of trees.

There is little evidence, however, for large-scale forest declines in the forests of North America due to acid rain precipitation. The cause of tree death in high-elevation areas continues to be under investigation and there is concern that, if deposition levels are not reduced in the south and western United States, adverse effects may develop in these regions in the future.

The impact of excess atmospheric nitrogen inputs to forests is now recognized as a serious pollution problem in Western Europe and the eastern United States (Vitousek *et al.*, 1997). Nitrogen inputs beyond normal background levels change ecosystem functions and make trees more susceptible to cold-temperature injuries, diseases and insects. Nutrient imbalances can result from over-fertilization, particularly of N. Magnesium deficiencies have been observed in conifers in areas of Europe receiving high atmospheric N inputs as a result of air pollution.

Emission reductions have significantly reduced precipitation acidity in the midwest, mid-Atlantic and north-east states in the United States, but as yet nitrate concentrations have not been reduced. Many impacted lakes and streams have responded positively, but not the Adirondack lakes. There is also concern that the gradual leaching of soil nutrients (especially Ca, Mg and K) because of the continued creation of acid rain could eventually affect the nutrition and growth of trees. How much tree growth will be reduced varies depending on the rate of cation or basic mineral depletion, soil cation reserves or storage capacity, forest age, weathering rates, species composition and disturbance history. Liming of forests has been implemented in Europe, particularly in Germany, to improve the nutrient status of soils. Changing the vegetation from conifers to hardwoods is also being implemented in Germany.

Influence of wars

The impact of human activities on forests and forest-dwelling animals can be positive or negative during a war. In contrast to forests being protected in many parts of Central and South America, many African forests were negatively impacted because forests were 'safer' places for urbanites to hide and to find food when war broke out. In these cases, refugees fleeing a war find forests to be locations where they can collect fuel wood and hunt animals for food, i.e. survive.

Positive aspects of war

War, unintentionally, can result in conservation of species and the prevention of forests from being over-exploited. Conservation of species is an unintended consequence of people staying out of forests because they were dangerous environments to go into because of guerrilla activity. This fear of losing one's life by entering a forest meant that forests were less extracted for timber and non-timber forest products, and less hunting occurred here. This scenario has been common in South and Central America, where forests were too dangerous for urban populations to flee into and hide during fighting.

Forests and their biodiversity have also been unintentionally protected due to governments setting up military training camps, using lands for military exercises or maintaining demilitarized zones between two fighting countries (e.g. military bases such as Ft Lewis, Washington, and Camp Pendleton, California, USA; lands being used for military exercises in Germany (Hopkin, 2005); the demilitarized zone between North and South Korea). The public is usually prevented from entering or collecting natural resources from these areas. These military training camps or the demilitarized zones have become locations where a higher diversity of many endemic species has been found. Many examples exist of forest areas remaining relatively intact and less impacted by human activities when people were prevented from entering them (e.g. lakes or rivers in forests being set aside to protect and provide clean water supplies for cities; the Panama Canal Zone had forests and the watersheds adjacent to the canal were set aside and protected to ensure sufficient water to run the locks in the canal; Corcoran, 1999).

Forests and forest-dwelling animals have also been unintentionally protected because war has made it difficult for people to venture into these areas to collect resources to sell in local, regional or global markets. How forests are unintentionally protected is nicely summarized by Le Billon (2002) when he wrote:

> Ironically, twenty years of war saved Cambodia's forests from the destruction associated with economic growth in the ASEAN [Association of South-east Asian Nations] region. Despite heavy US bombing and the murderous agrarian utopia of the Khmer Rouge,

forests survived the 1970s. Their exploitation during the 1980s remained limited, the result of continuing war and a trade embargo by the West.

In another example, biodiversity can be protected by having areas restricted to military exercises (Hopkin, 2005). Hopkin recounted how land areas used for military exercises in Germany have more endangered species present than even found in national parks. Two American bases in the German state of Bavaria comprised less than 1% of the land area of the state but contained 22% of its endangered species.

Another good example of warfare protecting forest-dwelling animals was reported on 28 September 2005 by Reuters News Service from Indravati, India, which read:

> The tiger population at a reserve in central India has increased after Maoist rebels who control most of the forest kept poachers away, a wildlife official said Tuesday. Tiger conservation became a hot button issue after reports in March that poachers had wiped out the entire population of 16–18 tigers at a leading sanctuary in western India. But poachers had kept away from Indravati tiger reserve in Chhattisgarh state for fear of the Maoist guerrillas who have banned all entry except once yearly visits by census officials. Authorities counted 39 tigers this year, up from 29 in 2002 when the Maoists imposed the ban, the state's chief wildlife warden said. 'The tigers are totally safe . . . we have had no report of poaching,' warden N.K. Bhagat said.
>
> The Maoists, who operate out of jungle bases in nine states, say they are fighting for the rights of poor peasants and landless labourers. They have killed politicians and policemen in a nearly 40 year revolt.
>
> Poaching is blamed for the drastic fall in India's tiger population from 40,000 a century ago to just about 3,700 today. Some conservationists put the number at fewer than 2,000.
>
> Trade in dead tigers is illegal but poachers still operate with impunity in much of the country. A single animal can fetch up to $50,000 as tiger organs, teeth, bones and penises fetch high prices on the black market, where they are bought for use in Chinese medicine.

Negative aspects of war

INCREASED NUMBER OF PEOPLE SURVIVING IN FORESTS THAT CAN ONLY PROVIDE SUBSISTENCE LIVELIHOODS   In some regions of the world, forests become refuge sites for less powerful people that are fleeing oppression, conflict and war. This situation has ramifications for conservation, food production and the over-harvesting of trees or non-timber forest products as people attempt to survive as extractors in the forest. Because of the dangers of being killed or maimed during a war, people flee into forests. This immigration of people to rural areas increases the population densities of these areas beyond its carrying capacity. For example, the eastern region of the Democratic Republic of the Congo (formerly Zaire) already had a high human population density (49 people per km^2) prior to the civil war. Rwanda had a population density of 293.3 people per km^2 in 2000 (FRA, 2000) which suggests that there should be no forests remaining, using the equation developed by Meyerson (2000) (see Fig. 3.1, this volume).

Using the Meyerson equation, Zaire should have 50% of its forest area remaining. These pre-war population densities suggest that a considerable portion of the forest lands was already threatened with deforestation and over-exploitation of forest resources. When war broke out in Rwanda and Zaire (e.g. the Rwandan civil war in

1994, Zairian civil war in 1996–1997), a significant influx of people from urban areas occurred in the forests of both countries (Sato *et al.*, 2000). During the Rwandan civil war alone, 1.5 to 2 million refugees entered the forests and needed it to survive.

Since Rwanda had only 12.4% of its land area in forests in 2000 (FRA, 2000), the civil war resulted in a significant influx of people into a small land area. A significant portion of the Rwandan population lived in rural areas in 1999 (93.9%), and people have to be able to rely on their ability to extract resources such as fuel wood from forests even during non-war times. Whether Rwanda will be able to use its forests as a safety valve during times of future wars is questionable. Between 1990 and 2000, FAO estimated that Rwanda also lost 15% of its forest cover, which means that less forest area will be able to provide a buffer for people during times of war.

Zaire contrasts sharply with Rwanda by having a larger portion of its land in forests (64.6%) and a smaller population of people living in the rural areas (38.3%) in 1999 and 2000 (FRA, 2000). So Zaire should be able to provide for the large influx of people that flee into the forests during a war. However, even Zaire is losing the capacity of forests to provide a safety valve for people fleeing a war since it lost 17% of its forest cover between 1990 and 2000.

NON-SUSTAINABLE FOOD PRODUCTION AND NO FOOD SECURITY   During wars, food production ceases in many places because of the safety concerns of farmers. Farmers are unable to grow crops because they are killed when tending their fields. So acquiring enough food becomes a serious problem.

OVER-HUNTING   When people flee into forests as sites of refuge, over-hunting is an inevitable consequence. This has been especially true in forests in Africa. Species that have conservation interest to international organizations are also hunted. Since bush meat is already an important protein source for diets in many parts of Africa, less availability of meat means that people's health deteriorates and they become more susceptible to diseases. The end result of all of this is the loss of food security, poorer human/animal health and hunting and fishing become more non-sustainable.

Even when wars are not occurring, wild meat is a significant source of readily available protein for landless, rural people throughout much of Asia, Africa and Latin America (Rao and McGowan, 2002). Much of this wild meat is hunted in forests. It has been estimated that wild meat provides more than half of the protein needs for many people living in tropical forests; some of the meat consists of crop pests like rats or small antelopes, which poor men as well as women trap (Wildlife and Poverty Study, 2002). In Liberia, 75% of the meat eaten is from wild sources (Rao and McGowan, 2002).

In many of the tropical forest regions, the population size of wildlife that could be hunted has been shrinking for several decades due to over-hunting. A war increases the pressures to further increase the hunting of animals for food. Since animals have been traditionally hunted to sell in urban markets, animals living in forests are an acceptable and important source of protein especially when few other options exist.

NO BIODIVERSITY CONSERVATION   Biodiversity conservation diminishes or ceases during a war (see Price, 2003, for examples of conservation occurring in areas of armed conflict). Nature reserves or parks established prior to a war are now seen as a source of readily available food on the hoof (i.e. bush meat) and their animals are killed to feed families. Nature reserves are not protected and are now used for subsistence survival both by people previously living in the forests and by people who have fled into the forests for security reasons. As more people survive by hunting wild animals for food (frequently from nature reserves or parks), many species being conserved are now seen as animals to be hunted for food. Since agriculture is difficult to practise in areas where fighting is occurring, the amount of available food is dramatically decreased, resulting in a greater dependence on hunting to survive.

At the same time, a demand for the same bush meat has also been occurring from the increasingly prosperous parts of Asia, which are 'increasing exploitation for booming commercial markets' (Rao and McGowan, 2002). These commercial markets for bush meat also provide an economic return from hunting when few other options exist for generating revenue. Commercial poachers hunt animals for bush meat to sell in the urban regions of Africa and most West African immigrants buy ape meat (Whitfield, 2003a). These commercial markets for bush meat mean that there is competition for hunting animals for bush meat for survival and for revenue generation.

Wars result in more contact between people and forest-dwelling animals. In Africa, this has increased disease outbreaks among apes because they are susceptible to many diseases transmitted by humans, which either weaken or kill them. In West Africa, both over-hunting and the Ebola virus have been dramatically decreasing the population of chimpanzees and gorillas (Whitfield, 2003a). Bush-meat trade is threatening their populations near towns while the Ebola virus is killing them in more remote areas (Ebola has killed up to 90% of the apes in some areas) (Whitfield, 2003a). Ebola spreads from apes to humans when apes are hunted for food and eaten or when humans come into contact with the infected body of a dead ape. In December 2001 and March 2002, an Ebola outbreak in Gabon killed about 50 people in the more remote areas (Whitfield, 2003a).

NON-SUSTAINABLE LOGGING PRACTICES   Wars result in the exploitation of forests for their economic value to pay for the war effort, as a source of income for some profiting from black marketeering of timber and also as a formula to pay for the recovery of economies after a war. Therefore, over-logging and unsustainable cutting of trees (much of it illegal) occur during and after a war.

Forests are exploited since they return a higher economic value than most other renewable resources (Le Billon, 2002). After the Second World War, Germany paid some of their retribution payments by cutting down trees because this was one of the few products they had with value in the global markets. This demand was fuelled by the major building boom that was occurring in the United States after the Second World War.

Similarly to Germany after the Second World War, after its two decades of warfare Cambodia used its forests to position itself in the global economic markets. In the post-war transition in the early 1990s, the focus was placed on transforming forests in the territory previously controlled by the Khmer Rouge into a major commercial timber centre. Timber was the most valuable and internationally accepted trade good for Cambodia (Le Billon 2002). Forests were the primary asset existing in Cambodia so the government took control over these resources to fuel their economic development (Le Billon, 2002).

COLONIZATION PROJECTS IN FRONTIER REGIONS TO MAINTAIN POLITICAL BOUNDARIES
Wars frequently result in the forced relocation of people because: (i) people are fleeing a war; or (ii) governments force people to move to border regions in order to maintain or establish the political boundaries of a country and to prevent neighbouring countries from occupying these regions. Forced relocation of people happened in Guatemala to establish ownership of lowland areas. People were transported from the mountain regions, which had volcanic soils, to the lowlands, which had calcareous soils. This move resulted in poor, non-sustainable agricultural practices occurring in the lowland areas because people did not know how to grow crops in these new environments.

HUMAN AND ANIMAL DISEASE OUTBREAKS   Wars increase the incidence of human and animal disease outbreaks because wars result in changing land-use practices, which increase the spread of disease vectors to landscapes where humans live. They also result in a greater frequency of contacts between organisms and disease vectors (see below – 'Disturbances in Forests and Human Health'). During wars, interior regions of forests are often exploited for timber or metals (e.g. gold) or diamonds, which increases the contact between humans and insect or mammal disease vectors. Not only are humans more susceptible to diseases but the increasing contact between humans and apes has also resulted in humans transmitting the Ebola virus to apes in parts of Africa (Whitfield, 2003a).

CHEMICAL CONTAMINATION OF SOILS AND PLANTS   Another factor that also exacerbates human disease outbreaks is the application of chemicals to forests to eliminate forest cover to make warfare easier. During the last half of the century, more toxic chemicals have been used in wars than historically. Many of these chemicals are suspected of being toxic to humans or to trigger the emergence of human diseases.

In the process of trying to eliminate forest cover, chemicals (herbicides) have been used to defoliate and kill trees (Warren, 1968; Bethel *et al.*, 1975). Agent Orange that was contaminated with dioxin was used in Vietnam. Dioxin can kill people when applied at large enough doses. Many of these chemical residues contain heavy metals, which are also toxic to humans. These chemicals have also left a residue in the soil, which persists for years in a toxic form and is also taken up by plants grown as food crops. A couple of decades after herbicide applications in Vietnam, Vtorova and Sergeeva (1999) detected high levels of lead (Pb) and nickel (Ni) in dead plant materials collected in South Vietnam.

Chemical contamination is also common in military training camps. These areas are highly contaminated from ammunition that has been left on the ground (including firing ranges and munitions storage facilities). Most ammunition is contained in receptacles that are made of heavy metals. These heavy metals eventually move into the soils and contaminate them. Heavy metals can cause animal deformities or birth defects, and they can kill plant species that are not tolerant of them.

ECONOMIC OPTIONS AND SUSTAINABLE LIVELIHOODS   The economic options pursued by people living in subsistence extractive economies are frequently lost during a war. This loss of economic options can occur in several ways:

- If ecotourism was the primary source of livelihood, tourists do not travel to war zones, so there is a loss of revenue. For example, during the fighting in Rwanda, ecotourism disappeared and many conservation groups who had been paying for the conservation efforts or helping communities to develop ecotourism projects left.

- Not only does the use of chemicals as defoliants to kill trees result in human health problems over long time periods, but trees that have economic value are killed when herbicides are applied to forests. This results in the loss of future economic revenue in global markets. For example, during the Vietnam War, largescale use of chemical herbicides occurred in defoliation missions.

> Eighty-nine percent of the herbicide chemicals were used in defoliation missions; that is, to remove leaves from trees, making enemy troops and the trails, camps, arms dumps, and other facilities visible from the air. The inland forests, with which this report is concerned, comprise about 62 percent of the land area of South Vietnam. These areas received about 72 percent of the herbicide sprayed.
>
> (Bethel *et al.*, 1975)

Wood loss due to herbicide applications in Vietnam was estimated to have resulted in the mortality of trees that totalled 5.6–11.9 million m^3 of merchantable tree volume (Bethel *et al.*, 1975). (Contrast this to the timber volume exported from Cambodia after its war, which was 1.1 million m^3 in 1998; Le Billon, 2002.)

Disturbances in Forests and Human Health

Many factors affect human health. The relationships between humans, the environment and disease can be summarized as follows:

- A poor environment will contribute to a poor diet and negatively affect nutrition (just as too much can also be bad).
- Poor nutrition may contribute to diseases and their emergence.
- Also social factors (governments, war, policy, economics, etc.) may affect nutrition and hence diseases.

How well humans are able to obtain nutrients needed for bodily maintenance and function has a significant impact on a person's susceptibility to diseases. The link between nutrition and disease is very strong, with the following diseases being more nutrition-related than genetic: iron-deficiency anaemia; vitamin deficiencies; mineral deficiencies; toxicities; and poor resistance to disease. So a malnourished individual will be more susceptible to infections and become sick more readily.

Therefore, inadequate food (which usually happens during a war) reduces human resistance to diseases. Since human subsistence and food availability commonly occur in or adjacent to forests, they provide an important source of food. Humans further impact their own health by how they interface with forests. As humans increasingly encroach into habitats where they typically have not lived and when a larger number of people try to survive as resource extractors in forests, they rapidly exhaust the available food supply. This means that as their nutritional intake is diminishing their immune systems become weakened and prone to illnesses. They also expose themselves to diseases they have no immunity against, such as malaria.

Several factors increase the susceptibility of humans to diseases, in addition to the susceptibility of any particular individual to illness (e.g. human nutrition). These factors typically increase the opportunities for humans to come into contact with disease agents or to increase the risk of transporting disease agents to regions where they have not previously been. These factors have been summarized as:

- High population densities.
- Settlers pushing into remote areas.
- Human-caused environmental change (use of fertilizers, pesticides, etc.).
- Speed and frequency of modern travel.
- Contact with water or food contaminated with human waste.

Poverty, war, corruption of governments and droughts are combining to make it difficult for sub-Saharan Africa to solve its disease problems. According to the World Health Organization, 30 new diseases have emerged in the past 20 years and there has also been a re-emergence of some old diseases (e.g. malaria and dengue, or breakbone, fever which are both transmitted by mosquitoes). For example, the spread of acquired immune deficiency syndrome (AIDS) throughout Africa is the result of a combination of factors, which makes it more difficult to control.

Not only do humans become exposed to more diseases when they move into forests but humans also transmit diseases to the animals living in forests (Adams *et al.*, 2001). At the same time, the transmittal of human diseases to wildlife also increases the population size of the disease vectors so that more humans can eventually become infected. Several outbreaks of disease believed to be of human origin have been reported among chimpanzees. In the late 1960s in Tanzania, an epidemic of a polio-like virus spread from a village and 15 chimpanzees were severely crippled or died. Recently, a polio-like virus occurred among chimpanzees and a flu-like epidemic killed 11 chimpanzees in the Democratic Republic of Congo. Documented cases of outbreaks of scabies, intestinal parasites, yaws (a syphilis-like infection in humans) and respiratory infections similar to measles have been reported for the great apes that were again suspected to be of human origin.

Humans do not appear to be very healthy when they travel and therefore are effective transmitters of disease to other animals (Adams *et al.*, 2001). For example, a survey of illnesses experienced by a village bordering the Kibale National Park, Uganda, showed the following results (% people showing symptoms): fever = 82%; coughing = 64%; respiratory distress = 26%; diarrhoea = 24%; vomiting = 24%; general illness = 22% (Adams *et al.*, 2001). The fact that tourists are not 'very' healthy prompted the International Gorilla Conservation Programme to establish rules of engagement between tourists and mountain gorillas. They changed the minimum distance between gorillas and tourists from 5 to 7 metres. To protect chimpanzees, they are also recommending that tourists should use face masks to reduce the risk of or prevent the transmission of airborne disease agents. They are also enforcing their right to refuse a tourist admission to the park if they are sick. Other measures include the mandatory washing of hands, use of disinfectant foot-baths for tourists prior to their visit and construction of adequate pit latrines to ensure proper disposal of human waste.

Specific environmental factors causing spread of contagious diseases

A primary effect of environmental change on the spread rate of disease is due to its impact on the breeding, development and proliferation of specific parasite species and their hosts. Some of these changes (Prothero, 1999; Patz *et al.*, 2000; Leonard, 2003) are summarized below.

Land-use changes

Deforestation is one of the most disruptive changes that can affect parasitic vector populations and the wildlife hosts (e.g. white-footed mouse needed by ticks to obtain their blood meal). Deforestation occurs for several reasons: (i) cleared tropical forests are converted to grazing land for cattle, small-scale agriculture, human settlement or left as open areas; (ii) expansion of existing human settlements; (iii) movement of humans into remote areas results in more area being cleared for food production. In the case of parasitic vectors, deforestation increases the spread of malaria because more breeding habitat is created for insect vectors, i.e. mosquitoes. In other cases, when wildlife hosts are needed to spread a disease like Lyme disease, deforestation can eliminate the forest habitat needed by the small mammals who are the wildlife hosts for the ticks (see Case 5.4 – Habitat Fragmentation and Disease Ecology: the Case of Lyme Disease). In this case, the removal of forest cover reduced both the population size of the wildlife hosts and the spread of Lyme disease to humans.

Deforestation can result in the spread of disease vectors when forests are replaced by crop farming and ranching. Small animals can create a supportive habitat for parasites and their vectors. Introducing cattle, pigs or chickens can increase or decrease the transmission of parasitic diseases. Parasitic disease transmission decreases if there is a reduction in the number of humans who are fed upon by parasites because they are feeding on the domesticated animals. In other cases, the presence of many domesticated animals can increase the incidence of a disease since the vectors can feed or obtain blood from so many animals (including humans).

Disease transmission can increase after deforestation depending on what vegetation regrows to replace the cut vegetation. Depending on which vegetation regrows after tree harvesting, there can be an increase in the presence of malaria. For example:

1. In Malaysia, during the 50-year period of repeat growing of rubber plants, there were cyclic malaria epidemics.
2. In Trinidad (in the 1940s following deforestation) the trees that regrew had a large number of bromeliads in their canopy. Since bromeliads collect water where leaves project out from the base of the plant, they have become the preferred breeding site for malaria parasites. When the bromeliads were removed, malaria prevalence returned to previous levels.

Standing bodies of water (no matter how small) are ideal habitats or breeding grounds for disease vectors. Forests tend to be heavily shaded with a thick organic layer that readily absorbs water, so they do not have standing puddles of water that could be used as breeding habitat. Cleared land is more sunlit and prone to form standing water puddles that are more neutral in pH and which favour the development of certain larvae. When forests are cleared to grow crops like sugarcane or rice, the farmers typically use a series of irrigation ditches, which provide good breeding habitat for disease vectors. New road construction into previously inaccessible forested areas can lead to erosion and create ponds that can become breeding sites for disease vectors. Roads allow construction workers, loggers, miners, tourists and conservationists to go to new areas that previously were forested and consequently less exposed to human intrusions.

Mining in forests also contributes to the spread of diseases. Gold mining in forest areas introduces chemical compounds such as mercury (which is toxic to humans and directly affects human health) into the environment. Mercury is used to help separate the gold from the rest of the rock material. Human health, however, declines due to mercury poisoning. At the same time, mining in forests takes humans deeper into remote regions so that they may come into contact with mosquitoes or eat animals that are diseased.

Moving into or through interior, remote forest areas

Humans are coming into closer contact with animals that are able to transmit diseases to them (e.g. eating infected monkey meat has spread the Ebola virus) or people travelling between country border regions in South-east Asia are getting bitten by malaria-infected mosquitoes. In some cases, the plasmodium agent for malaria is drug-resistant.

When deforested areas are replaced by crops, the pattern of human settlement changes and the migrants become reservoirs for transmission of parasitic diseases that were endemic to their former homes but foreign to where they have relocated. Indigenous forest dwellers are typically immune to forest-dwelling parasites but new settlers who enter for commercial activity in forests are not.

Climatic events

Climatic events are part of the problem in increasing disease transmission to humans (Epstein *et al.*, 1998). Disease vectors (e.g. arthropods) are sensitive to climate, which is why public health researchers recognize the importance of weather with regard to the timing and intensity of disease outbreaks. In 1987, the El Niño–Southern Oscillation event resulted in an increase in the mean night-time temperatures in the highlands of Rwanda, which coincided with significant increases in *Plasmodium falciparum* malaria. Minimum air temperatures of 15–18°C are needed for the development of the malaria parasite. A threshold temperature greater than 20°C is needed to set off an epidemic of malaria cases.

El Niño years have been correlated with malaria outbreaks in different parts of the world, especially in those areas that normally have higher amounts of rain (e.g. the tropics) (Epstein *et al.*, 1998). This does not mean that malaria is restricted to the tropical zone. During the movement of pioneers into the forests in North America during the 18th century, pioneers avoided the low forested areas because of the high prevalence of malaria (Williams, 1989). Since malaria either killed or debilitated the pioneers, they learned very quickly to avoid these locations.

The reason that malaria is not so commonly found in North America today is that control measures were effective at controlling it (Rosenberg, 2004). This is not the case for many parts of Africa or South America where malaria has been difficult to control and chemicals effective at controlling malaria are no longer acceptable to use (i.e. DDT) (Rosenberg, 2004). For this reason, climatic events have significant impacts on increasing the prevalence of malaria in regions where it has not been controlled. For example, the 1997/98 El Niño, which was the largest recorded in history, resulted in torrential rains in parts of East Africa and a malaria epidemic in the Uganda highlands.

Wars

Wars have significant impacts on the spread of human diseases. Food security decreases during a war and people's health in general declines. This makes people more susceptible to diseases and premature death.

Wars also result in the resettling of people or people spontaneously migrate, which exposes them to disease vectors that are unfamiliar to them. A good example of this occurred in Indonesia. People were resettled from the densely populated islands of Java and Bali to a more sparsely populated and densely forested outer island. These people do not have any immunity to malaria and on these outer islands the plasmodium thrives in conditions created by this new development project. Most of the people who have moved here are poorly educated and do not know how to protect themselves against malaria and other health risks.

Malaria, forests, environmental change and people

All information provided below is from Prothero (1999) and www.wehi.edu.au/MalDB-www/intro.html. Malaria parasites probably originated in Africa and 30-million-year-old fossils of the mosquito – the disease vector – have been found. The *Plasmodium* parasite is highly specific to humans, who are its only vertebrate host, and it has a long relationship with humans. *Anopheles* mosquitoes are the vectors. Today, malaria is generally endemic to the tropics and subtropics but it used to be found in the New England states of North America in the 18th century (Williams, 1989).

Today, travel by air is an important factor in transferring malaria to areas where it had been eradicated or was not normally found. In 1990, 80% of the malaria cases were in Africa with the rest of the cases found in India, Brazil, Afghanistan, Sri Lanka, Thailand, Indonesia, Vietnam, Cambodia and China (see Case 4.4 – Malaria and Land Modifications in the Kenyan Highlands). Today malaria is endemic in 91 countries.

Malaria is found in particular habitats in South-east Asia: forests and their fringes; lowlands and lower hill slopes largely cleared of forest that are more densely inhabited; and large urban places. In general, forests and forest fringes have most of the malaria cases because their ecological conditions favour the disease. *Anopheles* mosquitoes can breed under a wide set of conditions: water surfaces at stream edges, in rock pools, mine pits, bamboo stumps, decaying tree trunks and logs, or in the footprints of elephants used for logging in the forests. These areas have year-round rainfall and temperatures that allow mosquitoes to breed continuously. In these regions, this disease is commonly called forest malaria.

Compared with forests, the lowlands and cleared hill slopes plus the urban areas are relatively malaria-free. Forest people are relatively immune to malaria but economic development (mining and timber exploitation, water for hydroelectric plants and irrigation) is moving people who do not have immunity to malaria into forest regions.

Agricultural development in and adjacent to forests also affects malaria transmission. Development of fruit, rubber and forest tree plantations increases the incidence of malaria. Logging, gem mining, smuggling (drugs, arms) and other illegal activity also increase malaria transmission by increasing the possibility of contact for people not immune to malaria and the disease vectors. Gem mining in Thailand and adjacent countries produces mining pits, which are good breeding habitats for mosquitoes. The temporary ramshackle settlements around these gem-mining locations also have poor drainage and sanitary conditions. These poor sanitary conditions result in good breeding sites for mosquitoes.

A shipping corridor transporting both legal and illegal trade and commerce has developed through Myanmar, Thailand, Cambodia, Laos, Vietnam and southern China. This shipping corridor has resulted in the movement of large numbers of people through an area with a high risk of acquiring malaria (about a million cases are developing per year). Official border posts have estimated that more than 10 million people a year move through these regions. The malaria being contracted in this region is also now drug-resistant and today there is no known cure for it. Movement of people for agriculture or other economic reasons has strong implications for the transmission of this disease. During the last four decades, this region has seen a large movement of refugees because of wars and political disruptions.

The safest antimalarial drug (chloroquine) no longer works as well as it used to because the malaria parasite is becoming immune to this drug. Control measures that are used today include: spraying with dichlorodiphenyltrichloroethane (DDT), coating marshes with paraffin (to block *Anopheles* mosquito larvae spiracles), draining stagnant water and the use of nets sprayed with chemicals to kill the mosquitoes. In the developed world, funding organizations are not supporting the use of DDT, even though some are suggesting that this is the only way that malaria will be controlled on the African continent (Rosenberg, 2004).

CASE 5.1. KUDZU (SARAH REICHARD)

Kudzu (*Pueraria montana* (Lour.) Merr. var. *lobata* (Willd.)), is a fast-growing vine that has blanketed the south-eastern United States in just several decades, to the point that it has become almost symbolic for that region (see Fig. 5.1). Biologically it has many of the traits of an ideal invader, ecologically its impacts are several, and culturally it is integrated into the folklore of the region.

Biology

Kudzu is a climbing, semi-woody vine in the *Fabaceae*, the pea family. It is native to temperate Asia, including Japan and China. It primarily establishes in disturbed soil, but by growing up to 20 m in a growing season, often rooting at the nodes, it is capable of moving into less disturbed areas.

Fig. 5.1. Kudzu growing in south-eastern United States. Photo by Sarah Reichard.

While it is generally not tolerant of shade, it can invade forests by climbing up and over trees.

Kudzu has a number of attributes that contribute to its amazing success as an invasive species. First, as noted, it spreads vigorously by a high allocation of resources to shoot extension. By rooting at the nodes when in contact with soil, it allows new plants to establish as the rooted fragments become isolated through disturbance to the population or senescence of connected plant parts. Such vegetative spread appears to be the primary method of reproduction; seed production, viability and survival of juveniles all appear to be low, though these vary by location and levels of herbivores (Forseth and Innis, 2004).

Kudzu 'fixes nitrogen', which means that, through a symbiotic relationship with bacteria in its roots, it can colonize soils with little nutrition (Erdman, 1953). The bacteria change atmospheric nitrogen into the form used by plants. This ability allows it to colonize disturbed areas where organic material in the soil may be lacking, and then grow up and over adjacent vegetation for structure. Other aspects of the roots that contribute to its invasive ability include large succulent primary roots – some weigh over 180 kg and have a diameter of about a fifth of a metre, reaching 3 m deep (Blaustein, 2001). This high allocation of carbon to the roots allows the plants to store large amounts of starch, nitrogen and water – which permit plants to survive stressful conditions.

Leaves can quickly be reoriented to maximize photosynthetic ability. Like many legumes, the base of the leaflets, the pulvinus, experiences movement through changes in potassium levels, which then change the turgor, or osmotic pressure. The leaves move so that they are perpendicular to the midday sun, reducing excess radiation and therefore reducing leaf temperature and transpirational losses (Forseth and Teramura, 1987). Not only does this increase the efficiency of the individual leaf, but it also allows multiple canopy layers to form; the sunlight not intercepted by the upper canopy passes through the lower ones, reducing self-shading.

Ecological and Economic Impacts

There are surprisingly few quantitative studies documenting the ecological impacts of kudzu and its effect on biodiversity. Perhaps because it is such an aggressive vine, researchers felt that its harm was obvious. Certainly the behaviour previously discussed results in superior competition for light and for soil resources, such as water, resulting in a monoculture of kudzu. The competition may be worse for later successional woody plants as opposed to herbaceous spring ephemerals, which complete their growth cycle before the kudzu leafs out (Forseth and Innis, 2004).

Another potential impact that is not fully understood is the introduction of large amounts of nitrogen through the fixation in the root nodules. Recent reviews (Vitousek *et al.*, 1997) have found that humans have increased the amount of nitrogen added to the biosphere nearly twofold. This increased nitrogen finds its way into streams and lakes, contributing to eutrophication, and can impact native plant species not adapted to the elevated nitrogen. Kudzu has the potential to add 235 kg N/ha per year (Forseth and Innis, 2004), compared with up to 200 kg N/ha per year from industrial activities (Compton *et al.*, 2003).

The economic impacts are a little better understood. Blaustein (2001) reported an estimate of 'greater than $500 million per year' in losses due to damage and decreased property values and lost productivity in forests. Blaustein also found that power companies in the south-east spend about $1.5 million annually preventing kudzu from bringing down electrical wires and poles.

Kudzu does have some beneficial uses as well. While its utility as animal fodder is sometimes questioned, livestock are often seen feeding on it. Chinese medicine has used it to treat effects of alcohol consumption and a recent study found that it reduced alcohol use in heavy drinkers by allowing them to achieve an effect with fewer drinks (Lukas *et al.*, 2005). And, of course, its reasons for introduction, discussed below, include soil stabilization and ornament.

History of Introduction

Kudzu is unusual among invasive plants because it has become extremely integrated with the culture of an entire region, the south-east United States, in a short amount of time. It was first introduced into the United States in 1876 as an ornamental plant featured in the Japanese display at the Centennial Exposition in Philadelphia. Its rapid growth and pretty pink flowers made it an attractive vine that would quickly cover and cool a porch on a hot summer's day – hence its early common name of 'porch vine'. In fact, one attendant at the Exposition wrote to her sister in Florida, 'Of all the marvelous things there, a vine which the Japanese call Kuzu [*sic*] astounded me the most ... Knowing how you suffer from that awful heat there, I am going to try to get one of the vines for you' (Hoots and Baldwin, 1996).

The rapid growth also made it attractive to the Soil Erosion Service and Civilian Conservation Corps. In the early 1930s, parts of the country experienced drought years, which removed much of the vegetative layer and left the soil exposed and vulnerable to erosion by high winds – an era referred to as the 'Dust Bowl Era' and memorialized in John Steinbeck's *Grapes of Wrath*. Over 85 million plants were provided for landowners by the Corps to quickly cover the soil and hold it in place (landowners were paid $19.75 per hectare to plant it). It did this very well; by 1946 over 1.2 million ha were covered with it (Everest *et al.*, 1991). In 1953 the federal government recognized the aggressiveness of kudzu and removed it from its list of preferred species for planting. While it is no longer planted, it is now an icon of the south-east, celebrated in festivals, comic strips, crafts, poems, songs, folklore, food products and many aspects of everyday life.

Kudzu continues to expand its range beyond the south-east. Its primary range is currently from New York in the north to Florida in the south and to Texas and Oklahoma to the west. Three small populations have been found in Oregon and one in Washington, but control efforts were immediately implemented and it is no longer believed to be in those states.

CASE 5.2. ACID RAIN, AIR POLLUTION AND FOREST DECLINE (JOHN L. INNES)

In 1967, newspaper reports published in Sweden drew attention to the acidification of Swedish lakes, blaming acid rain as the cause. The subject created a political awareness and at the 1972 United Nations Conference on the Human Environment in Stockholm several papers addressed the issue of acidification. This was followed by a report of the Organization for Economic Cooperation and Development (OECD), published in 1977, which established the imports and exports of sulphur between European countries. The environmental and political awareness that was generated by water acidification provided the backdrop to reports in the late 1970s that forests were dying in parts of Germany, Poland and Czechoslovakia.

As forest decline in the 'Black Triangle' formed by East Germany, southern Poland and northern Czechoslovakia had been documented for some time as being closely associated with air pollution, the inference was readily made that the decline of trees seen in the Black Forest, Bavarian Forest, Harz Mountains and elsewhere in Germany was also attributable to air pollution. Although doubts were raised at the time (e.g. Binns and Redfern, 1983), there was a widespread belief that acid rain was at the root of the problem, and claims were made of the impending demise of forests throughout Europe.

Surveys of Forest Health

The decline observed in some trees was characterized by a number of phenomena, including loss of density in the crown (allowing more light to pass through the crown) and foliage discoloration. Initial surveys undertaken by scientists, foresters, environmental campaigners and others quickly revealed that trees all over Europe had evidence of thin crowns, and this was assumed to be evidence of a massive and almost universal decline of forests in Europe. Those involved also visited North America and identified similar problems there, particular in the Appalachian Mountains. As a result, under the auspices of the Convention on Long-Range Transboundary Air Pollution, an international programme was set up to look at 'forest health' across Europe, using a systematic grid of sampling plots. At each site, a specific number of trees (usually 24) were assessed annually for evidence of loss of density or discoloration. Assessments were made from the ground, using binoculars, and in most cases no attempt was made to undertake standard pathological investigations of trees (Skelly, 1990; Innes, 1998).

The surveys revealed that many trees had transparency levels in excess of 10% (i.e. more than 10% of the crown profile was transparent). This figure increased in successive inventories during the 1980s, but gradually stabilized in the 1990s. Many inconsistencies were identified in the survey methodology, the manner in which the data were analysed and the interpretation of the results (e.g. Innes, 1988a, b) but, led by Germany, these were largely ignored. For example, major differences in transparency levels tended to follow national boundaries, suggesting differences in the standards used by observation teams. A key factor was whether or not absolute or local references were used. The absolute references were defined in guidebooks (e.g. Innes, 1990), whereas local references were taken as the best tree in the area of the sample plot. When Great Britain switched from an absolute to a local reference system in 1992, a marked improvement in 'forest health' was recorded.

Research

The surveys of forest health were accompanied by major research programmes into acidification and forest decline, such as the French DEFORPA programme (Landmann and Bonneau, 1995). Very large amounts of research funding were made available to universities and forest research institutes, and a substantial proportion of European forest research capacity was devoted to the issue. The research looked into the details of the mechanisms by which acidification might affect trees, often using highly unrealistic experimental conditions or modelling scenarios. Although symptoms could be induced, these were generally at levels well above ambient and on sensitive seedlings. Modelling studies helped to identify areas where impacts might be occurring, although often field assessments in those areas failed to reveal any indications of a problem.

What was Really Happening?

There is no doubt that acidification of streams and lakes occurred in areas sensitive to acidic deposition (areas where the soils had a low buffering capacity). In some areas, individual trees and some stands of trees (but not forests – Skelly, 1989) were clearly suffering from environmental stress, although drought appears to have been considerably more important in inducing the observed symptoms than air pollution. In some areas, pollution appears to have been important, but the nature of this pollution and its impacts were very variable. For example, in the Black Triangle, the problems were clearly associated with sulphur dioxide emissions associated with the burning of brown coal. In parts of the Netherlands, exceptionally high levels of ammonia were a problem, caused by the use of agricultural wastes as fertilizers. Around particular point sources, such as smelters and brickworks, a range of different pollutants could be important, including sulphur dioxide, hydrogen fluoride and heavy metals.

A pollutant that was initially dismissed in Europe was ozone. This had previously been recognized as the cause of decline in eastern white pine (*Pinus strobus*) in eastern North America and of several species, particularly ponderosa pine (*Pinus ponderosa*) in California. However, studies undertaken in southern Switzerland, northern Italy and Spain revealed that symptoms typical of ozone were present on a variety of species (Skelly *et al.*, 1999). Using the classic principles of pathological research described as Koch's postulates, cause–effect research demonstrated that some, but not all, of the symptoms were indeed caused by ozone, enabling the publication of an atlas of ozone symptoms (Innes and Skelly, 2001). Ozone has since been shown to be more widespread than previously recognized and to cause injury to the foliage of many species at concentrations below those previously thought capable of inducing injury.

Lessons Learnt

The forest decline debate cannot be considered to be one of the more credible periods in European forest science. Many scientists were too ready to accept hypotheses as facts, to ignore evidence that falsified those hypotheses and to use alarmist tactics to raise research funding levels. The credibility of forest science was badly damaged and, in Germany in particular, forest science suffered major cutbacks in the period immediately following the debate. Poor scientific methods were frequently used, such as the lack of proper pathological investigation of symptomatic trees, and there was an unwillingness to engage in the costly and time-consuming cause–effect research required by Koch's postulates.

There was also an unwillingness to consider alternatives – although much of Europe had suffered major consecutive droughts in 1975 and 1976 and many symptoms developed in 1977 and 1978, this fact was largely ignored. Interestingly, today the argument is being made by survey advocates that the surveys provide an important mechanism to assess the deleterious effects of climate change – an argument that will probably do little else than damage the credibility of the climate change research community. We have since learnt that European forests were not in a state of terminal decline (Spiecker *et al.*, 1996), that the phenomenon was not new (Kandler, 1992) and that, overall, our forests are acceptably healthy (from a traditional pathological viewpoint).

CASE 5.3. WILDFIRE IN THE BOREAL FORESTS OF ALASKA (ROBERT A. WHEELER)

The Role of Wildfire in Boreal Forests of Alaska

The interior boreal forest of Alaska covers a broad geographical area, estimated at about 332 million acres, of which about 116 million acres are forested. Of these estimated 116 million acres of forest land only about 22 million acres are estimated to be commercial forest land. Although the boreal forest covers a large portion of the state, the forest overstorey composition is based upon only a relatively few tree species, which includes white spruce, black spruce, quaking aspen, Alaska paper birch

and balsam poplar. These forest stands can occur as either pure-composition or more commonly as mixed-species stands. The impact of wildfires, insects and diseases and other agents of forest change has resulted in the creation of a vast mosaic of stands of different age and species composition.

Wildfires in interior Alaska are more commonly controlled by changing weather patterns than by direct human intervention. The fire season typically begins in May with the drying of grassland areas. A review of fire statistics from the past 40 years indicates that for the boreal forest about 38% of fires are lightning-caused. Even though lightning causes

Major Fire History of Alaska

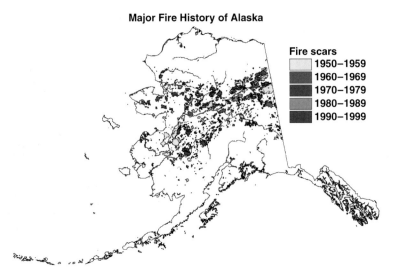

Fire scars
1950–1959
1960–1969
1970–1979
1980–1989
1990–1999

Fig. 5.2.   A visual presentation of the history of the major fires in Alaska from 1950 to 1999.

a smaller proportion of the fires, they account for about 83% of the area that is burned in forests. The other 62% of the forest fires are human-caused but they only account for 17% of the burned forest area.

During a Forest Health conference held in Fairbanks, Alaska, in 1998, Dr R.A. Werner stated that:

> Boreal forests are under stress in much of interior and south coastal Alaska because of changes in forest conditions or pest problems. These stress factors have caused a reduction in tree growth and increased mortality. Severe insect infestations, high incidences of root and stem decay and high fuel loads with related extreme fire hazard, have combined to cause a general condition of poor health in forests of Alaska.

The conference concluded that warming trends will probably lead to more frequent outbreaks and more extensive regions of impact for both insects and pathogens and regional increases in frequency and intensity of wildfires (Fig. 5.2).

Insect activity contributes to the forest health crisis

During the 10-year period 1990–2000, it was estimated that more than 3 million acres of

spruce-dominated forest were devastated by heavy mortality caused by the spruce bark beetle in south central Alaska (Fig. 5.3). Infested stands will typically exhibit more than 90% mortality of all overstorey trees. This mortality has led to heightened concerns for severe risk of wildfires in rural communities and within the urban Anchorage complex.

Unfortunately, the 2005 fire season in the coastal zone of south central Alaska has seen a dramatic increase in the occurrence of thunderstorms and lightning strikes, resulting in several destructive wildfires, which have threatened rural homes and communities in the region. As of the end of July 2005 13 wildfires had burned more than 1 million acres, where there had only been 12 wildfires reported for the previous decade.

The accumulated mortality associated with the spruce bark beetle outbreak has dramatically increased the risk of associated wildfires. Even stands that were thinned down to 4 inches' diameter at breast height have had high levels of mortality, which is unusual susceptibility for small-diameter trees in thinning operations.

Evidence of regional warming

Until recently, an average of about 2 million acres of forest land burned each year. However, recent trends in fire area and severity have dramatically increased

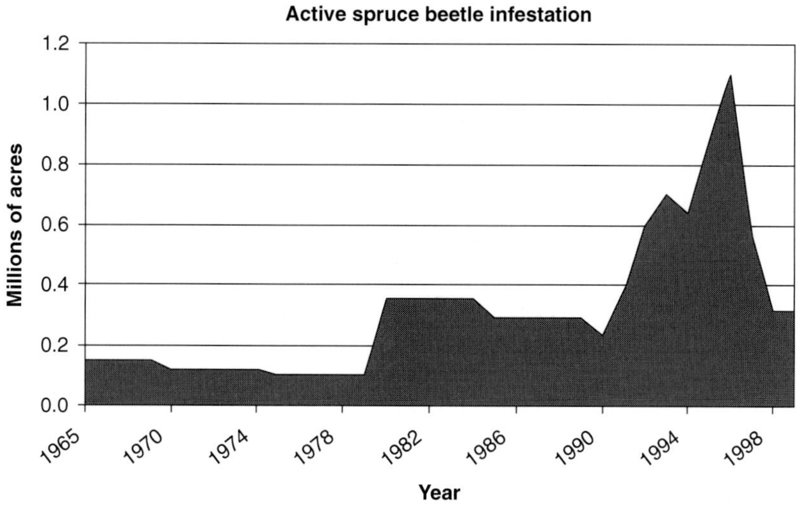

Active spruce beetle infestation

Fig. 5.3.   Spruce bark beetle-impacted acres as reported in the annual Forest Health Protection Report for Alaska produced by the US Forest Service.

the annual acreage burned. During 2004, nearly 7 million acres of forest land burned and by the end of August 2005 more than 3 million acres had burned, derived primarily from lightning stikes from the severe thunderstorms that have frequented the region.

A review of the number of frost-free days per year for the period of 1900 to 2000 indicates that there has been nearly a 50% increase in the average number of frost-free days (Whitfield, 2003b). The increased number of warmer days and extended dry or droughty conditions, with frequent summer lightning storms, have all contributed to larger than normal wildfire acreages. Also, the impact of increased fire frequency on forest areas of the boreal forest has resulted in more frequent release of sequestered carbon in both above- and below-ground carbon pools.

The Effects of Wildfire Smoke on Trees and Vegetation

The wildfire season for 2004 was historic both from the standpoint of number of acres burned and in the extent and levels of air pollution experienced by communities throughout the interior of Alaska. Extremely high levels of wood smoke were observed throughout the interior region, with visibilities less than three-quarters of a mile and hazardous particulate levels sometimes exceeding 500 µg/m^3 (Fig. 5.4).

For the 10-year period of 1992–2001, it is estimated that the total emission of carbon dioxide from wildfires in Oregon alone would be about 65 million metric tons of CO_2 released from a total of 2,006,336 acres that were burned. This amount of carbon dioxide is equivalent to the carbon found in more than 740 million barrels of crude oil. For the estimated 6.7 million acres of forest land burned during the summer of 2004 in Alaska, if we assume a similar relationship of acres burned to carbon dioxide released, then it is estimated that these fires released approximately 217 million metric tonnes of carbon dioxide or the equivalent of carbon found in 2 billion 470 million barrels of crude oil or about one-third of the annual consumption of crude oil in the United States (estimated at 19 million barrels per day).

Wildfire smoke varies depending upon the fuel source but is composed mostly of carbon dioxide, water vapour, carbon monoxide (a poisonous gas), particulate matter, nitrogen oxides and other hydrocarbons and minerals.

Particulate emissions from wildfires can be especially troublesome as a health hazard for humans. The particulates are less than 1 µm in diameter;

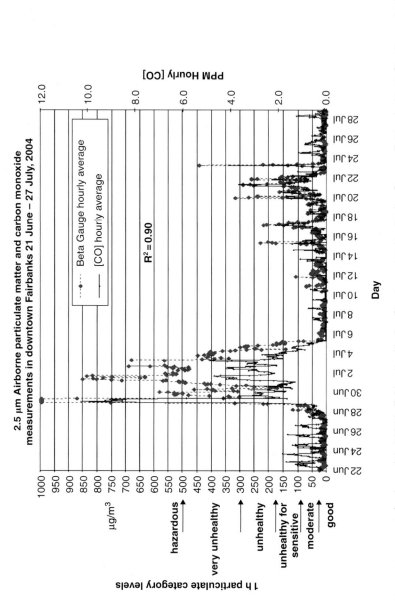

2.5 μm Airborne particulate matter and carbon monoxide
measurements in downtown Fairbanks 21 June – 27 July, 2004

Beta Gauge hourly average

[CO] hourly average

$R^2 = 0.90$

Fig. 5.4. Air particulate measurements for June–July 2004 in Fairbanks, Alaska. Graph provided by Dr Jim Conner, Fairbanks North Star Borough, Alaska.

Fig. 5.5. Smoke in downtown Fairbanks during the summer of 2004. Image provided by Dr Jim Conner, Fairbanks North Star Borough, Alaska.

this size is near that of the wavelength of visible light and so the particulates are very effective at dispersing light and reducing visibility (Fig. 5.5). The combination of air pollutants and reduced light levels is likely to have had some effect upon plants and trees in the interior of Alaska.

Nitrogen oxides (NO and NO_2) are known as plant toxicants and can cause damage to leaves, resulting in interveinal necrosis in hardwoods (aspen and birch) and tip burning in spruce and other conifers. These nitrogen oxides typically reside in the atmosphere for only a few days. Rain or wet conditions will accelerate their filtering from the atmosphere.

Long-term exposure of plants to nitrogen oxides found in air pollution from combustion of fossil fuels was found to have some impact on forest growth. The deposition of the nitrogen oxides can result in increased availability of nitrogen as a fertilizer and result in some initial increase in growth but will probably result in lowering soil phosphorus levels, leading to growth reduction, increased levels of plant physiological stress, reduced carbohydrate storage in roots and heightened susceptibility to insect and disease attack. Given our current semi-drought-stressed forest condition, this could contribute to further heightening concerns over forest health and landscape-level management issues to reduce fire risk and maintain forest vigour.

CASE 5.4. HABITAT FRAGMENTATION AND DISEASE ECOLOGY: THE CASE OF LYME DISEASE (RICHARD S. OSTFELD)

Background and Historical Issues

Lyme disease (LD) is one of the so-called emerging infectious diseases. This zoonosis was first described in the mid-1970s, was listed as a 'notifiable' disease by the US Centers for Disease Control and Prevention (CDC) in 1991 and has undergone explosive growth in incidence, recently exceeding 20,000 reported cases per year in the United States (Centers for Disease Control and Prevention, 2003). LD is not a notifiable disease for health authorities in the European Union or Russia, but incidence in some parts of Europe and Asia, particularly Austria, Germany and Slovenia, might equal or exceed that in highly endemic zones in the north-eastern United States. On both sides of the Atlantic Ocean, strong evidence suggests that LD existed for centuries or millennia before its discovery, but there is little doubt that LD incidence in people rapidly increased in the late 20th and early 21st centuries (Stanek et al., 2002). This increase in LD incidence at both national and global scales includes both escalation in local incidence and geographical spread to previously unaffected areas (Ostfeld, 1997). These features meet criteria for designation as an emerging infectious disease (Daszak et al., 2000).

The aetiological agents of LD include several genotypes of the spirochaete bacterium *Borrelia burgdorferi*, which is transmitted to humans and among wildlife by ticks in the *Ixodes ricinus* species complex. In eastern North America, the predominant tick vector is *Ixodes scapularis* (the black-legged tick). Larval ticks typically hatch from eggs in mid-summer uninfected with spirochaetes and take a single blood meal, lasting 2–5 days, from a vertebrate host. After the meal they drop off, moult into a nymph and remain inactive until the following late spring or early summer, when they again seek a host from which to take a blood meal. Larval ticks that feed from an infected host may acquire a *B. burgdorferi* infection, in which case they can transmit the spirochaetes to their host during the nymphal meal. When the infected nymph transmits *B. burgdorferi* to a wildlife host, this serves to perpetuate the enzootic cycle by providing later-feeding larvae with a source of infection. When the infected nymph transmits *B. burgdorferi* to a human, the result can be a case of LD. Following the nymphal blood meal, the tick drops off the host and moults into the adult stage, which seeks a host during mid-autumn. Like nymphs, adults are capable of transmitting *B. burgdorferi* to their hosts, but spirochaete transmission from adults is relatively unimportant to both the enzootic cycle and LD epidemiology, because adult ticks tend to bite 'dead-end' hosts (e.g. white-tailed deer and humans), which do not serve as important sources of infection to later-feeding larval ticks (LoGiudice et al., 2003).

It seems likely that black-legged ticks and *B. burgdorferi* spirochaetes were widely distributed occupants of eastern North American deciduous forests in pre-colonial times, but that both species were nearly extirpated by the middle of the 19th century, when > 90% of these forests were cleared and replaced with agricultural fields. Black-legged ticks and spirochaetes apparently survived in small, uncleared refugia on both the northern and southern shores of Long Island Sound, but remained sufficiently scarce to be undetected. Following the completion of the Erie Canal in 1825, farmers of New England and New York began to abandon their marginally productive farmland in favour of deeper, more fertile soils of the Ohio Valley and the Midwest. Massive farm abandonment continued throughout much of the 19th century, leading to the natural reforestation of the north-eastern United States over the next several decades. Small, forest-dwelling mammals, such as white-footed mice, eastern chipmunks and shrews, as well as larger mammals, such as white-tailed deer, probably began to reoccupy the land as reforestation proceeded, re-establishing the community of hosts for ticks and spirochaetes necessary for LD maintenance and transmission. Apparently, however, dispersal by ticks and their commensal spirochaetes from presumed refugia throughout the region proceeded much more slowly, with re-invasion of many north-eastern regions lagging many decades behind. This re-invasion appears to be continuing in the north-eastern and mid-Atlantic regions of the United States (Waller et al., 2006).

The Problem

A fascinating paradox of LD ecology is that, even though reforestation of the landscape was apparently necessary to create the emerging LD problem in the north-eastern United States, subsequent forest destruction and fragmentation appear to be strongly exacerbating the LD problem and elevating human risk. A wholly deforested landscape cannot support the species responsible for maintaining LD. Yet, once LD is established in a largely forested landscape, fragmentation of the forest exacerbates the health threat to humans.

Both the tick vectors and the wildlife hosts of *B. burgdorferi* predominantly occupy deciduous forests, with some spillover into residential lawns and ornamental vegetation (Ostfeld, 1997). Apparently, most human victims of LD are exposed to infected ticks on or near their own property while

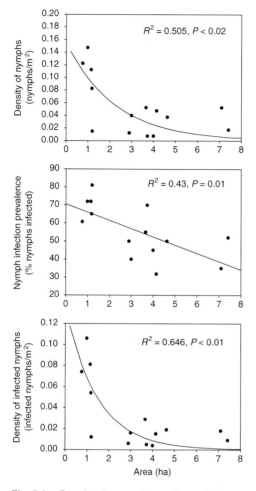

Fig. 5.6. Results of research on effects of forest fragment size on tick variables in Dutchess County, New York State. Small, isolated fragments have significantly higher proportions of infected ticks and higher densities of ticks, and thus higher DIN. Figure reproduced from Allan *et al.* (2003) with permission from the Society for Conservation Biology.

residences, they also comprise the riskiest landscapes with regard to LD risk.

Recent research has demonstrated that forest fragmentation strongly increases risk of human exposure to LD via its effects on both tick abundance and spirochaete infection prevalence. In Dutchess County, New York, a county that often leads the nation in both total numbers of cases and per capita cases of LD, forest fragments less than about 2 ha supported densities of infected nymphs on average *c.* four times higher than those in fragments of 2–15 ha (Fig. 5.6; Allan *et al.*, 2003). This heightened human risk is not associated with changes in human behaviour, but rather appears linked to changes in the diversity and species composition within the community of vertebrate hosts for ticks.

The Players

The immature stages (larvae and nymphs) of *I. scapularis* are highly non-specific in their distribution on vertebrate hosts, having been reported on > 80 species of mammals, birds and reptiles (Lane *et al.*, 1991; LoGiudice *et al.*, 2003). These vertebrate hosts differ dramatically in the probability that they will infect a feeding tick with the LD spirochaete. In the eastern and central United States, the white-footed mouse (*Peromyscus leucopus*) is the principal natural reservoir for the LD spirochaete, because *c.* 90% of larval ticks feeding on free-ranging white-footed mouse acquire the spirochaete (Schmidt and Ostfeld, 2001). These ticks then become capable of infecting subsequent hosts during their nymphal and adult stages. A few other mammalian and avian hosts (e.g. eastern chipmunks (*Tamias striatus*) and short-tailed shrews (*Blarina brevicauda*) are moderately effective reservoirs, but most others are incompetent reservoirs of LD (Mather *et al.*, 1989; LoGiudice *et al.*, 2003).

Two ecological metrics of disease risk are relevant to the epidemiology of LD. The first is the density of infected nymphs (DIN) occurring in areas that people use recreationally or peridomestically (Barbour and Fish, 1993). This metric determines the probability that a person will encounter a tick capable of infecting him or her, given that the person enters tick habitat. However, DIN incorporates two factors, density and infection prevalence, which may be determined by different biological processes. Density of nymphs may be a complex

gardening or during recreation (Barbour and Fish, 1993). Counties in the USA with the highest incidence of LD are characterized by moderately to densely populated suburban communities within a largely forested matrix. Such a landscape provides ample zones of contact between forest-dwelling vertebrate hosts for ticks, the ticks themselves and humans. Thus, although suburbs with ample forest cover are highly desirable locations for function of: (i) composition of the community of hosts for larvae, which determines average feeding success for larvae and therefore survival to the nymphal stage; (ii) abundance of white-tailed deer, which is the predominant host for adult black-legged ticks (Wilson, 1998) and which is therefore critical for maintaining a tick population (Wilson *et al.*, 1988, 1990); and (iii) abiotic (climatic) variables that influence tick survival when off hosts (Bertrand and Wilson, 1997; Fig. 5.6).

The second ecological risk factor is nymphal infection prevalence (NIP) (the proportion of nymphs infected with the disease). NIP appears to be largely independent of the population density of ticks and should be a function of the distribution of tick meals on the host community. The distribution of tick meals is crucial, owing to variation among host species in reservoir competence. NIP is an important risk factor because variation in NIP will strongly influence the probability that a given tick bite will result in a case of LD.

A third ecological risk factor has only recently been discovered. Populations of *B. burgdorferi* maintain extremely high, stable levels of polymorphism at a particular outer surface protein (OspC) locus. Fifteen OspC genotypes co-occur at stable frequencies within local populations of the pathogen. Brisson and Dykhuizen (2004) have carefully examined the distribution of OspC genotypes among the four most commonly captured species of tick host at our Dutchess County study sites (white-footed mice, eastern chipmunks, grey squirrels and short-tailed shrews) and found that each host favours a distinct subset of OspC genotypes and apparently eliminates all others via an immune response. Only four of the 15 OspC genotypes are infectious and therefore pathogenic to humans (Seinost *et al.*, 1999); thus, the frequency of pathogenic genotypes within tick populations is also a key determinant of ecological risk. This frequency is a function of the distribution of tick meals on vertebrate host species.

All three of these factors determining ecological risk of LD are strongly influenced by the composition of the vertebrate host community. Previous research suggests that the composition of the host community is largely determined by landscape structure and habitat configuration. For example, studies in the Midwest have shown that forest fragmentation increases the population densities of white-footed mice (Nupp and Swihart, 1996;

Krohne and Hoch, 1999), the most competent reservoir for the LD bacterium. This effect is probably caused by the loss of predators and competitors of mice in small fragments (Rosenblatt *et al.*, 1999). Thus, in small fragments, more tick meals are probably taken from mice because alternative vertebrate hosts are absent, and these mice are the most competent reservoir for the Lyme bacterium (Ostfeld and Keesing, 2000); the net result is an increase in the density of infected nymphal ticks (DIN; Fig. 5.6).

Lyme Disease, Forests and Society

Although a vaccine for LD was available for several years in the late 1990s, this vaccine is no longer produced and new vaccine development appears years away. Consequently, avoidance of bites by infected ticks and prompt removal of embedded ticks (pathogen transmission begins 24–48 h after tick attachment) are the only direct preventive measure available. Thus, in contrast to many infectious diseases, much responsibility rests on individuals, rather than on health-care providers or the government, for avoidance of disease. Avoidance strategies include: avoiding tick habitat (forests and forest edges) during early summer (nymphal activity peak); wearing protective clothing; using insect repellents; adopting landscaping practices that discourage tick entry or survival; and daily autogrooming or allogrooming to inspect for and remove ticks. The recent research linking forest fragmentation to increases in LD risk and ongoing studies asking related landscape-level questions are likely to highlight the possibility of landscape management as a tool to reduce the burden of LD on society.

Local governmental planning or zoning boards are already paying attention to the finding that small forest patches pose higher risk than do larger (> 3–5 ha) patches (R.S. Ostfeld, personal observations). Real and perceived risk of exposure to LD, as well as to other newly emerging tick-borne illnesses, are likely to be strong motivators to act on research findings that link landscape structure with risk. Rigorous, replicated studies assessing these landscape effects on epidemiology are rare but vital.

Beyond its direct effects on human health, LD profoundly impacts society via both perceptions of risk and fear of illness within LD-endemic zones. The potential exists for the real and perceived threat of exposure to LD to affect societal attitudes towards nature in general and forests specifically. Clearing of forests might be considered more acceptable to those who perceive a strong health threat lurking therein. However, knowledge that deforestation that creates small forest patches in a suburban landscape is likely to increase risk of exposure to LD could influence attitudes towards forest and landscape management. Surprisingly little is known about the effect of variation in human knowledge about LD ecology and risk factors on people's attitudes towards nature and their protective and controlling behaviours concerning ticks, tick habitat and LD (Orloski *et al.*, 2000; Hayes and Piesman, 2003). Research at the interface of landscape ecology, infectious disease biology, sociology and psychology will probably be necessary in order to understand feedbacks between actual LD risk, perceptions of LD risk and human-induced changes to the landscape that affect LD risk.

Take-home Message

Lyme disease is an enormously important epidemiological and sociological problem within portions of the United States, Europe and Asia. Ecological risk, defined as the abundance of infected ticks or the proportion of ticks infected, is determined largely by variation in the abundance of specific forest-dwelling vertebrates as well as by the composition of the entire community of tick hosts. Vertebrate communities, in turn, are apparently affected by the composition (types of patches) and configuration (relative positions of patches) of forest-dominated landscapes.

Urban sprawl and other causes of deforestation can, under some circumstances, increase risk of human exposure to LD by creating species-poor vertebrate communities dominated by highly competent LD reservoirs, such as white-footed mice. Knowledge about what types of landscape changes are likely to increase or decrease LD risk is likely to influence both the ethics and the politics of forest and landscape management.

6 Forests and the Carbon Cycle

DANIEL J. VOGT, RAGNHILDUR SIGURDARDOTTIR, DARLENE ZABOWSKI AND TORAL PATEL-WEYNAND

Introduction

Carbon is one of the most important elements needed for life as we know it. It is the basic building block of all life on earth. Carbon occurs in all organic life and is in the food you eat, in the clothes and cosmetics you wear and in the petrol you use for your car. Probably the most recognized form of carbon in the world is a diamond. Carbon also has the interesting chemical property of being able to bond with itself and a variety of other elements, forming nearly 10 million known compounds. When it bonds with two oxygen molecules, it forms carbon dioxide (CO_2), which is vital to plant growth. Combined with both oxygen and hydrogen it can form numerous groups of compounds, including fatty acids, which are essential to life. Bonding of carbon and hydrogen produces hydrocarbons in the form of fossil fuels (e.g. coal, petroleum and natural gas).

Any stopping, slowing down or speeding up of the availability or movement of carbon from one form to another can have repercussions not only locally but also globally. For example, if large areas of forests are cut down (such as extreme clear-cutting practises) and not replanted with trees, forest assimilation of carbon dioxide from the atmosphere will be decreased, which could lead to increases in global temperatures. Or, if large areas of forest are burned and not replanted or naturally regenerated with trees, the combustion will have a net effect of increasing the atmospheric carbon dioxide concentrations. The resulting change in CO_2 levels can lead to increase in global temperatures. It needs to be emphasized that the replacing of trees with other types of plants will not provide the same carbon dioxide assimilation capabilities. Per cubic centimetre there is not another plant species on earth that can perform this job as well as a tree.

For all practical purposes, the amount of carbon on earth is constant. Processes that use carbon must get it from somewhere and it must be disposed of in some way. The paths that carbon follows in the environment are called the carbon cycle (Fig. 6.1). An extremely simple example of this process would be something along the

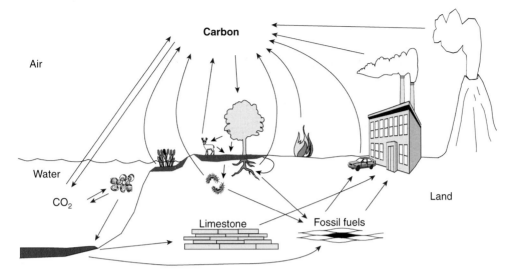

Fig. 6.1. Carbon cycle in the lithosphere (land), hydrosphere (water), atmosphere and biosphere (organisms). Major pools are shown along with arrows indicating fluxes between the pools.

following lines. Plants take carbon dioxide from the environment and use it to grow in size (create biomass). Animals eat some of this biomass. When animals consume biomass, they exhale CO_2. At the same time that this simple process is going on, more complicated cycles are taking place all over the globe.

All of the carbon on earth can be organized into five primary categories, called pools. These five pools in order of magnitude are:

1. Lithosphere (Earth's crust).
2. Oceans.
3. Soil (e.g. soil organic matter – SOM).
4. Atmosphere.
5. Biosphere (consists of all organisms, both living and dead).

To understand how human activities impact the carbon cycle, it is important to have some familiarity with these pools or storage sites for carbon and the fluxes of carbon contained in them. Human activities do change how carbon is stored and cycled at local, regional and global scales. Since carbon can also exist as a gas, there are few barriers to its movement from local to global scales, so local impacts can and do have global effects.

Since forests are a large repository of carbon (i.e. in tree biomass) and trees in the forests can be managed, it is forests that can help to mitigate and restore carbon cycles that have been disturbed by human activities and/or by climate change. For example, forests can substitute for fossil fuels in the energy sectors or be used to increase the sequestration of carbon from the atmosphere in tree biomass. When forests are used in these ways, they contribute towards managing the global carbon cycle and could help to restore some of the carbon cycle functions altered by human activities. Since these changes can affect human and ecosystem health, understanding what triggers changes in the carbon cycle is crucial.

Carbon Cycle

Plants are autotrophic and obtain their carbon from CO_2 in the atmosphere while synthesizing organic compounds. Animals (and humans) are heterotrophic and require carbon-based organic compounds, which they break down to obtain their energy. This breakdown process releases CO_2. Also burning fossil fuels (e.g. oil) will release CO_2 into the atmosphere, similar to what happens when forests and other plants burn. Even differences in land use (e.g. forestry, agriculture, grazing, urban development) have different effects on carbon pools and fluxes. In fact, it is important to know what the carbon form is within each pool of the carbon cycle in order to know how fast it may cycle from one pool to another. Anything that can affect one part of this carbon cycle could have a dramatic influence on other parts of the cycle. We also need to know about carbon cycles and how they are influenced so that we can better understand and manage forests to eliminate the possibility of negatively impacting human and ecosystem health locally or globally.

The carbon cycle interconnects global processes

By understanding some of the components of the carbon cycle, one will probably be able to realize the limits and opportunities that exist in managing carbon globally. This basic understanding could also allow us to determine whether certain human activity could be detrimental to the carbon cycle and perhaps make our environment unsustainable. For example, the ability to determine potential impacts on the global carbon cycle and the use of forests as a tool to help in mitigating these impacts can be strengthened by knowing the answers to some of the following questions:

- How much of the global carbon is found in forests and how easily can this carbon be managed?
- Does carbon move around and, if so, where does it move? What causes the carbon to move?
- Does it make any difference if carbon moves and whether it accumulates or diminishes in one pool versus any other?
- How much should tissue-chemistry differences be included in the calculation of the carbon storage capacity of a forest?

Knowing the answers to any of these questions helps us to better understand our environment and the interconnectivity that exists among the biosphere, hydrosphere, lithosphere and atmosphere of our world.

When the availability of carbon-based resources is reduced due to over-exploitation and over-consumption (Diamond, 2005; see Chapter 1, this volume), the effect can be devastating in carbon-based economies. Carbon is the key mineral that integrates and connects many of the resources that society is dependent upon globally. When humans modify the storage and cycling of carbon, these changes could have broad ramifications for the quality of life as we know it and they would dramatically impact the quality of life for plants and animals. In the past, there are many historical examples of civilizations collapsing (see Chapter 1, this volume) due to their mismanagement of carbon-based resources (e.g. over-cutting trees contributed to the collapse of the Sumerian civilization).

Now humans are modifying the terrestrial lands, the oceans and the atmosphere at an unprecedented rate because of our use of fossil fuels (a high carbon-based compound) as an energy source. Fossil fuels are originally derived from biomass but the process of converting biomass to fossil fuel requires millions of years to occur (Fig. 6.2). Since many scientists have linked the high rate of fossil fuel combustion to climate change, some suggest that we need to move away from the use of carbon-based compounds. However, relative to other sources of energy, carbon-based compounds provide energy extremely efficiently so it will be difficult to entirely eliminate their use unless something can be economically substituted that has similar characteristics.

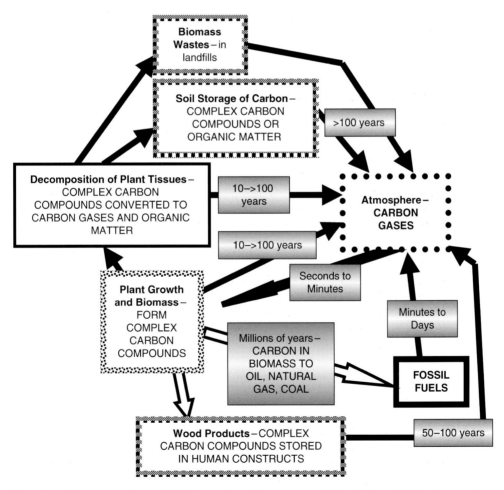

Fig. 6.2. A conceptual model of the pools and interconnectivities in the global carbon cycle.

New breakthroughs in the technology of converting biomass – a carbon-based compound – are starting to provide a mechanism to take advantage of the carbon molecule in an environmentally friendly manner.

The carbon cycle interconnects the terrestrial, aquatic and atmospheric parts of the globe, which means that changes in one part of the cycle will reverberate through the other parts of the cycle. For example, if the CO_2 levels increase in the atmosphere because humans are combusting carbon-based fuels, temperatures are predicted to increase, which can cause global warming. The increase in temperature is also predicted to result in different rainfall patterns across the landscape, which may alter the current ability of inhabited landscapes to provide for people. Furthermore, increases in temperature can change an environment sufficiently for plants and animals to be unable to compete or survive in their current habitats.

When plants are unable to fix enough carbon from the atmosphere, they become less able to grow and compete for space. Not being as competitive can result in plant mortality and the transfer of their tissues into the decomposition part of the carbon cycle.

Animals are also affected if their food source, also carbon-based, is either no longer available or what is available is less digestible because of changes in plant chemistry (e.g. producing more lignin in tissues). When plant tissues become less digestible, animals have to eat more food to obtain the nutrients/minerals required to survive or are unable to obtain enough food and die.

Carbon in soils tends to be more stable and less impacted by environmental change and human activities. However, land-use activities that expose deeper soil profiles can result in decreased storage of organic matter – carbon high in lignin – in the soil. Agriculture has decreased the organic carbon contents of soils by more than half throughout human history.

The simplified global carbon model (Fig. 6.2) demonstrates that carbon accumulates slowly in terrestrial landscapes but it can be rapidly converted to its gaseous form, especially when fossil fuels are combusted for energy. This figure also shows the interconnectivity of the carbon cycle in the globe and how changes in the terrestrial or aquatic realms or the atmosphere can feed back to affect the other parts of the carbon cycle. Ultimately, human survival is dependent on how well the carbon cycle is managed or manipulated. An overview of the global carbon cycle is presented next to show where the dominant pools of carbon are found globally.

Carbon pools in the global carbon cycle

The pool sizes of carbon are provided in Table 6.1 for the globe. This table shows only the stored pools of carbon and not the amount of carbon that moves from these pools annually. Carbon pools can be broadly separated into those that are found in the atmosphere, the hydrosphere, the lithosphere (the earth and soil) and the biosphere (e.g. plants, animals, microorganisms). The amounts of carbon vary dramatically among those pools.

An uneducated guess might assume that most of the global carbon would be found in the biosphere (specifically the vegetation) because about a third of the terrestrial land area is in forests, 23% is in pastures and rangelands, 10% in agriculture and

Table 6.1. Estimates of the carbon pools in various global reservoirs (modified from Siegenthaler and Sarmiento, 1993; Sundquist, 1993; Schimel *et al.*, 1995). (Gt = 10^{15} g = billion metric tons.)

Global pools of carbon (or reservoirs)	Amount of carbon (billion metric tons)	Percentage of total (%)[a]
Atmospheric	750	1.8
Oceanic	39,000	93.4
Vegetation	550	1.1
Soils	1,500	3.7
		100.0
Recoverable fossil fuel	> 4,000	–
Geological substrate (marine sediments, sedimentary rocks)	> 65,000,000	–

[a]Exclusive of recoverable fossil fuel and geological substrate.

only 2% in built land or human habitation (e.g. cities) (Fig. 2.1). But, in fact, as shown in Table 6.1, the vegetation pool is the smallest pool of carbon in the categories presented in this simplified global diagram.

The greatest pool of carbon is found deep in the earth (e.g. fossil fuels, like coal and oil, and carbonate rocks, like limestone), followed by the ocean (e.g. bicarbonates or carbonates, like those mixed with calcium and magnesium in calcite and dolomite, which we commonly use in our gardens to increase the pH of the soil). Another large pool of carbon is the soil (a thin, live skin of the earth that contains live organisms and detritus or dead organic matter such as humus). The soil has more than twice the amount of carbon that is found in the atmosphere (e.g. carbon dioxide and methane). The lowest carbon pool is the vegetation (e.g. carbohydrates or complex carbon compounds). So carbon in vegetation and soils is a very small part of the total carbon stored globally (Table 6.1).

If the recoverable fossil-fuel and geological pools are excluded from this discussion, the ocean contains the highest proportion of the global carbon (93.4% of the total) (Table 6.1). At this broad scale of analysis, the vegetation would be considered unimportant in the global carbon cycle, since only 1.1% of the global carbon is found in this category. Similarly the atmosphere contains 1.8% of the global carbon pool, while the soils would have a higher per cent of total carbon (3.7%). The higher percentage of carbon in the soil compared to the vegetation is a result of the greater mass or volume of soil compared to the mass or volume of vegetation. Despite these low percentages in the latter three categories, these are the pools that are important because these pools can be managed and this is where carbon levels can be manipulated. The fact that these pools provide invaluable services to maintain the quality of life on the planet is above all the key reason for their sustainable management.

The recoverable fossil-fuel and geological pools have accumulated over millions of years and therefore are pools that are difficult to manipulate in the short time scales of a human lifespan. The recoverable fossil-fuel pool is also the pool that is being rapidly collected by humans and combusted for energy production. Combustion of this pool is releasing significant amounts of CO_2 into the atmosphere.

Forests as Interconnectors of the Carbon Cycle

It has been estimated that carbon found in forests around the world contributes 90% of the carbon that is found in the total vegetation pool (Houghton, 1995). This high percentage of vegetative carbon in forests explains why any disturbance that significantly reduces forest area becomes a concern. Knowledge of how trees store and release carbon contributes information on how to manage forests to increase their carbon storage capacity. Forests interconnect with the carbon cycle as live plants as well as after the trees die.

Most carbon enters the soil as organic detritus or litter. Litter includes a variety of materials, such as leaf litter, woody debris, dead roots and bodies of various organisms. Once litter is in the soil, it can undergo decomposition. The rate and type of decomposition depends on the chemical quality of the litter (see below – 'Decomposition (and composting): converting complex carbon compounds into their molecular forms'). As decomposition proceeds, successive organisms breaking down the litter can produce humic substances or simple organic compounds. Humus is one type of humic substance: a complex mixture of blackish, amorphous, organic substances resulting from decomposition and synthesis that has no specific chemical formula. Humus gives soil good physical properties, nutrients and the ability to store nutrients, it holds water and it helps bind soil particles together to prevent erosion.

Simple organic compounds (such as oxalic acid, sugars, etc.) are other possible products of decomposition, as are the more complex organic molecules. If decomposition is complete, the C in the litter is completely converted to CO_2. Simple organic compounds such as sugars and proteins are often completely decomposed, but humus and other humic substances are more resistant to further decomposition and can persist in the soil for long periods of time. These materials, particularly insoluble forms, are likely to accumulate in soils.

Carbon can also enter soil through the weathering of sedimentary rocks containing carbonates. This is a source of inorganic carbon. When soil pH is low, carbonates can be converted to carbonic acid and CO_2.

Soil organic matter (SOM) is produced and accumulates when dead plant and animal matter is deposited in or on the soil surface and when the rate of plant and animal detritus production exceeds its rate of decomposition (Johnson, 1995). Through natural decomposition processes, these once identifiable materials are physically and chemically broken down into smaller unrecognizable pieces (Swift *et al.*, 1979). A portion of the original organic material is lost as part of microbial respiration during decomposition and leaves the soil system as CO_2. Another portion is incorporated by the decomposers in their own tissues as cell walls, proteins and other compounds and eventually becomes the substrate for other decomposers. Some of the original organic material, however, becomes immobilized or biologically unavailable either on soil mineral surfaces or flocculated by inorganic cations (Johnson, 1995).

Carbon sequestration, however, is not a simple accounting approach that catalogues the amount of live tissues produced each year and the amount of dead tissues decomposing in the forest. Carbon sequestration needs to be balanced with maintaining a healthy forest (e.g. the right number of both live and dead trees capable of supporting the unique habitats for many species that are dependent on these structures for their survival) and a forest that provides sustainable livelihoods while being used to sequester carbon (see Chapter 7, this volume).

To better understand these interconnectivities and how carbon fixed at one location becomes part of the carbon cycle, it is worth discussing how carbon is taken up by trees and how it is stored and cycled in forests. All these interconnectivities have been and can be managed or modified by humans.

Trees as carbon interconnectors in ecosystems

Trees accumulate or 'fix' carbon during photosynthesis when they take CO_2 and convert it into six-carbon sugar molecules. Trees release CO_2 back to the atmosphere through respiration when these sugar molecules are used to maintain the functions of a tree and to build more complex molecules to produce plant tissues and organs. Trees can use the fixed carbon to build different tissues, such as leaves, branches, stems and roots. The chemistry of these tissues varies considerably and is comprised of chemicals called hemicellulose, cellulose and lignin. As the tree grows, more of these chemicals are used to make the different plant tissues and the tree becomes larger.

Eventually plant tissues or even the whole tree can senesce and die. For example, the leaves on deciduous trees may die after one growing season or they die after several years on evergreen trees, and then fall to the ground as litter fall. The whole tree may die when outcompeted for the growing space by trees or plants, or from a disease, insects or other natural or anthropogenic disaster, or just from old age! At this point, with the mortality of the tree or tree part, this structure and its tissues of fixed photosynthate will be mostly oxidized through microbial decomposition back to CO_2, which can then return to the atmosphere, beginning its cycle over again (Fig. 6.2). It has been suggested that the life cycle of a carbon dioxide molecule (going through all four spheres – atmosphere, hydrosphere, biosphere and lithosphere) is hundreds of years relative to 2000 years for an oxygen molecule and 2,000,000 years for a water molecule (Laporte, 1975).

How carbon is cycled within an individual tree needs to be discussed next since disturbances modify carbon cycles. When a human or natural disturbance shifts the balance between the carbon-capturing process of the plant and how much carbon is needed to maintain plant tissues, a plant may become uncompetitive and not be able to continue growing at the site. When plants are unable to fix sufficient carbon to maintain their tissues, the carbon cycle shifts towards having more dead tissues (and more food for the microbes or detritivores) than live tissues (see Chapter 4, this volume). An ecosystem characterized by having more carbon moving through the decomposer community than in growth and maintenance of live tissues is considered to be unhealthy. Some other vegetative community will typically replace these decomposer-dominated ecosystems since the existing vegetation is not competitive at accessing the resources they need for growth.

In forests, human activities are able to directly shift the balance between how much carbon is stored and how much is released. Forests are very amenable to having their carbon cycles manipulated even though how much of a change will occur for any given site has been difficult to predict. Even the process of fixing carbon by individual plants has been modified by humans when:

- they select particular species of plants to grow in nature that are more efficient at fixing carbon (i.e. have higher photosynthetic rates) or respond more to fertilization applications; or

- plants are selected for planting that convert a higher proportion of their biomass into the part of the tree desired by humans as wood products (e.g. the bole or main stem of a tree versus allocating carbon to roots).

How trees fix CO_2 from the atmosphere will be discussed next.

Photosynthesis and respiration: carbon dioxide and trees

While trees are alive, they grow and accumulate biomass as they age. This accumulation of biomass is achieved through the fixation of carbon by a process called photosynthesis. Photosynthesis is the process of combining carbon dioxide (CO_2) from the atmosphere and water (H_2O), generally from the soil, into simple sugars ($C_6H_{12}O_6$) by utilizing the energy of light. Photosynthesis can be expressed simply in the following equation:

$$6CO_2 + 6H_2O + \text{solar radiation} \rightarrow C_6H_{12}O_6 + 6O_2 \qquad \text{[Equation 6.1]}$$

Besides solar radiation, the tree also needs chemical energy to grow. This energy is produced within the tree by the process of respiration, in which some of the sugars produced from photosynthesis are broken down back into carbon dioxide and water. So respiration is the use of oxygen (O_2) to break down organic compounds metabolically for the purpose of releasing chemical energy:

$$C_6H_{12}O_6 + 6O_2 \rightarrow 6CO_2 + 6H_2O + \text{chemical energy} \qquad \text{[Equation 6.2]}$$

Respiration is similar to the process of burning or combusting of organic matter, in which organic material when exposed to oxygen and heat can burn and release CO_2 to the atmosphere.

The sugars produced by photosynthesis can then move around the plant and be used to produce a multitude of tissue components that are needed by the plant to help build the various plant structures.

Measuring the amount of carbon found in forests is challenging because of our inability to measure the amount of carbon taken up by different forests and how this carbon is mobilized within the plant to grow new tissues or to maintain plant tissues already formed. The current method for cataloguing carbon is so labour-intensive that alternative approaches have been used to estimate the carbon cycle within individual trees. These individual tree data are then transferred to a stand of trees and across regional landscapes. How carbon is measured (biomass is a surrogate for carbon) is discussed in the next section since this is a difficult value to obtain. New techniques are evolving using remote sensing and satellite imagery to determine the amount of carbon in a forest (see Case 6.1 – Measuring Carbon in Forests Using High-resolution Digital Imagery). It is critical that these new techniques are developed because of the difficulty of assessing forests on a global scale and the fact that not all countries have the resources to quantify their forest area.

Net primary production: the amount of carbon produced annually by a forest

A common unit used by ecologists to measure plant growth is net primary production (NPP). It is important to understand how and what is included in NPP measurements. A brief review of NPP and how it is measured is presented next.

NPP is the total amount of plant material produced by a forest per unit area of land (typically this is a hectare) per year. This measure does not express forest productivity in terms of harvestable products, and thus has more value for forest carbon sequestration calculations than for commercial forest management. Furthermore, NPP is a useful measurement that allows comparisons among regions. The measurement of NPP is also valuable to obtain information on factors influencing productivity by observing how trees allocate growth on various sites and how different species allocate their growth on a particular site.

NPP can be expressed as:

$$NPP = GPP - R_a \hspace{4cm} \text{[Equation 6.3]}$$

where GPP is the gross primary productivity or total photosynthetic assimilation for a unit of time (usually a year) and R_a is the total metabolic respiration for the same year. NPP has long been recognized as the only complete measure of ecosystem dry-matter production (e.g. Assmann, 1970; Whittaker, 1975). Essentially, NPP is the amount of organic matter produced per unit area per year minus the organic matter lost from this production annually.

Since the terms GPP and R_a are difficult to measure, NPP is usually expressed as:

$$NPP = \Delta B + D + G \hspace{4cm} \text{[Equation 6.4]}$$

where ΔB is biomass increment or any increase in dry weight of living green plants over a year, D is annual production of detritus, including litter fall, fine-root turnover and tree and shrub mortality, and G is annual consumption of plant material by herbivores during that year.

NPP may be expressed either in calories, as the amount of energy required to fix a given amount of carbon or as the mass of organic matter produced as described above (i.e. mass per unit area per year). On average, 1.0 g of carbon is equivalent to about 2.2 g of organic matter or about 9.35 kelvin calories (Whittaker, 1975).

Biomass increment is the component of NPP that most concerns forest managers. Biomass increment is the annual increase in dry weight and volume of wood, bark, foliage and coarse roots. Biomass increment occurs most rapidly in young stands, and can approach zero in over-mature stands (Grier and Logan, 1977). This fact is one of the reasons that over-mature stands do not sequester carbon at the same high rates that can be found in younger stands.

Another major component of NPP is termed 'detritus production'. Detritus production incorporates mortality, litter fall and the turnover of the herbaceous understorey, fine roots and mycorrhizae, which develop, mature and die in one growing season. The amount of detritus produced by a forest varies with stand age and stand developmental stage or the structural and functional changes that a forest goes through with time (Oliver and Larson, 1995). Actively growing young stands in the stand initiation stage of stand development have relatively low production rates of detritus, with reported values ranging from 15 to 25% of annual NPP (e.g. Fujimori *et al.*, 1976; Turner, 1977). When a stand reaches the stem exclusion stage of stand development (when the competition between individual trees increases sufficiently for many trees to die), detritus production increases significantly. Grier *et al.* (1981), for example, reported that detritus production was 25% of NPP in a 23-year-old *Abies amabilis* (Pacific silver fir) stand. This means that, of the total amount of carbon fixed by this stand on an annual basis, a quarter was lost from the plant as dead plant tissues.

The importance of detritus production also increases with stand age. Net production in several old-growth Douglas-fir stands was entirely detritus (Grier and Logan, 1977). When detritus production becomes so dominant in a stand, this means that most of that ecosystem's energy is transferring through the decomposer community. At this stage, the existing vegetative community will be outcompeted by other plant species and high mortality of the overstorey trees will result. The ecosystem will not survive in this landscape because the formerly dominant trees are no longer healthy. Trees are no longer healthy due to their inability to fix enough carbon to maintain their physiological processes or to protect themselves against insects and pests (see Chapter 5, this volume).

Grazing (or herbivory) is the final term in the NPP equation, and the one that is often the most difficult to measure. In forests, grazing generally accounts for only a small portion of net production (Grier *et al.*, 1987). Under extreme conditions, however, grazing not only becomes an important term in estimating NPP but can also reduce future productivity and cause considerable mortality of the grazed species. When grazing by insects or mammals increases above the normally low 2%, the forest is unhealthy (Chapter 5, this volume).

Below-ground production is a component of ecosystem productivity that is often ignored. Despite a considerable body of information on above-ground productivity, information on below-ground production is exceedingly sparse (Vogt *et al.*, 1986, 1996). Below-ground productivity, however, is highly variable and can range between 30 and 70% of the total forest net primary productivity (Grier *et al.*, 1981; Vogt *et al.*, 1996). This variability can be attributed to differences in site quality (e.g. the amount of water and nutrients available for plants) and stand age. For example, a higher proportion of NPP is allocated below ground as site quality decreases (Keyes and Grier, 1981) and stand age increases (Grier *et al.*, 1981). Plants allocate more of the fixed carbon to the root system in order to enable them to grow more extensively, thus increasing their ability to search for nutrients and water when these resources are less available for plant growth. When more below-ground resources are available to the already existing root system, plants allocate less carbon to maintaining a fine root system (Gower *et al.*, 1992).

In summary, NPP varies with stand development and is highest at the time of canopy closure in young, newly established stands (Vogt *et al.*, 1983). During the stand initiation stages of stand development, NPP is largely a function of how much biomass is added by trees to its supportive structure (i.e. the main stem or the bole of a tree). Once a forest becomes older, the amount of plant tissues lost to detritus production increases as a greater proportion of fixed carbon is used in maintaining older tissues and little goes into producing new tissues. The proportion of NPP in detritus production reflects an inability of a plant to maintain its tissue functions, which results in a greater proportion of the plant tissues dying (i.e. higher rates of dropping foliage as litter fall or branches to the ground as dead plant parts).

How data are collected for determining NPP

Several different methods or approaches with differing efficiency and reliability have traditionally been used to estimate the different parts of the tree needed to estimate

NPP, using the equations given above. The most direct method of estimating NPP is the harvest technique. This is a very labour-intensive approach to determining NPP since plants need to be harvested during the peak growing season or several times during the growing season and separated into all their different tissue types (Newbould, 1967; Woodwell and Whittaker, 1968). This information is used to develop regressions that allow the same type of data to be collected in another forest for comparative purposes.

Regression equations are based on direct measurements of sizes and weights of plants and plant parts. Allometric equations are regression equations that describe the growth (weight and volume) of the total plant in response to changing proportions, such as diameter of the tree stem taken at breast height (dbh) and the height of the tree. A problem with the regression analysis approach is that it does not account for changes in growth rates and growth forms caused by changes in age, site or management (Grier *et al.*, 1984). Allometric equations tend, therefore, to be site- and stand-specific.

Forest age, tree species and carbon sequestration

An understanding of the pattern of forest growth with stand development and how different tree species differ in their carbon sequestration potential is fundamental for determining the role of forests as sources and sinks in the global carbon budgets. A source is where the carbon is initially obtained from, such as the atmosphere (e.g. trees take up atmospheric CO_2 during photosynthesis). A sink is a carbon pool where the carbon accumulates and is used to maintain organisms or to produce biomass (e.g. tree biomass is a sink for atmospheric carbon). Forest stands have different growth rates, depending on their age and the nutrient supply capacity of the soil. In an even-aged forest (or in forests where the trees are of the same age), growth and biomass accumulation decline after reaching a peak relatively early in a stand's life (Assmann, 1970; Vogt, 1991). These decreases in growth generally coincide with the peak in the stand's leaf area or the photosynthetic area of the plant.

The age of a forest determines how much carbon is sequestered. If only live trees were used to sequester carbon, the solution to increasing carbon sequestration would be to convert most of our forests into young stands and to eliminate older forests. You would arrive at this conclusion since young forests take up more carbon annually than an old forest (Table 6.2). Carbon sequestration, however, should not just include live trees but also dead tissues of plants. If these dead tissues were included in the amount of carbon sequestered, old-growth forests are ideal locations to sequester carbon in tissues that were formed some 200 years ago (Table 6.3). It is important to

Table 6.2. General summary of forest carbon storage and uptake of carbon annually in a young and an old forest (from Vogt *et al.*, 1996).

	Storage (Mg/ha)	Uptake annually (Mg/ha/year)
Young forest	50–100	5–10
Old forest	400–1000+	~1.0

Table 6.3. Biomass data (Mg C/ha) for a 60-year-old Douglas-fir stand and a 450-year-old Douglas-fir/hemlock stand in Oregon (from Harmon *et al.*, 1990).

Component	60-year-old stand	450-year-old stand
Tree	194.1	432.1
Fine and coarse woody debris	10.9	123
Soil carbon	56	56

remember that the carbon found in an older forest has taken several centuries to accumulate and is beyond the time scale of the typical human lifespan.

The literature shows that this decline in growth rates as forests become older always occurs, but the age at which tree growth slows down does vary (Ryan *et al.*, 1997). In general, more productive stands (i.e. those that accumulate biomass at a faster rate early in the growth of a stand) reach a peak or optimal growth at a younger stand age (e.g. 20–30 years old) (Ryan *et al.*, 1997). If the site has less nutrients and water available for plant growth, tree growth rates may not slow down until the forests are 40 to 50 years old.

Some scientists suggest that, if we want to increase the amount of carbon stored in forests, forest landscapes should be comprised of more young stands. The rationale for having more young stands in forest landscapes is due to the fact that tree growth is higher in younger forests and slows down when trees are 20 to 40 years old (Table 6.2). If no consideration is given to the age of trees in a forest, old-growth forests do have a high amount of total carbon stored (mostly in dead wood) but this carbon accumulates in the forest over a 200-year-plus time period. If one wants to store more carbon during a human lifespan, growing younger stands of trees is appropriate; however, the carbon sequestration potential of these forests is negligible if these lands were previously occupied by older forests that were cut to make room for the younger forest. As part of the Kyoto protocol (more later in this chapter), carbon credits can be obtained for afforesting previously degraded pastures or agricultural lands but not if forests existed on that land during the last 50 years and were cut down to afforest that land.

The species of tree that is growing at any site will also determine how much carbon will accumulate in any given site. Many examples exist of exotic or non-native species being planted around the world because they have faster growth rates. These faster growth rates mean that they sequester more carbon in biomass during their early years of growth. Many plantations are comprised of exotic species because of their higher NPP when compared with native tree species.

It has been difficult to measure the environmental impacts of native versus exotic species, but in Iceland a well-documented history has allowed comparisons to be made on their impacts (see Case 6.2 – Carbon Sequestration in a Boreal Forest in Iceland: Effects of Native and Exotic Species). In this Iceland case, the introduced species sequestered higher levels of carbon than the native birch. The exotic species also appeared to be more effective at acquiring nutrients from the soil horizons, raising questions about the long-term impacts of introducing plants into new environments. The exotic species also changed the understorey composition in the Iceland forests, which has implications for maintaining the biodiversity in native Iceland forests.

Why carbon accumulates in the soil

Soil organic matter can be divided into three different categories: (i) simple substances, such as small-chain phenolic compounds, amino acids, sugars and other small molecules; (ii) identifiable high-molecular-weight compounds, such as polysaccharides, proteins and lipids; and (iii) humic substances, a series of high-molecular-weight compounds formed by condensations of smaller organic radicals by secondary synthesis reactions (Fisher, 1995). The first of these groups of compounds provides a ready source of energy for soil organisms. The second group is decomposed to yield members of the first group and to provide building blocks for the third group. It is this latter group of compounds that gives soil organic matter most of its unique properties.

The humic substances are defined as naturally occurring, biogenic, heterogeneous organic compounds, which can generally be characterized as being yellow to black in colour, of high molecular weight and refractory (Aiken *et al.*, 1985). Humic substances are higher in lignin, phenolic-type compounds and waxes. Humic substances are important in the soil for several reasons. They provide buffering over a broad range of soil acidity, since much of the soil exchange capacity may be due to colloidal humic substances. Humus greatly affects soil physical properties, such as aeration, aggregate stability, water-holding capacity and permeability (Stevenson, 1982).

Soil organic matter is further separated into two types of carbon pools: the soil organic carbon (SOC, henceforth referred to as soil organic matter = SOM) pool and the soil inorganic carbon (SIC) pool (Lal *et al.*, 1995). The SOM pool is mostly concentrated near the soil surface, usually within the top 1 m of the soil, and has been estimated to be about 1550 Gt globally (Eswaran *et al.*, 1993). The SIC pool is composed of mainly calcium carbonate ($CaCO_3$) or caliche and is often deeper in the soil and in sediment layers (Lal *et al.*, 1995). The SIC pool has been estimated to contain about 1700 Gt of carbon (Post *et al.*, 1982; Schlesinger, 1991).

Both external and internal factors have been found to have an effect on the accumulation and stability of organic matter. These external factors include temperature, precipitation, vegetation and disturbance (Greenland *et al.*, 1992). The internal factors or intrinsic soil properties that affect the accumulation and stability of soil organic matter include:

- the amount of aluminium and iron in the soil;
- the kind and amount of clay-sized minerals;
- soil pH; and
- the abundance of base cations, in particular Ca^{2+} (Johnson, 1995).

It is assumed that, when other internal and external factors stay comparable, soil organic matter increases with cooler soil conditions, moister soil conditions and improved soil fertility but decreases with disturbance and aeration (Jenny, 1980). The amount of variation in soil temperature, rather than average temperatures, has been found to be of greater importance for determining how much carbon is sequestered in the soil (Buol, 1991). For the same mean annual temperature, soils that have less than a 5°C difference in winter and summer soil temperature at a 50 cm depth tend to accumulate more carbon (Buol, 1991).

Soil carbon has generally been considered a stable resource and, when changes do happen, they occur very slowly through time (e.g. Schlesinger, 1990; Aradóttir *et al.*, 1992). Recent research, however, has shown that this is not necessarily the case. Various plant species, or plant types, seem to have different inherent characteristics that affect carbon sequestration in the soil (Vogt *et al.*, 1995, 1996).

The rate of accumulation in newly developing soils characteristic of primary successional systems can, however, be very low. Schlesinger (1990) reported a long-term soil carbon accumulation rate for newly formed land surfaces (e.g. mudflows, retreating glaciers) of 2.4 g C m^2/year. Higher accumulation rates have been recorded in soils that have legacies remaining from the previous vegetation communities that grew on the site. For example, Alexander *et al.* (1989) reported soil accumulation rates for forested soils in south-eastern Alaska ranging from 29 to 113 g C m^2/year. In managed tropical systems, accumulation rates as high as 120 g C m^2/year have been reported (Lugo, 1991).

The amount of organic matter that accumulates in soils is highly variable and depends on many factors. As a generality, soils rich in organic matter tend to be more productive than organic-poor soils. Despite soils higher in organic matter being considered more productive, there is seldom a strong correlation between total organic matter content and crop yield (Fisher, 1995). In wetter soils, growth often declines as soil organic matter increases because of anaerobic conditions, which limit decomposition and the availability of O_2 in the root zone; a condition commonly found in bogs. In drier soils, the reverse is true and, if organic matter is added, it increases the water- and nutrient-holding capacity of the soils.

Today managing the carbon storage pools in vegetation and soils is insufficient to allow forests to provide services and goods for society

Information on the carbon storage pool sizes for forests is useful to have as a measure of the carbon sequestration potential of any given forest. This information can be used to estimate how much of the carbon emitted during fossil-fuel combustion can be mitigated by growing trees and to calculate how much more area should be planted in trees. This information does not, however, address other aspects of the carbon cycle that are important to know.

Silvicultural manipulation of forests for carbon alone, for example, will not allow you to determine how long wood will sequester carbon in a forest when its tissues are dead (e.g. decaying large logs). To obtain this information requires an understanding of the chemistry of plant tissues, since the rate at which biomass decomposes depends on its chemistry. Some of these chemicals are broken down very quickly (e.g. sugar compounds break down in a matter of minutes), while others are more difficult to break down and may even persist for several hundred years (lignin persists in a decaying log from years to several centuries). The importance of managing global carbon cycles at the plant chemistry level is discussed next. Managing carbon pools in forests alone is important but forests can have an even greater impact on restoring the imbalances that currently exist in global climates when wood is used as a renewable resource. This can be done by managing plant chemistry for carbon-based products that can substitute as one alternative for carbon-based fossil fuels.

The Importance of Managing the Carbon Cycle and Mitigating Climate Change by Managing Plant Chemistry

It is the presence of complex chemical compounds (e.g. cellulose, lignin) in wood that is providing forests new opportunities to help mitigate CO_2 emissions. The chemistry of wood allows an amazing number of different products to be manufactured that can substitute for fossil fuels. Since wood and fossil fuels are all carbon-based, they can be used to produce similar products.

Prior to explicitly discussing the new ways that plant chemistry is being used to mitigate CO_2 emissions into the atmosphere from fossil-fuel combustion, it is worthwhile to briefly discuss the complexity of carbon compounds that can be found in trees. Wood can be transformed to many different compounds because of its chemistry. The different chemical compounds found in wood or biomass can be converted to different products because of the ease with which they can be broken down to their molecular constituents.

Tissue components: complex carbon compounds used in structure and function

Most plant tissues, especially the structural components, are mainly composed of carbon compounds, such as hemicelluloses, celluloses and lignin. The proportions of these components in plants vary and are dependent upon the type of plant species, the structural part of the plant and its maturity (Table 6.4). Trees have in general more lignin in their tissues, while many agricultural crops have less lignin. The high sugar and lipid contents of agricultural crops are the reason that they have been used to produce ethanol or bio-oils.

Only a small fraction of the total plant chemistry, however, has been used to produce liquid products. For example, sugarcane has been used to produce ethanol but

Table 6.4. The chemical composition of wood and agricultural crops as a per cent of the total biomass (from K. Vogt, unpublished; www.fpl.fs.fed.us/documents/pdf1998/han981.pdf; www.wsu.edu/~gmhyde/433_web_pages/ 433Oil-web-pages/Soy/soybean1.html; www.soyohio.org/edu/facts.cfm; www.ctic.purdue.edu/Core4/StoverNCNU.pdf; www.chemeng.lth.se/exjobb/025.pdf).

Chemical component	Wood – conifers	Wood – deciduous	Soybeans	Maize stover	Wheat
Cellulose	37–47	35–51	17	38	29–51
Hemicellulose	15–27	15–21	–	32	26–32
Lignin	25–29	23–29	5	20	16–21
Lipids	–	–	20	–	–
Sugars	–	–	7	5–11	–

its extraction efficiency was only 10% of the sugarcane biomass (only the sugar compounds can be converted to ethanol). In the past, the cellulose in plant tissues has been used to make paper products but not liquid products. This inability to use cellulose has reduced the efficiency at which waste products from agriculture or forests could be converted to new products. Now genetically modified organisms are being developed to convert cellulose to sugars so that more of a plant's biomass could be efficiently converted to other liquid products. Historically, the different chemical components of plant tissues were individually extracted and converted to products but not the entire plant tissue because the other chemicals could not be converted and they impeded this conversion. So the chemical composition of biomass is important in determining what products can be produced from biomass.

All these chemical compounds are found in wood at the same time. Functionally hemicellulose is a carbohydrate or sugar compound found in the cell walls of herbaceous and woody plants along with cellulose and lignin. Cellulose is a carbohydrate that strengthens the cell walls of most plants and is the principal component of wood. Lignin then acts as a binder for the cellulose fibres in wood and certain other plants and also adds strength and stiffness to the cell walls. So hemicellulose and lignin form the matrix in which the cellulose fibres are embedded.

In the past, manufacturing products from biomass required each of these compounds to be selectively separated from one another. Separating compounds so that one chemical component of the biomass could be selectively extracted has been difficult and used to require the use of strong acids or bases. On the other hand, the presence of different compounds is also what allows so many different products to be formed from wood (e.g. cellulose to make paper or to form rayon; lignin is the main component remaining after biomass wastes are composted).

An understanding of the different types of compounds found in wood helps to explain much of the carbon cycling that occurs in forests. For example, sugar-based compounds are readily used by microbes as a food and therefore disappear quickly when they become available in an ecosystem. This contrasts with compounds like lignin, which require specialized organisms to break the bonds holding these compounds together. These compounds can persist for many decades or centuries in a forest and contribute to the carbon sequestration that is found here (e.g. logs persist in old-growth forests and their slow decay rate is due to their lignin content). Coniferous or evergreen tree species have higher lignin contents in their tissues. These forest types are also found to sequester higher amounts of carbon in their tree biomass (Vogt *et al.*, 1996). Other properties of lignin enable it to be used to control insects and pests. The polyphenolic ring structure of lignin is ideal as a pesticide since it persists in environments long enough to kill the target organisms.

The characteristics of hemicellulose, cellulose and lignin will be briefly discussed next. This overview demonstrates the unique characteristics of each and explains why they were historically used to produce particular products.

Hemicellulose

Hemicelluloses are short, highly branched chains of five- and six-carbon sugars and uronic acid. Hemicellulose is a polymer of five different kinds of sugars, in contrast to cellulose, which is a polymer of only one type of sugar (i.e. glucose). The branched nature of hemicellulose renders it amorphous and relatively easy to hydrolyse (a chemical reaction in which water is used to break down a compound) into its constituent sugars.

When hydrolysed, the hemicellulose from hardwoods releases products high in xylose (a five-carbon sugar), whereas the hemicelluloses contained in softwoods yields more six-carbon sugars (e.g. DOE, 2004). Because these simple hemicellulosic sugars are easy to break down, they do not need complex enzymes to break them apart like the other more complex compounds, such as cellulose and lignin.

Cellulose

Celluloses are a series of glucose sugar molecules linked by bonds between the first and fourth carbons of two adjacent glucose molecules. However, unlike hemicellulose, cellulose does not easily dissolve and only then in strong alkalis and some acids. Celluloses are resistant to chemical attack because of the high degree of hydrogen bonding that can occur between the adjacent chains of cellulose. Hydrogen bonding between these adjacent cellulose chains makes the polymers more rigid and imparts a three-dimensional micro-fibril structure to cellulose. Cellulose, however, can be broken down by heat as well as by some bacteria (such as those found in the gut of cattle and horses) and fungi.

One type of fungi that can utilize the cellulose and hemicellulose portions of wood but are incapable of decomposing lignin are called brown-rot fungi, which produce a rotting wood characteristic of a blocky, crumbly, brown-coloured appearance. When organic matter is composting or decomposing, the cellulose is broken down into sugar molecules by bacteria and fungi using a special cellulase enzyme. These separated sugars are then used as a carbon and energy source for the microorganisms.

Industrially, lignin can be removed from wood chips, leaving behind the cellulose, which can then be used to make paper. When making paper, strong bases or acids are used to separate cellulose from the other chemical compounds. The need to use these strong acids or bases has resulted in environmental pollution of waterways when the waste products were dumped into rivers. This practise occurs less frequently now and new and improved approaches have been developed to separate cellulose from other compounds (e.g. hydrogen peroxide).

Wood is composed of 40% to over 50% cellulose, depending upon the tree species and structural part. In fact, wood may have as high, or even higher, concentrations of cellulose as many crops; however, there are some crops, such as cotton, which can have cellulose comprise up to 98% of its tissue chemistry (Table 6.4).

Using all this information about cellulose, you can understand why your blue jeans could easily get holes in them when you are in a chemistry lab. Remember that most jeans are made of cotton, which is primarily composed of cellulose. And cellulose is a compound made up of chains of sugar molecules that are acid-soluble. So the acid that your lab partner spills on your jeans can then break up the bonds and dissolve the cellulose – resulting in the hole in your trousers and possibly a sensitive spot on your leg.

Because fibres of cellulose are found everywhere in nature, it is not surprising to find that it was used to make some of our first synthetic polymers (e.g. cellulose nitrate, cellulose acetate and rayon) (PSLC, 2003). For example:

- Cellulose nitrate, also nitrous cellulose or nitrocellulose, was originally used to make plastics (an early imitator of ivory), motion picture film (but highly explosive and caused lots of fires in cinemas) and clear lacquer coatings (e.g. a protective coating for furniture, musical instruments and many other wood objects).

- Cellulose acetate is not explosive and has now replaced cellulose nitrate in motion picture film. It is also used to make film negatives, print film and clear plastic sheets.
- Rayon was originally a fibre made from cellulose nitrate and therefore very flammable. However, rayon is now made from cellulose xanthate, which is much safer and less flammable. Rayon is extensively used for fabric material, since it has qualities of natural plant fibres. It has a smooth texture and is shiny like silk (initially rayon was an inexpensive substitute for silk).

Lignins

Lignins are based on a six-carbon ring structure, as found in benzene, and is composed of complex polyphenolics (similar to what is found in pesticides). Lignin is what makes plants woody and allows trees to grow tall (for example, cellulose would not make a plant woody). Lignin is what is removed from wood biomass when making paper.

Lignin is highly resistant to microbial attack and comprises a significant portion of what remains in the soil after microbial decomposition of plant materials (i.e. organic matter). Lignin is also what is left after composting plant waste materials.

The structure and composition of lignin make it very resistant to decay. In fact, only very specialized organisms with complex enzymes are capable of breaking down lignin. One type of fungus that is capable of utilizing lignin, as well as cellulose and hemicellulose, as a food source is the white-rot fungus. So lignin is not only involved in the structural support of the plants, but it is also part of the plants' chemical defence against insects and most fungi.

Decomposition (and composting): converting complex carbon compounds into their molecular forms

In plant succession, there is generally a progressive accumulation of organic matter (carbon) and energy as the ecosystem matures from the pioneer stage to a climax stage (see Chapter 4, this volume). In the soil ecosystem, this accumulation of carbon and energy is reversed, where organic matter and energy become depleted as decomposition proceeds through a succession of microbes. These contrasting functions highlight the importance of decomposition to the carbon cycle.

As mentioned in the preceding section, different tissues have different degrees of resistance to microbial decomposition. And, as there may be a succession of plants in an ecosystem, there may also be a succession of microbes during the decomposition of the organic material (Richards, 1987). There is a certain succession of microbes that occur during decomposition, but generally it is dependent upon the environmental conditions (e.g. moisture, temperature, pH), the tissue quality (e.g. complex tissues, available nutrients) and the microorganisms present. Other conditions that may affect the rate of decomposition would certainly include the presence of organisms (e.g. mites, springtails, earthworms) that fragment the organic tissues into smaller pieces, creating larger surface areas on which the microbes can work.

Initially organic material is decomposed by the saprophytic 'sugar fungi' and bacteria, which utilize sugars and other carbon compounds simpler than cellulose (Richards, 1987). These fungi are then generally followed by cellulose decomposers and associated secondary sugar fungi and then finally by the lignin decomposers. Bacteria are, of course, present during all these stages but are mostly limited to the breakdown products of the fungi. However, bacteria may play an even more dominant role in decomposition in the more alkaline environments, such as in grasslands or hardwood forests, versus the more acidic conditions found in the coniferous forests.

Decomposition is a natural process affected by biological and non-biological (e.g. temperature, water, nutrients) conditions. Humans and some other organisms (e.g. leafcutter ants) have learned to mimic decomposition but have speeded up this process when garden wastes or food wastes are composted. Composting is the rapid (6 months compared with > 1 year) conversion of organic materials (e.g. leaves, branches, grass cuttings) to organic matter (humus), CO_2, water and nutrients. The conversion of the organic materials is mainly changing or converting the cellulose of the decomposing material to its more basic components (e.g. sugars), but without changing the lignin. When the composting process is finished, the material remaining is mainly comprised of lignin compounds. Composting occurs at very high temperatures, which are above 60°C during the first 10–20 days, and generally bacteria are responsible for transforming organic materials to compost.

Natural Patterns of Storing and Cycling of Carbon by Forest Soils and Vegetation

Natural patterns of carbon storage by forest climatic types

By summarizing global carbon budgets for forests, it becomes apparent that some parts of the world have a greater capacity to sequester carbon. The total amount of carbon sequestered in a forest over time is also not the same thing as the amount of carbon that a plant is capable of fixing annually (Table 6.5). This apparent discrepancy between the amount of organic matter sequestered in a forest and the annual amount of carbon fixed is due to the fact that fast-growing trees contribute a greater proportion of NPP to detritus (i.e. a larger proportion of plant tissues dies in a given year). For example, the NPP, or the amount of carbon fixed annually, is extremely high for subtropical needle-leaved evergreen forests but the total organic matter found in these forests is only half or less of what occurs in some other forests. These forests are good sources of timber supply since trees grow rapidly and can be harvested at a shorter time scale. However, they are not good sites to sequester carbon.

So, if the goal is to use forests to sequester carbon, forests in the cold temperate and tropical climatic zones have the greatest potential to sequester carbon compared with other vegetative communities and forests in other climatic zones (Table 6.5). In particular, forests comprised of the cold temperate needle-leaved species (e.g. coniferous forests in the Pacific North-west United States) and the tropical broadleaved evergreen forests are ideal locations to sequester carbon. These forests typically grow on poorer soils and may grow more slowly each year

Table 6.5.   Total living biomass, total organic matter and total NPP by forest biomes by climatic zones (from Vogt *et al.*, 1995).

Forest climatic biome	Total living biomass (metric tonnes/ha)	Total organic matter (metric tonnes/ha)	Total NPP (metric tonnes/ha/year)
Boreal broadleaved deciduous	78	322	7.5
Boreal needle-leaved evergreen	93	247	–
Cold temperate broadleaved deciduous	245	458	15.4
Cold temperate needle-leaved evergreen	367	585	16.2
Warm temperate broadleaved deciduous	170	357	14.8
Warm temperate needle-leaved evergreen	103	242	14.4
Subtropical needle-leaved evergreen	204	259	21.4
Tropical broadleaved semi-deciduous	359	437	24.5
Tropical broadleaved evergreen	332	666	14.5

(note that the NPP of both of these regions is only half of what was recorded in forests with the highest NPP, i.e. tropical broadleaved semi-deciduous) but accumulate more carbon in the long term.

If one is interested in using forests to sequester carbon, particular forest types are not good locations to pursue this goal (Table 6.5). For example, the boreal forests, deciduous forests in the cold and warm temperate climatic zones and needle-leaved evergreen forests in the subtropics may not be the ideal locations from a carbon sequestration viewpoint alone to grow trees because these forests do not sequester as much carbon in their standing biomass as other forest types.

Despite having lower total living biomass in the boreal zones, the higher total organic matter recorded here shows the important role of detritivores in these forests. Decomposition rates are slow in these cold climates, which means that detritus requires a long time to be decomposed, so that the total organic matter found here is higher than in other areas with more conducive climates for decomposers. Because of the lower decomposition rates, the soils also tend to have lower organic matter than what is found in warmer climatic zones (Vogt *et al.*, 1995). Whatever carbon does accumulate in the boreal climatic zone occurs over century time scales, so land-use activities that decrease the carbon stored in these forests will require long time scales to re-establish them to levels prior to the land-use activity. This is an important consideration for forests in Siberia, which are now being cut to provide timber for global markets.

Natural patterns of carbon storage capacity in soils

Organic carbon accumulates in soil if the rate at which C enters the soil exceeds the rate at which it leaves. In general, if litter fall is high and decomposition is low, organic matter accumulates. Some specific factors can increase the accumulation of humus and humic substances in soils by preventing further decomposition or movement. For example, clays can chemically bind to soil organic matter (SOM) and protect it from microbial decomposition by preventing enzymes from accessing binding sites. Calcium and other cations can also stabilize humus and humic acids by chemically binding to the SOM in a similar fashion to that of clays. High or low soil pH will slow decomposition, as these extremes are generally harmful to organisms and will decrease their numbers and activity. Similarly, if any highly toxic substances are in the soil, they will harm the microorganisms and decrease decomposition rates. Too little water or oxygen will also slow decomposition, as microorganisms need both materials to reproduce, thrive and be effective decomposers. Although there are organisms that can decompose without oxygen (anaerobic decomposers), their decomposition rates are very slow. This is the reason why carbon accumulates in waterlogged soils and wetlands. Conversely, any conditions that favour high decomposition rates (e.g. favourable soil temperatures and soil moisture, plenty of soil oxygen) will decrease C accumulation in the soil.

Overall, since most organic matter enters the soil near the surface, where most organisms occur, most organic matter is stored in soil near the surface. The major types of soil horizons, which are called the 'O', 'A', 'E', 'B' and 'C' horizons (horizons are soil layers that have distinctly different properties, such as quantity of SOM, colour, clay content, etc.) are shown in Fig. 6.3. 'O' horizons are typically found in forests and are composed almost entirely of litter (they are often called the forest floor). 'A' horizons are mineral soil that is high in organic matter (often called the topsoil); they are usually darker in colour, as SOM will darken the mineral material.

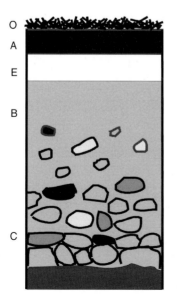

Fig. 6.3. Major soil horizons. O and A horizons are typically high in carbon.

'E' horizons may have many roots and some SOM, but usually are a zone where organic matter (OM) is not high. Most soils do not have an 'E' horizon, but they can be found in some forest soils. Some 'B' horizons may be high in SOM if they: (i) accumulated the SOM lost from an 'E' horizon (as soils are porous, soluble SOM can leach down into the 'B' horizon); or (ii) are high in clay or cations. 'C' horizons are typically at the bottom of the soil and contain little organic matter and few roots. Figure 6.4 shows some examples of different types of forest soils with the total amount of carbon in each.

Carbon can leave the soil in several ways. First, it can leave as CO_2. One of the products of decomposition is CO_2, which is a gas. Since soils are porous, the CO_2 released by decomposition can migrate up through its pores and escape to the atmosphere. This microbial respiration, in conjunction with root respiration, is the major pathway for the return of terrestrial carbon to the atmosphere. Second, erosion can also remove SOM and transport it into rivers, lakes or oceans. Since most SOM is located near the surface of the soil, this can be an important mechanism for loss of soil C. Third, a minor loss of soil C occurs when soluble forms of C (e.g. bicarbonate (HCO_3^-); carbonic acid (H_2CO_3) or soluble organic acids) are able to percolate through the entire soil profile and enter the groundwater. These soluble forms of C can subsequently flow into bodies of water.

All of these processes can result in highly varying amounts of carbon stored in soils. Globally, major differences occur in soil C depending on climatic factors and vegetation types. Generally, most forest soils have less C than grasslands or agricultural soils. Even among forest soils, there are large differences in the amount of carbon stored (see Fig. 6.5). Many moist tropical forest soils are low in carbon, even though large amounts of litter may enter the soil. This low carbon content in soils in the moist

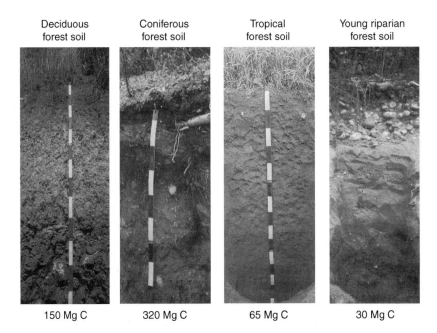

Deciduous forest soil	Coniferous forest soil	Tropical forest soil	Young riparian forest soil
150 Mg C	320 Mg C	65 Mg C	30 Mg C

Fig. 6.4. Soil profiles and the changes in soil organic matter for different soils.

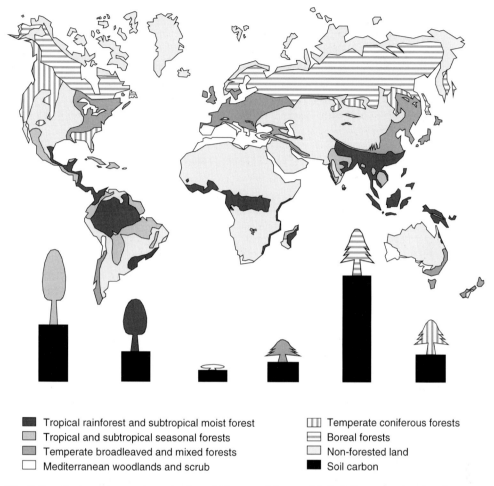

Tropical rainforest and subtropical moist forest

Tropical and subtropical seasonal forests

Temperate broadleaved and mixed forests

Mediterranean woodlands and scrub

Temperate coniferous forests

Boreal forests

Non-forested land

Soil carbon

Fig. 6.5. Carbon storage in major forest biomes of the world. Relative quantity of carbon in each forest (biomass C) and soil is indicated by the height of the tree or soil icon. Non-forested land includes grasslands, wetlands, deserts, savannahs, tundra and bare land. Carbon pools adapted from Lal *et al.* (1995). Approximate biome regions taken from WWF (2005). See Table 6.6 for specific values of each pool.

tropics is due to the moisture and the warmer temperatures favouring high decomposition rates. Conversely, boreal forests can have very high soil C contents even though litter-fall rates are low. In this case, decomposition rates are much lower, so that organic matter accumulates in the soils found in the boreal climatic zones.

The Altered Carbon Cycle due to Global Climate Change and the Connection to Forests

Global climate change has been classified as the major environmental issue of our times. During the past 100 years, the mean temperature on earth has increased by

Table 6.6. Carbon pools in forest biomass and soil carbon by major forest biomes. Percentages of each pool relative to total global C in biomass and soil, respectively, are also included. Data are taken from Lal *et al.* (1995).

Forest biome	Forest C (Pg)	% of global biomass C	Soil C (Pg)	% of global soil C
Tropical evergreen and woodlands	122	21%	82	6%
Tropical seasonal	169	29%	125	9%
Temperate evergreen	81	14%	68	5%
Temperate deciduous	48	8%	49	3%
Boreal	105	18%	241	17%
Temperate woodlands	7	1%	18	1%
Global total for all biomes	592	–	1431	–

about 0.5°C and the concentrations of greenhouse gases have increased by more than 30%. These changes have been attributed to human activities, which are implicated in changing the natural cycles of greenhouse gases in the atmosphere (Schlesinger, 1991; Walker and Steffen, 1999; IPCC, 2001). The two main causes of the global warming have been identified as: (i) burning of fossil fuels, responsible for 75% of the increase in greenhouse gas (GHG) concentrations in the atmosphere; and (ii) land-use change, particularly deforestation, which is responsible for about 25% of the increase in GHG emissions (Houghton, 1995).

Greenhouse gases are so named because of their ability to trap the outgoing infrared radiation from earth after absorbed solar energy warms the earth's atmosphere and surface. This trapping of heat is generally referred to as the greenhouse effect. Despite these concerns about climate change or global warming, it is important to remember that the greenhouse effect is a natural phenomenon and has been going on for millions of years. In fact, the habitability of the planet depends on these greenhouse gases never falling too low or rising too high. Without them, the average Earth surface temperature would be –6°C. Comparatively, at current levels of greenhouse gases, the earth's average surface temperature is about 15°C.

Important greenhouse gases in the atmosphere include water vapour (H_2O), carbon dioxide (CO_2), methane (CH_4), nitrous oxide (N_2O), oxides of nitrogen (NO_x), ozone (O_3) and chlorofluorocarbon (CFC). Carbon dioxide is the most abundant of these gases and global warming is to a considerable extent attributed to the substantial and steady increase in the anthropogenic emissions of CO_2 since the onset of the industrial revolution (Houghton, 1995; Lal *et al.*, 1995). The global warming potential of the different greenhouse gases is usually represented as a carbon dioxide equivalent (CO_2e), where carbon dioxide is the reference gas against which the other greenhouse gases are measured.

There are many sources of carbon that are increasing atmospheric carbon contents. Some of the following contribute to increasing the amount of carbon in the atmosphere:

- When fossil fuels are combusted, a significant amount of carbon is emitted into the atmosphere since fossil fuels are high in C (70–80% C).

- When forest fires occur, trees combusted during the fire release a significant amount of carbon into the atmosphere (50% of the biomass of a tree is carbon).
- Deforestation of large areas of forest land will increase the evolution of CO_2 into the atmosphere (FAO estimated that deforestation globally emitted 1.6 Gt of C per year; FAO, 2001).
- Volcanoes add CO_2 to the atmosphere during volcanic eruptions – this is a natural source of carbon.
- Caliche or calcium carbonate deposits in desert areas today volatilize CO_2 when rainfall that is acidic falls on the deposits.

How can Society Manage Carbon in Forests?

If the emissions of CO_2 into the atmosphere are to decrease, many of the sources of this emission have to be curtailed. The continuing high global deforestation rates are decreasing the carbon sequestration capacity of existing forests (Table 6.7). Since trees soak up or sequester more carbon than they release, forests are classified as being carbon sinks ('sink' is used for the processes that remove CO_2 from the atmosphere; UNFCCC, 1992). This sequestration of carbon is denoted as the capture of CO_2 in a manner that prevents it from being released into the atmosphere for a specified period of time.

If the carbon sequestration rates of forests are to increase, deforestation rates have to decrease. As already discussed in Chapter 2 and will be discussed in Chapter 7 of this volume, forest exploitation (i.e. deforestation) is occurring in those regions of the world where forest biomass is higher because these areas were not as hospitable for human survival in the past.

Other approaches to sequestering carbon in forests result from planting forests in areas where they have not existed for at least 50 years. Since afforestation, reforestation or management practises can increase the amount of carbon stored in vegetative biomass instead of the atmosphere, these approaches are effective in increasing the 'carbon sink'. Afforestation is defined as the establishment of forest plantations in areas not previously containing forests for the past 50 years, such as on bare or cultivated land. Reforestation is defined as the establishment of forests

Table 6.7. Above-ground biomass (including trees, shrubs and bushes) and the per cent change in forest cover between 1990 and 2000 for different regions of the world (from FRA, 2000).

Region	Above-ground biomass (t/ha)	Forest cover change 1990–2000 (%)
Africa	109	−0.78
Asia	82	−0.07
Oceania	64	−0.18
Europe	59	+0.08
North and Central America	95	−0.10
South America	203	−0.41

(e.g. through planting or seeding) after a temporary loss of the forest cover. Even though afforestation is seen as an important approach to increasing the sequestration of C in forests to decrease atmospheric CO_2 levels, the amount of land area converted into a forested condition has not reached its potential. The continued loss of forests to wildfires continues to decrease the amount of land in forests and afforestation has not been able to replace even the amount lost in these fires.

Any discussion of the global sources and sinks of carbon has to evaluate the success of current practices to mitigate the emissions of CO_2 into the atmosphere. A focus on only sequestering carbon will be unable to mitigate CO_2 emissions into the atmosphere if current rates of energy consumption and deforestation rates continue. Today, using forests to sequester carbon will only mitigate a small fraction of the carbon released during fossil-fuel combustion. In the United States, forests and land use mitigated 19% of all the CO_2 emitted during the combustion of fossil fuels in 1990 (Fig. 6.6). However, in 2002, forests and land use mitigated only 12% of the CO_2 that was emitted during the combustion of fossil fuels. So the ability to use carbon sequestration in forests and land use is decreasing in the United States.

Much of the effort in the European Union has been towards supplementing or substituting biofuels, methanol or ethanol for petrol or diesel versus direct sequestration of carbon in forests. The European approach has been driven by the fact that the transportation sector produces such a high proportion of the CO_2 being emitted into the atmosphere.

It is important to recognize that the discussion should not be over whether to choose carbon sequestration projects over producing biofuels to replace fossil-fuel-based transportation fuels. All of these approaches will be needed if human attempts at mitigating the effects of greenhouse gases are likely to succeed.

The opportunities existing with using biomass to produce environmentally friendly energy will be discussed next since this area has tremendous potential to reduce the emissions of fossil-fuel-based greenhouse gases. A discussion of the industrial sectors that produce greenhouse gases will introduce the topic of why biomass is ideally suited to these efforts.

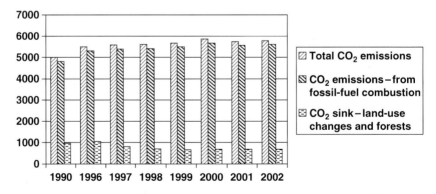

Fig. 6.6. United States CO_2 emissions and the proportion that land-use change and forestry provided as a sink in 1990 (19% of total) and 2002 (12% of total) (EPA, 2004).

Fossil-fuel combustion significant source of atmospheric carbon emissions and why climate-friendly forestry has a strong future

Principal sources of anthropogenically induced carbon dioxide emissions have been identified as the release of carbon from the non-circulating reservoir (e.g. burning of fossil fuel) and the degradation of the terrestrial reservoir due to deforestation and land use (Houghton, 1995). Burning of fossil fuels releases about 5.4 Gt C annually to the atmosphere, while the deforestation of tropical rainforests and other land uses have been estimated to emit about 1.6 Gt carbon annually (Lal *et al.*, 1995). Major sinks for this released carbon dioxide are: the atmosphere, where 3.2 Gt of carbon are annually added, and the ocean, which absorbs an additional 2.0 Gt of carbon per year (Lal *et al.*, 1995). The remaining 1.8 Gt/year, which is often termed 'the missing carbon', is unaccounted for, but is hypothesized to be absorbed by terrestrial eco-systems (Sundquist, 1993).

Three-quarters of the carbon emissions into the atmosphere are from fossil-fuel combustion. Fossil fuels are fuelling and maintaining the development of industrial-ized countries. They have allowed industrialized countries to obtain cheap energy and to maintain a high standard of living. Today, however, there is not enough oil to continue to support all of its uses. In the United States, 65.5% of the oil supply is imported (DOE, 2003). The United States has not been self-sufficient in natural gas since the late 1980s. Annually, 64% of the United States petroleum use is for trans-portation fuels and the other 36% is used for industrial chemicals and materials (manufactured plastics, paints, paper, textiles, pharmaceuticals, building materials, etc.) (DOE, 2003).

Of the energy used in the United States in 2003, only 9% was from renewable resources and only about half of that (4.1%) was from biomass (DOE, 2003). Biomass for energy is mostly derived from the wastes of agriculture and forests. In the United States, little biomass is used to produce energy, in contrast to the European Union, where biomass is viewed as the dominant renewable resource to use for energy production.

The advantage of biomass is that, when transformed using new technology, it is a more consistent and reliable source of energy than wind or solar power. For exam-ple, tree growth is not restricted geographically so biomass can be used in almost all regions of the world. Trees are also effective at capturing solar radiation at the times when it is available and can be harvested at any time.

Trees, therefore, have fewer limitations than some of the other renewable energy resources. For example, solar photovoltaic systems are limited by the need for sun, and high-latitude areas do not have much sun during the winter months when energy needs are high. Windmills have been encountering problems because some old windmills are attracting endangered bird species that have been killed by their blades. In addition, some people do not want large windmills located in their neighbourhood.

Using wood to produce energy is not a new idea; humans have long been burn-ing wood in open fires. In the 20th century, interest in bio-energy has increased whenever the supply of fossil fuel has been reduced (e.g. during the Second World War, in the 1970s and 1980s, and in 2005). Today we are still burning wood to produce steam energy in direct fire and cogeneration plants throughout the

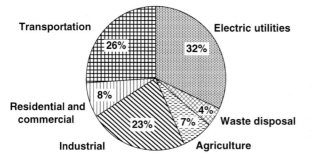

Fig. 6.7. The total greenhouse gas emissions from fossil-fuel combustion by energy-use sector for the United States in 1998 (EPA, 2001). Transportation and electricity combined produce > 50% of greenhouse gas emissions from fossil-fuel combustion.

United States. In fact, since the year 2000, biomass has been the leading source of renewable energy (this is not the total energy consumed) in the United States. In 2003, biomass provided 47% of all renewable energy but only 4% of the total energy produced in the United States (DOE, 2003). In some regions of the world, such as the Pacific North-west United States, there is some recalcitrance to use any wood for energy production because a segment of the population is against cutting any trees. This view has been changing since the fire risk has increased in these forests.

When considering the use of biomass to produce energy, the total amount of greenhouse gases emitted from the combustion of fossil fuels is important to discuss. The different energy-use sectors in the United States and the proportion each contributed to the total United States emissions in 1998 are shown in Fig. 6.7. From the combustion of fossil fuels, the electrical utilities contribute 32% of the annual greenhouse gas emissions, followed by the transportation sector at 26%. It is obvious why those pursuing solutions to reducing or mitigating greenhouse gas emissions are focusing on these two sectors (Fig. 6.7). Since much of the fossil fuels are used in the transportation sector, it produces the greatest amount of greenhouse gas emissions globally. The European focus on producing alternative biofuels is obvious. Globally, the transportation sector emits 22% of the global CO_2 emissions while consuming 25% of the world energy (Azar and Rodhe, 1997; Azar *et al.*, 2003).

Biomass can be converted to all the products produced from oil since they are all carbon-based materials. Therefore, biomass can be used to substitute for fossil fuels used in both the electrical and transportation sectors.

New role for forest materials to produce energy and chemical commodities

If forests are to continue to be a dominant part of our landscapes, new economic values have to be produced for forests. More value will have to be given to forests so that they will be kept in their present state, rather than being converted to other uses (e.g. urban development). Whatever new products come from forests have to be environment- or climate-friendly while contributing towards sustainable livelihoods of rural communities. New products from forests will only be accepted by a broad segment of the population if they are climate-friendly.

It is possible to manufacture new products from forest material. These new products typically use forest or urban wood wastes that have been traditionally taken to landfills and have a high fire risk (Vogt *et al.*, 2005). Wood can be converted to by-products that:

- Expand markets and suppliers of renewable energy systems by including forest-dependent communities as energy providers.
- Improve overall energy system reliability from interconnecting biomass to other energy sources.
- Provide sustainable and environmentally friendly energy production systems.
- Enhance renewable resources transformed environmentally.
- Decrease loss/conversion of forest lands (e.g. urban sprawl) by providing market return from non-timber forest products.
- Obtain conservation values and environmental services from forests by decreasing the pressures to over-cut forests.
- Provide electricity end-users with more energy choices and less volatile prices and costs.

Forest materials can be used in the creation of several products that will diversify and create new environmentally friendly markets in rural areas for: (i) electricity production using hydrogen fuel cells; (ii) selling methanol to chemical companies, using it as an industrial chemical commodity; (iii) using wood as pharmaceutical precursors; and (iv) biofuels for the transportation sector (see Chapter 7, this volume for more details). By diversifying the product options available from forests that are environmentally based, people dependent on forests for their survival should be more in control of their economic opportunities by developing local/regional markets that will be less controlled by global economic markets.

When rural community livelihoods come from sustainably managed forests, international markets for forest products are likely to also develop. The ability to produce environmentally friendly products should also help to decrease the conversion of forests to other uses when there is little or no economic return from forests.

Global Management of Carbon

Global institutions are actively working at managing human uses of resources so that they can decrease the emissions of carbon into the atmosphere. Most of the efforts in forests initially focused on using forests to sequester carbon. Today, the international community is pursuing many efforts, which also include the use of renewable resources for energy production as an alternative to fossil-fuel use. Some of these efforts will be discussed here briefly as this is part of the globalization of forest uses, which was introduced in Chapter 3 of this volume – 'Globalization of Forest Management and Uses'. These efforts reflect the importance of the carbon cycle to society globally.

UNFCCC and the Kyoto Protocol

Following international acceptance of the global-warming theory, the United Nations decided that international measures were needed to halt the anthropogenic accumulations of greenhouse gases before they reached levels that were hazardous for earth systems. This was an attempt to restore global carbon cycles so that changes in this cycle did not occur at a faster rate than the earth's systems were capable of evolving to cope with (e.g. food production capacity would not be endangered, economic development could evolve sustainably in the face of a changing supply of resources). An international agreement to address these issues was reached in the United Nations Conference on Environment and Development (UNCED) held in Rio de Janeiro, Brazil, in 1992. This conference resulted in the formation of a voluntary multinational agreement, the United Nations Framework Convention on Climate Change (UNFCCC), which has been ratified by 170 countries.

To address these pressing issues and the increase in GHG emissions and to address the issue of meeting targets, the international community developed the Kyoto Protocol. The Kyoto Protocol establishes legally binding commitments for Annex I or industrialized countries to collectively reduce GHG emissions by more than 5% below 1990 levels by 2008 to 2012. The Kyoto Protocol was adopted at the Third UNFCCC held in Kyoto, Japan, in December 1997 and entered into force in February 2005. Although party to and active under the UNFCCC, the United States, Australia and several other countries have not ratified the Kyoto Protocol and have sought to explore alternative options through other alliances to address the problem.

To help meet the emission-reduction targets, the Kyoto Protocol developed mechanisms for countries unable to reach their commitments and to finance projects elsewhere where the realized carbon benefits would be higher. The Kyoto Protocol set of mechanisms to help reduce GHG mechanisms include Joint Implementation (JI), where an industrialized country can finance projects that reduce net emissions in another industrialized country, and the Clean Development Mechanism (CDM). The CDM was designed to help developing countries address sustainable development and to help developed countries meet their emission-reduction commitments. The rationale behind both JI and CDM projects is that emissions know no borders and, for global purposes, reducing emissions in any one country is as good as reducing it in the country where the greenhouse gas is produced. These projects included bio-energy or other renewable energy projects and growing trees in plantation settings for carbon sequestration purposes.

How successful these mechanisms will be in helping to mitigate CO_2 emissions into the atmosphere is not clear at this stage. These projects are adapting and evolving as the constraints and opportunities become clearer. The CDM, as a tool to reduce emissions of greenhouse gases, can in theory be effective in sequestering carbon from the atmosphere. The CDM is one of the solutions proposed by the international community to develop sustainable economies in the tropics (see Chapter 3, this volume). For CDM to be successful, explicitly linking sustainable livelihoods of indigenous communities to afforestation and reforestation efforts will be important.

CASE 6.1. MEASURING CARBON IN FORESTS USING HIGH-RESOLUTION DIGITAL IMAGERY (SANDRA BROWN)

Interest in implementing land-use change and forestry projects for mitigating carbon dioxide emissions has increased interest in measuring the carbon content of forests and other land uses accurately and precisely and yet cost-effectively. Precision is the key term for projects because investors and the regulatory arena want the changes in carbon stocks measured to high degrees of precision – generally with 95% confidence intervals of 10% or less of the mean. For heterogeneous forests, for example, this could mean measuring tens to hundreds of plots over large areas of hundreds to thousands of hectares (Brown, 2002).

Ground-based methods for estimating biomass, and thus carbon, of the tree component of forests are well known. Typically, lots of ground plots are installed and the diameter and often height of all trees in the plots measured. Then allometric equations are used that relate biomass as a function of a single, e.g. diameter at breast height (dbh), or a combination of tree dimensions, such as dbh and height, and even dbh, height and wood density (Brown *et al.*, 1989). A potential way of reducing costs of monitoring the carbon stocks of forests is to collect the key tree data remotely. As remote methods would collect data from above the forest, the question is what data are needed to accurately estimate the biomass of forests. Based on an extensive experience of research on forest biomass, to accurately estimate biomass carbon stocks with high precision from 'overhead' imagery data requires a measure of the crown area and possibly height of individual trees from a known area in combination with allometric equations based on these dimensions, similar to the ground-based approach.

The available aerial and satellite data collecting systems are comprised of an array of methodologies with varying accuracies and amenabilities for use in carbon studies. Most of the satellite systems collect data that are too coarse in resolution to identify individual tree crowns. Even the recent availability of higher-resolution data from sensors such as IKONOS and QuickBird has met with mixed success in attempts to use them to measure crown area or crown diameter of tropical forest trees (e.g. Asner *et al.*, 2002; Clark *et al.*, 2004). Another remote-sensing tool is a scanning laser system – lidar – but at present this is deployed from aeroplanes only. Tests of this system in temperate conifer and deciduous forests and tropical forests have shown its ability to produce high correlations between the sensor metrics and above-ground biomass (Lefsky *et al.*, 1999; Means *et al.*, 1999; Drake *et al.*, 2003).

Experience with biomass estimation tells us that what is needed is a system that could collect high-resolution digital imagery (≤ 10 cm pixels) from which individual trees or shrubs could be distinguished, making it possible to identify them to plant type (e.g. broadleaved, needle-leaved, palm, shrub, etc.) and measure their height and crown area. In essence, a virtual forest would be created that could be used to collect tree measurements. The measurements would then be used to derive estimates of above-ground biomass carbon for a given class of individuals, using allometric equations. Biomass can thus be measured in the same way that ground plots are measured to achieve the same accuracy and precision, but with potentially less investment in resources. In addition, the data can be archived so that, if needed, the data could be re-evaluated or used for some future purpose.

The system that we have developed, the multispectral three-dimensional aerial digital imagery (M3DADI) system, is composed of two digital cameras for data collection at different scales, a profiling laser, an altitude and heading reference system, a Global Positioning System (GPS), and a hard-drive storage unit (Fig. 6.8). The system can be loaded onto just about any small aircraft that can fly at low altitudes (1000 feet above ground level) and at relatively slow speeds for image acquisition (a single-engine Cessna can be used). The M3DADI system can be flown under cloud cover, flown at high temporal frequency and viewed as automatically georeferenced strips and stereo pairs in a standard computer. Spacing camera exposures for 80% overlap provides the stereo-image pairs while the profiling laser, altitude and heading reference system, and GPS provide georeferencing information to view and measure the stereo imagery within a common three-dimensional (3D) space of geographical coordinates.

A series of systematically spaced parallel transects, at approximately 1 km intervals and 200 m in width, with a random start, are flown over the entire study area to collect the imagery (Fig. 6.9). These data are woven into continuous strips and bundle-adjusted blocks of stereo imagery viewable in 3D and calibrated against the laser profiler data. This coverage, combined with real-time differentially corrected GPS data and a pulse laser, provides a detailed and accurate profile of the canopy. The bundle-adjusted blocks of imagery reconstruct a 3D virtual forest that can be analysed at a computer station, using ERDAS-Imagine Stereo Analyst (Leica Geosystems, Atlanta, Georgia) and Arcview 3.3 (ESRI, Redlands, California).

Data Collection from the Imagery

A series of images can be selected along the transects in a systematic manner. Thus each sample image is a 3D digital representation of the forest, from which an analyst makes measurements of

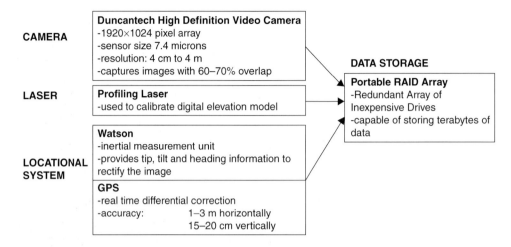

CAMERA

Duncantech High Definition Video Camera
-1920×1024 pixel array
-sensor size 7.4 microns
-resolution: 4 cm to 4 m
-captures images with 60–70% overlap

DATA STORAGE

LASER

Profiling Laser
-used to calibrate digital elevation model

Portable RAID Array
-Redundant Array of
Inexpensive Drives
-capable of storing terabytes of
data

LOCATIONAL
SYSTEM

Watson
-inertial measurement unit
-provides tip, tilt and heading information to
rectify the image

GPS
-real time differential correction
-accuracy: 1–3 m horizontally
                         15–20 cm vertically

Fig. 6.8. Diagram of an overview of the multispectral three-dimensional aerial digital imagery system (top) and the attachment of the camera system in a pod to an aircraft door (bottom). The whole system runs off the aircraft's 12 V power system.

the vegetation. To efficiently measure each sample image, a series of nested plots can be 'installed' around a plot centre that was located at the image centre, just as one would do in the field. The digitization process begins with the creation of the nested plot circles in ArcView 3.3 (Fig. 6.10). The plot circles are imported into ERDAS Imagine's Stereo Analyst, where the interpreter creates polygons around the crown of each vegetation type (Fig. 6.10). After each polygon is created, heights are determined (in the z plane) from measurements of

the highest point on the tree crown and a point in the vicinity for ground height. The resulting databases associated with each plot and vegetation type are then exported into Excel for analysis.

Collection of Ground Data

To convert the measurements from the imagery (crown area and height) to estimates of biomass carbon, a series of allometric equations between

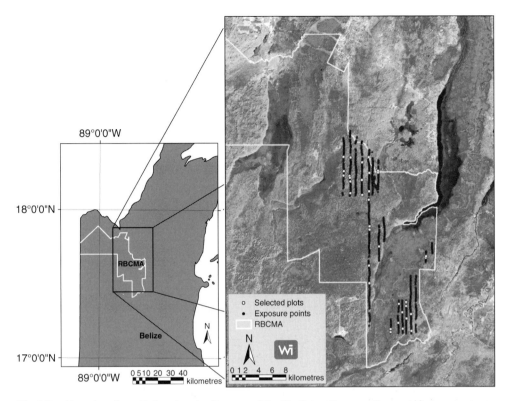

Fig. 6.9. Map of northern Belize showing the area of the Rio Bravo Conservation and Management Area (RBCMA) and areas of pine savannah (darker areas) embedded in the broadleaved forest landscape (lighter areas). The black lines running north and south are the paths of the flight-line transects over the broadleaved forest; the white dots along the lines represent images that were selected for analysis.

biomass carbon and the imagery measurements need to be derived. Some examples of such equations are given in Table 6.8 for both broadleaved species and pine species and for two species of palms in the tropical forests of Belize.

Allometric equations such as those in Table 6.8 are applied to crown area and/or vegetation height data per individual obtained from the analysis of the imagery. Estimates of biomass carbon per individual are summed on a plot basis and then scaled to a per hectare basis, in just the same way field data would be analysed.

Application to Tropical Broadleaved Forest in Belize

A total of 117 km of transects were flown over the broadleaved forest area in Belize over a period of 2 days in May 2003 (Fig. 6.8). Thirty-nine images were systematically distributed along these transects at approximately 3 km intervals and represented about 1.5% of the total images collected. Two nested plots were 'installed' on the images – a 20 m radius plot for all broadleaved trees and a 10 m radius plot for all palm trees. Analysis of the data collected from the 39 image plots resulted in a mean carbon stock in trees and palms of 117.3 t C/ha ± 8.7 (mean ± 95% confidence interval). The 95% confidence interval for the total carbon stock is equal to just 7.4% of the mean.

Conventional measurements taken in 101 permanent ground plots in the same forest type in 2000 resulted in a mean estimated carbon stock of 124.4 t C/ha ± 10.6 (mean ± 95% confidence interval). The confidence interval (CI) is equal to 8.5% of the mean. This value is from the digital-imagery-based estimate (t-test; $P > 0.05$).

Fig. 6.10. A screen-grab showing the image-analysis screen. A pair of nested plots are shown with trees delineated in yellow and palms (in the small plot) in pink (the red circle denotes a palm that is actually in the small plot but looks outside in the larger image on the left).

Table 6.8. Examples of equations relating factors measurable from aerial imagery with biomass carbon for tree species of closed forest and pine savannah in Belize (from Brown *et al.*, 2005; Pearson *et al.*, 2005).

Plant type	Equation	R^2	Max. value
Tropical broadleaved species	Biomass carbon (kg C) = 0.7619 × (crown area × height)	0.62	17,297 m^3
Tropical pines	Biomass carbon (kg C) = 0.49 × (crown area × height)	0.95	1,995 m^3
Tropical palms	Biomass carbon (kg C) = 0.5 × (0.182 + (0.498 × height) + (0.049 × height^2))	0.94	10 m
	Biomass carbon (kg C) = 0.5 × (10.856 + (176.76 × height) − (6.898 × height^2))	0.93	16 m

It can be seen from this example that the imagery and field results are not significantly different from each other and that the 39 image plots resulted in a more precise estimate (95% CI of 7.4% of the mean) than the 101 field plots (95% CI of 8.5% of the mean).

Other Applications

This approach was used in the pine savannah system of Belize (Brown *et al.*, 2005) and the very high spatial variability in this system meant that, for the very large number of plots required, analysis of aerial digital imagery was more cost-effective.

For the mixed bottomland hardwood forests in Delta National Forest in Mississippi, relatively few plots were required for either the field-based or the digital-imagery-based approaches but use of the aerial approach removed the complications of access in the flooded topography (Pearson *et al.*, 2005).

Cost Comparison

For the three case studies mentioned above, the carbon stocks in trees were measured using both the M3DADI system and conventional field approaches. Data were collected only for the time (in person-hours) involved in each of the various steps in preparing, acquiring, interpreting and entering the data into spreadsheets for the final step in the analysis. The overall goal of the cost comparison is to compare the total person-hours needed for the two approaches at the three sites to collect the same set of data to achieve a 95% CI of ±8% of the mean based on the sampling error only. It is recognized that different steps in both approaches require different skill sets – e.g. for the digital-imagery approach the image interpreter will require a background in geographic information system (GIS) and image software, whereas for the field approach the technician will need skills in forest mensuration. These differences in required skills will clearly affect overall cost in dollars.

In all three study sites, the time cost for the field approach is about 2.5 to 3.5 times longer than for the digital-imagery approach to achieve the same precision level (Brown and Pearson, 2005). The main reasons for the difference in time costs for the M3DADI approach versus the field approach is caused by the shorter time needed to measure an

image plot versus a field plot and the lower coefficient of variation in carbon stocks among image plots versus field plots, resulting in fewer plots needed for the M3DADI approach to achieve a given precision target. However, the M3DADI approach took much more time to prepare and process the imagery for interpretation compared with the conventional field approach. The M3DADI approach requires time to prepare the equipment and load it onto the plane, the flight time, the time for downloading data, processing the imagery into the 3D block files and selecting the images to interpret. The field approach requires organizing and maintaining the field equipment, checking supplies, processing the field data after completion of the fieldwork, including quality-control checks, collating all the field sheets, checking them for completeness and consistency, and preparing them for entry into a spreadsheet. The overall time involved in the data collection phase using the M3DADI approach represented 20–80% of the total time compared with the field approach, which took about 5% or less of the total time.

The image plots are presently interpreted and measured manually – a relatively time-consuming step in the process. However, due to the recent advances in high-resolution imagery and sophisticated software, automating these processes may be possible. After testing four of the commercially available products on images of complex tropical forests, the existing software products are unable to automatically delineate tree crowns and heights at present.

The research reported here was supported by a Conservation Partnership Agreement between the Nature Conservancy and Winrock International (prime award DE-FC26-01NT411511 between the Department of Energy and The Nature Conservancy, Bill Stanley, principal investigator).

CASE 6.2. CARBON SEQUESTRATION IN A BOREAL FOREST IN ICELAND: EFFECTS OF NATIVE AND EXOTIC SPECIES (RAGNHILDUR SIGURDARDOTTIR)

Introduction

Iceland is an ideal site to study the effects of planting exotic or non-native boreal tree species on ecosystem processes and how their carbon sequestration

potential differs from the native birch found growing here. In contrast to most forests found around the world, Iceland has soil conditions that are relatively uniform and has a low diversity of tree species. This low diversity of tree species simplifies the ability to measure the impact of a specific tree species on ecosystem processes. If many generations of different tree species have been growing at different times at one site, it becomes difficult to determine the impact of just one species on the ecosystem.

In Iceland, mountain birch (*Betula pubescens*) is the single forest-forming tree species in the country. This condition has persisted for at least the last 10 thousand years or even just before the Ice Age glaciers covered the country for the last time. Icelandic forests have minor constituents of mountain ash (*Sorbus aucuparia*) and tea-leaved willow (*Salix phylicifolia*). All coniferous tree species and other broadleaved deciduous species are relatively recent introductions and are growing in Iceland on a first-rotation basis (i.e. that species has yet to be harvested in that location). Therefore, a clear history of land uses allows one to separate the impacts of individual tree species (e.g. the many exotic species that have been planted in Iceland compared with native species). The low biodiversity and the absence of native conifers in Iceland are probably caused more by Iceland's isolation from continental areas rather than any ecological constraints.

Previous studies elsewhere have indicated that the most severe threats to conservation globally are from the introduction of exotic species (Mooney and Hobbs, 2000). In Iceland, introducing exotic tree species has not yet threatened conservation efforts. This may be due to the fact that these plantings are limited spatially, they are young or the environment is constraining.

The first exotic trees were introduced into Iceland in 1899, but these planting efforts were small in extent and rare until after 1950 (Loftsson, 1993). Since that time, forest stands of exotic or introduced tree species have increasingly been planted. Some of these plantings were stimulated by the reductions in conventional farming practises. Less farming has left large areas of degraded pastures that needed to be restored. Much of these afforestation efforts have led to plantings of exotic tree species.

The interest of the Icelandic government to afforest these former agricultural lands is evidenced by its passing of a law to increase the forests (Act 56, 19 March 1999). This Act stipulated the planting of trees to convert these former agricultural lands into forests. The government of Iceland set a goal of covering at least 5% of the land area below 400 m elevation in forests within the next 40 years.

The major incentive for enacting this new law was to stimulate planting of forests to provide stability and economic returns for rural communities.

Whether the government's goal for restoring Iceland's forested area will be met remains to be seen, but one of the rationales behind the project is to reverse the deforestation and decrease the erosion which has taken place in Iceland for the last 1100 years. When the Viking settlers came to Iceland, it has been estimated that about 25% of the country was forested (Arnalds, 1987). The Vikings soon started to clear the land for agriculture, until now only about 1% of Iceland's land area is in forests. Reforesting Iceland will be a major challenge since so little of the original forest remains.

The most significant challenge will be to determine what the composition of the future forests should be. Will managers in Iceland have to plant more exotic, non-native species because those are the trees that sequester the most carbon, or should new forested areas have a mix of native and exotic species? How much land area should be expanded to include the only native tree species (i.e. the birch) that forms forests in Iceland? How much are birch forests needed to provide Icelandic forests buffers from human and natural disturbances? It is too early to state what approach should be used in Iceland but the decisions need to be made in a context that considers all the constraints and opportunities that each provide.

Introduction to the Species and How Carbon Sequestration was Determined in Iceland

The case examines the carbon sequestration potential of two exotic tree species compared with the native birch in Iceland. In addition, the impact of planting exotic tree species on understorey species composition and biomass was also examined as a way of assessing whether exotic plantings would start modifying plant succession.

The tree species

The species selected for the study contrast with each other in terms of leaf form and longevity. As part of this study, the carbon sequestration potential was determined for a native birch forest and two forests planted with exotic or introduced species of larch and pine. The native mountain birch (*B. pubescens*) is a broadleaved deciduous species, while the Siberian larch (*Larix sibirica*) is a deciduous needle-leaved conifer and the lodgepole pine (*Pinus contorta*) is an evergreen conifer.

The study site was located in Hallormsstadur forest in eastern Iceland, where the largest continuous forest area is found. This forest covers an area of about 1850 ha (Blöndal, 1995). To date, all stands of exotic tree species are currently being grown on their first-rotation basis so there are no complicating factors such as having several species growing on the site at different times. The introduced tree species were about 45 years old at the time of the study, while the native birch forest was about 60 years old.

How biomass was measured

Above-ground net primary production was determined on the sites by measuring change in plant biomass and monitoring tree litter fall. Carbon budgets were produced at the ecosystem level, which included tree biomass (above and below ground), understorey biomass, ground-surface organic and soil organic matter. Nine to ten trees of each species were taken down, dissected and measured to obtain the relationships among branches, stems, bark, foliage and roots. Allometric relationships were then calculated for each of the species, which allowed the biomass of stands to be calculated. By adding all the different components together, the total amount of carbon in each of the different ecosystems was found.

If Carbon Sequestration is the Only Goal of Icelandic Forest Management, Exotic Tree Species Sequester More Carbon than the Native Birch

The introduced coniferous species, both larch and pine, had a significantly higher amount of carbon sequestered in tree biomass after growing for about 45 years in areas previously occupied

by the native birch forest. Of the two exotic species, lodgepole pine had a greater capacity to grow and produce biomass. This is partly due to lodgepole pine being somewhat more shade-tolerant than larch and therefore being able to grow as a denser stand than Siberian larch. Also, lodgepole pine is an evergreen conifer and maintains its foliage year-round. This allows lodgepole pine to photosynthesize and sequester carbon for a longer period each year. In contrast, larch is an evergreen species but drops its foliage similarly to a deciduous species.

If a policy is being pursued to maximize carbon sequestration in Icelandic ecosystems, planting introduced conifers would be an obvious choice over the more slowly growing native birch. For example, after 45 years, the pines had accumulated 242 tons of biomass per hectare. Larch had the next highest accumulation of biomass during this 40-year period, with a biomass of 172 tons per hectare. The native birch forest accumulated a total of 102 tons of biomass per hectare during a 60-year period.

Based on these data, it appears that Iceland would need to increase its planting of exotic tree species if the goal is to increase the amount of carbon being sequestered by Icelandic forest lands. However, there are many other effects of introducing exotic species that need to be considered before making the decision to expand the area of forest in Iceland using introduced species. Several of the considerations are discussed below.

A Metric Based on How Much Carbon is Sequestered Neglects whether Ecosystem Processes and Disturbance Regimes are Altered

Too much detritus?

Despite significant differences in the annual amount of carbon that is sequestered in biomass, the exotic and native forests all produced similar amounts of detritus (dead parts of the tree that fall to the ground). However, the composition of the detritus varied greatly among the different forests. For example, the larch and pine forests had 14 and ten times more dead branches, respectively, attached on the trunks of the trees than the native birch. Larch also has a large build-up of dead branches on the ground. This build-up of dead, dry branch material is likely to affect the combustibility or wildfire risk of these forests in the foreseeable future.

Until now, forest fires have been very rare in Iceland. Forest fires were associated almost exclusively with forest clearing by the Viking settlers 1100 years ago. More recently, infrequent and rare fires that were not controlled were associated with the burning of straw in early spring in nearby grasslands.

With the introduction of exotic conifers to Iceland, the previous fire scenario is expected to change since the conifers are likely to convert forests to those that are fire-prone systems. This change in the fire-proneness of Icelandic landscapes is not surprising since some of the introduced tree species' (e.g. lodgepole pine and larch) native habitat is fire-dominated (see Chapter 5, this volume). So the higher carbon sequestration potential of the pines has to be balanced by their greater susceptibility to being lost in a fire at some future time. When managing to maximize the carbon sequestration potential of forests by selecting certain fast-growing species, the fire risk of the systems should be factored into the decision-making process. If slower-growing trees with less ecosystem accumulations of carbon (e.g. birch) are less fire-prone, then they may in fact be a more stable sink for atmospheric carbon.

A complete conversion of forest planting to the native birch has also to be pursued with caution and may not be the logical solution for Iceland. The birch forests are showing signs of more damage due to insect herbivory than most exotic tree species in Iceland. For example, insect herbivores (e.g. *Epinotia solandriana* and *Operophtera brumata*) were introduced into Iceland in the 20th century. Planting native tree species is an ideal solution as long as there is sufficient genetic diversity within the birch to withstand disturbances such as the introduction of these invasive insects. In the case of larch, there are no invertebrate pests that constantly devour its foliage (Halldórsson *et al.*, 2002).

The absence of pests on exotic tree species does not suggest that insect pests will not attack these tree species in the future. History has shown that tree species may grow well for several decades until a certain pest is accidentally introduced.

For example, Scots pine (*Pinus sylvestris*) used to be one of the most commonly planted tree species in Iceland from 1948 to 1960. However, the introduction of the Eurasian pine adelgid (*Pinus pini*) has almost completely eradicated it from Iceland.

Despite faster growth rates and greater ecosystem accumulations of biomass, some of the introduced species may become stressed in the variable climates of the Icelandic environment. This may result in trees appearing to grow well in Iceland but then having significant mortality after several decades of good growth. For example, warm temperatures in late winter and late spring frost spells can be a disturbance that exotic species are not adapted to withstand in Iceland. In the 1960s, climatic variations resulted in the eradication of nearly all of the introduced *Salix viminalis* trees in Iceland. Similar climatic fluctuations in 2003 severely damaged a large number of larch trees in Iceland. Before massive plantings of an exotic tree species occur, those sensitive to such climatic disturbances should be eliminated from consideration.

When reforesting 5% of Iceland, selection of trees to plant will also have to consider the ecosystem impacts of individual species. For example, understorey plants native to Iceland are not found growing under the canopy of all tree species. This fact means that selecting overstorey trees to plant has also to consider their impacts on the conservation of the understorey species. This will be discussed next.

Introduction of Larch and Pine to Ecosystems and their Effects on the Understorey Plant Communities

Forty-five years after lodgepole pine and Siberian larch were introduced into the Hallormsstadur forest for the first time, considerable changes in understorey biomass, productivity and biodiversity were recorded. These changes in the understorey composition and structure are compared with what is typically found in a native birch forest.

The change in the understorey species is especially obvious in the lodgepole pine forest. The ground in the pine forest is covered with a thick mat of dead pine needles, in which only a few plants are capable of growing. The pine forest therefore does not support native understorey plant communities under its canopy.

Visually, the understorey of the larch forest appears quite luxuriant, with noticeably taller vegetation composed of grass species, a few forbs and *Equisetum* spp. Larch therefore does not have the same effect on the understorey as lodgepole pine. Even though the larch understorey appears more luxuriant, the biomass of the understorey is similar to or less than that recorded in the native birch forest.

Native understorey shrub species are found less commonly growing under the canopy of both the pines and the larches. The native birch forest has the greatest biodiversity of plants and greatest biomass, especially of understorey shrubs (such as blueberries, *Vaccinium uligosum*, and willow, *Salix phylicifolia*) and mosses. In fact, compared to the birch forests, the biomass of moss in the larch and pine forest was only half of what was found in the native birch forest.

Take-home Message

Introduction of exotic tree species as part of management objectives should consider how these species will modify or influence the type of disturbances that may occur in the future (e.g. changing the fire risk of a forest) and how they influence the diversity of other native plant species that coexist with trees in a forest. The management objectives have to be balanced with long-term repercussions for ecosystem functioning, since these environments have a high probability of being altered when exotic species are planted.

This case study showed that selecting fast-growing tree species to accumulate or sequester more carbon in biomass does result in higher levels of carbon sequestration as long as the disturbance regime is not altered. The slower-growing native species do not sequester as much carbon as the introduced species but may provide a more stable sink for carbon if the disturbance regimes of the introduced species become a 'normal' part of the forest disturbance cycle (e.g. fire in pine forests). In this study, the introduced species produced more dead branches, which increases the risk of fire in these forests. If the fire pattern of Icelandic forests changes, the carbon sequestered there may be a less stable carbon sink.

This study also showed that introducing new species into an area will modify the presence of other understorey species that are typically found growing in the native forests. In the case of the pine, a very limited amount of understorey plants were able to grow under the dense canopy and the deep detrital layer that is formed by dead needles. If pine were to become a larger proportion of the forest area, the capability of understorey plants to persist as part of Icelandic forest landscapes would be reduced. The larch did stimulate the growth of some understorey species also found in native birch forests. This stimulation of growth may be associated with the direct or indirect effects of larches in assimilating nitrogen in the system.

The selection of tree species to plant therefore does have ramifications and repercussions on the vegetative dynamics of the entire forest landscape. Since Iceland has a major programme to afforest its lands, careful consideration will have to be made as to which species to plant since these plants will be impacting the environment for decades.

7 Emerging Issues in Forests

Kristiina A. Vogt, Toral Patel-Weynand,
Gretchen K. Muller, Daniel J. Vogt, Jon M. Honea,
Robert L. Edmonds, Ragnhildur Sigurdardottir
and Michael G. Andreu

Continuing Challenge for Sustainability: Linking Social and Natural Sciences and Codifying Indicators

Developing a natural and social science for sustainable management of forests

The tools to measure the natural science side of forest sustainability have been developed relatively independently of the more recent research that assesses how people's behaviour and patterns of land use have impacted forest health. Previous work in the natural science realm is now helping to link the natural and social science of forests for measuring sustainability at the ecosystem level (which includes the conservation of a suite of multiple organisms living in a forest). To understand the links that bridge the two areas, there is merit in exploring the evolution of knowledge in the natural sciences that has produced the principles of forest sustainability that are currently in practice.

Many of the emerging issues in understanding natural forest science involve changing paradigms on how to manage forests and how to quantify sustainable forestry principles. Since indicators of sustainable forestry are derived from how these terms are defined, measuring sustainable forestry is a significant challenge.

The organizing principles of sustainable forestry have evolved over time. Definitions of sustainable forestry of 20 years ago would have made measuring forestry sustainability much easier since it primarily consisted of understanding the silviculture of the trees being harvested. Except for economic analyses, which were the driving force in defining sustainable management, society at that time was rarely part of the equation, and sustainable forestry comprised a not very complex web of interactions, which were geared to ensuring sustained yield (Fig. 7.1).

The goal of sustainable forestry in the past was to achieve a sustained yield of timber (Franklin *et al.*, 1981; Franklin, 1989). Today sustainable forestry incorporates a multiple-use strategy conscious of the complexity of interactions at many different levels, where forests are managed so that future timber extraction is possible, while

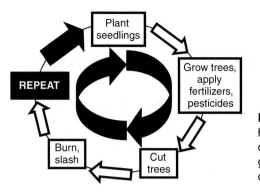

Fig. 7.1. A conceptual diagram of how sustainable forestry was defined in the past when it was geared to ensuring sustained yield of timber.

maintaining biological diversity in forests and protecting watersheds and habitats. A case study from Australia demonstrates how forest management has changed and how today restoration of forests is an important element of sustainable forestry (Case 7.1 – Restoration of Degraded Rainforest Sites in the Wet Tropics of North-east Australia).

Today potable water, recreation, aesthetics and wildlife are as important as the sustained yield of wood. What human societies want from forests today, at the very basic level, can be summarized as:

- Sustainable flow of wood products.
- Recreational opportunities.
- Wildlife habitat.
- Sustainable flow of clean water.

Understanding sustainable forestry as a natural science

Any discussion of sustainable forestry has to include a discussion of the terminology used in today's forest management (Box 7.1). Since sustainable forest use does not include practises commonly used in the past (e.g. clear-cutting), these forest management practices will not be discussed here. Sustainable forestry has been fundamentally refocused to include all components of the ecosystems and landscape in management decisions. These forest management decisions must take into consideration both human and natural disturbances and how they affect forest health.

One management approach that allows for careful preservation of environmental services from forests while managing them for timber is called variable retention harvesting (VRH) (Franklin, 1992). As one alternative to clear-cutting, VRH focuses on managing forests for ecological values and to ensure that some biological legacies (e.g. old, large-diameter trees) are left behind after trees are cut. Quite simply, VRH harvesting practises specifically shy away from removing all trees from a stand (trees left can be scattered evenly or be clumped in distribution or both) and encourage maintaining the existence of all species present to some degree, so that biological legacies are left behind.

Box 7.1. Terms used in defining sustainable and non-sustainable forestry (Franklin, 1992).

The following terms are also important for the reader to know since they are used to define whether sustainable forestry was being practised:

- Legacies – remnants, live or dead, in ecosystems that are important in determining how ecosystems function and their resilience or recovery after a disturbance.
- Clear-cutting – cutting all trees in an area, no trees left for regeneration or for habitat (also referred to as clear-cuts).
- Biodiversity – the variety of life in the world or in a particular habitat or ecosystem.
- Lifeboating of species – structures or functions (e.g. coarse wood) that allow species to survive or persist after a major change in the system or a landscape matrix.
- Fragmentation – changing landscape matrix of tree distribution so that trees persist in the landscape in smaller patches and therefore have more edge environments and greater disconnects between forest areas.
- Structural enrichment – managing forests so that structures required by species remain as legacies in younger systems and can provide habitat, such as owls' roosts and nests in younger forests when a few old trees are left.
- Maximum wood production – focusing on obtaining the maximum growth of trees for harvesting without considering all the other services that could come from the forest.

The following terms are considered positive and important elements of sustainable forestry: variable retention harvesting, legacies, biodiversity, lifeboating of species and structural enrichment. Society has also defined the following terms but these have negative connotations and are a detriment to forest sustainability: clear-cutting, loss of biodiversity, increasing fragmentation, maximum wood production.

Implementing VRH practices is not easy because of the level of social and natural complexity and competing uses that must be considered. Questions that help to determine how to implement VRH practices are (Franklin, 1992):

- What should we leave?
- How much should we leave?
- What spatial pattern do we want to attain?

The reason these questions need to be asked is to determine whether different habitats that are desirable in the landscape can survive after the trees have been cut and are maintainable. An ideal situation is one where a variety of ecosystem conditions are present in the landscape, from reserves at one end of the spectrum to cut-over areas at the other end. The spatial pattern desired in the landscape varies depending on what the landscape is being managed to preserve (e.g. wildlife biologists like aggregated patterns, hydrologists like the dispersed pattern; see Hunter, 1990, for more information).

Indicators of sustainable forestry

Biodiversity alone not a good indicator of forest sustainability

In the 1950s, wildlife biologists considered the natural forests of the Pacific North-west in the United States to be 'biological deserts' because the larger animals, the charismatic megafauna, were missing. However, in the 1980s, scientists found high biodiversity in these forests, which consisted mostly of small organisms called detritivores – organisms that break down organic material and thus help in the decomposition process (see Chapter 4, this volume). Detritivores were found to be important in maintaining the productive capacity of these forests. Detritivores tend not to be visible with the naked eye and therefore are easy to ignore (e.g. fungal tissues are finer than human hair) when developing the sustainability equation.

Even if all the species of detritivores in any forest could be listed, this number is not indicative of their importance to the system since it is their functional diversity that reflects whether or not sustainable forestry is being practised. The variety of chemicals (e.g. cellulose, sugars, lignin, waxes, etc.) that comprise plant materials require specialized organisms that have the enzymes capable of breaking down each of these chemicals. Since most detritivores are efficient at utilizing one plant chemical efficiently, a healthy decomposer community has to be functionally diverse. In this case, measures of biodiversity numbers alone are not particularly useful by them-selves because the functional diversity of many of the species that make up a particu-lar ecosystem may be key but not accounted for in a biodiversity count, a case in point being the detritivores.

At present, there is little consensus among scientists on how to measure the func-tional diversity of detritivores in an ecosystem. However, tools for measuring their functional diversity are being developed, but at present they tend to require sophisti-cated equipment and knowledge, which are not readily available to most institutions and agencies managing forests.

In the past, recording higher species diversity in a forest was regarded as indica-tive of a healthy ecosystem. However, high biodiversity numbers that are comprised of many weedy and undesirable species do not contribute to long-term sustainability. It is also a well-established fact that a higher number of species are found in clear-cuts after harvesting compared with those found in old-growth forests. We know that many of the species that initially establish in clear-cuts are weedy or non-native spe-cies, which further demonstrates the point that biodiversity numbers alone tend not to be accurate indicators for sustainability. Biodiversity is a complex term and has to consider:

- the type of species present and their functional role in a forest (e.g. blackberry bushes growing in a forest may not be desirable since they dominate the site and compete with native species);
- where the diversity is located (e.g. there exists more diversity in detritivore or decomposer communities when compared with the diversity of trees in an old-growth Douglas-fir forest in the western United States);
- what relationships exist between species (e.g. in the Pacific North-west part of the United States, several species are important components of old-growth Douglas-fir forests but are also part of the food chain for the species that make

up the biodiversity for that area – spotted owls feed on flying squirrels, which feed on below-ground fungi).

The importance of dead trees

Today, we recognize that not only live trees but dead trees as well have an important ecological role in forests (Franklin, 1989, 1993). In terms of sustainability, this means that sustainable forest management requires that dead trees remain on a site after harvesting in order to maintain the many complex ecosystem functions, listed below, for that site.

Live-tree functions include:

- root systems to control soil erosion;
- branches and stems that provide habitat;
- subsistence for many small mammals that dig up and eat the fruiting bodies of fungi that are linked in symbiotic associations with tree roots.

Dead-tree functions include:

- decay of the dead tree, which results in the release of nutrients needed for the growth of many other species in the forests;
- large pieces of dead wood (either a snag or an upright or downed log) provide habitat for many small mammals, birds, amphibians and insects;
- large pieces of dead and decomposing wood (called nurse logs) are locations where seedlings start to grow so that regeneration of plant species occurs.

The importance of disturbances in sustainable forestry

How forests respond to disturbances has shown us some other significant components of forests that need to be considered when practising ecologically based, sustainable forest management (Franklin and Forman, 1987; Franklin, 1989, 1992, 1993). Some of the factors mentioned below are useful indicators to determine whether sustainable forestry is occurring:

- Cumulative effects. Recognizing that forests are impacted by multiple disturbance events and that those disturbances are a normal part of a forest's development (Chapter 5, this volume) is important. These disturbances do not occur in isolation or in a vacuum and each disturbance will probably modify how another disturbance impacts any given forest (e.g. clear-cutting followed by snow or rain can result in large erosion events).
- Legacies. Structures that remain after a disturbance will mitigate or determine how future disturbances impact an environment because of the role that they play in the recovery (e.g. dead stems of trees after the eruption of Mt St Helens provided habitat for surviving organisms and hastened the recovery of these areas).
- Survivors are important in the recovery process of forests and should be used to define what needs to be left in a forest (e.g. both living and dead trees) for their sustainable management. Some disturbances, such as fire or big storms that occur at decadal time scales, have important roles to play in ecosystems, and how we manage forests should mimic these natural forest processes.

- Animals respond to structural legacies. It used to be thought that spotted owls could only be found in old-growth forests in the Pacific North-west United States. Now we know that owls can survive outside old-growth forests as long as the younger forests have the structural characteristics of old-growth forests.
- Landscape ecology. There is a need to look at the cumulative effect of our activities at a landscape scale since fragmentation, with isolated pockets of forests, does not provide good habitat for forest-dependent species and the edges of forests are more susceptible to disturbances (e.g. wind blowing over trees growing along the edges of forests that are adjacent to other land uses; see Franklin and Forman, 1987).

Start of forest certification: difficulty of codifying acceptable indicators of sustainability that integrate the social and natural sciences

One tool that incorporates natural and social science indicators in implementing sustainable forest management practises that is being widely used is forest certification (Vogt *et al.*, 1999). Forest certification uses social and natural science indicators and criteria to evaluate whether sustainable forestry is occurring. It has been extremely difficult for scientists to identify a few simple indicators to measure whether a system would degrade as a result of implementing a particular management practice.

Simple indicators, like a canary in a mineshaft, do not exist for assessing forest health or whether sustainable forest practises are occurring. Ecosystems are complex and need to include both social and natural science indicators to determine the impact of cumulative effects. How and what type of past disturbances and land-use activities have influenced the landscape and how the landscape matrix affects forest health (see Chapter 5, this volume) are important considerations for sustainability (see Case 7.2 – Forest Informatics: Need for a Framework for Synthesis, Data Sharing, Development of Information Infrastructure and Standards for Interoperability to see how organizations are dealing with large data sets and the need to share information).

Forest ecosystems are similar to a bowl of clam chowder. You can identify and recognize all the ingredients that went into the making of the clam chowder but when you taste the soup no one ingredient stands out as being more significant than any other (Fig. 7.2). When you taste the soup, it is difficult to tell what really went into it or if something critical was left out. But, if all the ingredients are there and in the correct proportions, the soup tastes good. Our ability to construct or restore an ecosystem is still rudimentary and, like an inexperienced chef trying to make clam chowder for the first time, we know what needs to be included (the indicators) but not how the different parts are linked or what the correct proportions of those parts are.

The time it takes for an impact to manifest itself and the time lag before an impact is measurable in an ecosystem creates great difficulties in credibly verifying that a particular practise or management activity is responsible for a demonstrated negative result. When insect and pest outbreaks are observed in forests, they are generally the secondary agents that are responding to an already weakened forest and are not the primary agents causing unhealthy forests (Chapter 5, this volume). The boiled-frog analogy emphasizes the difficulty in identifying the causal factor due to the time lags before a response can be detected (Fig. 7.3). As an analogy, if the water

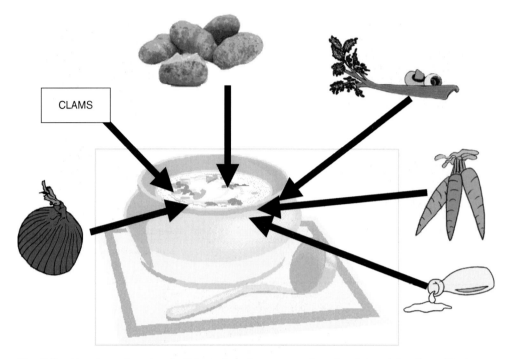

CLAMS

Fig. 7.2. How to make a bowl of clam chowder – an analogy for determining the critical indicators of an ecosystem to show that it is functioning normally.

in which the frog is sitting is slowly heated, the frog will not immediately jump out and there is a strong likelihood that it will become lethargic and be cooked. If the water is heated slowly, the frog will not detect it is in danger of being cooked.

Similarly to the 'frog analogy', if forest health is deteriorating due to some human or natural disturbance (see Chapter 5, this volume), the lack of visible symptoms does not allow us to recognize that the health of the forest is changing. It is not until trees are being killed by insect and/or pest outbreaks that a forest is recognized as being unhealthy. Insect and pest outbreaks are not the primary indicators of decreasing forest health and by the time an outbreak occurs it is too late to respond to or to even recognize the other causal factor that has been slowly contributing to the decline in forest health (e.g. pollution, drought). The insect and pathogen pests come in for the 'kill' after pollution, drought or mismanagement has predisposed plants to be more susceptible to them. Pest infestations are good indicators of an already existing debilitating condition, but unfortunately they are not an early warning indicator of an impending problem.

Many have explored the use of animals as indicators of ecosystem health and sustainability and there are some success stories in identifying indicators (Azevedo-Ramos *et al.*, 2003). For example, mining companies in Australia have successfully tracked the types of ants they find near abandoned mines to see if the sites have been fully restored. In other cases, waste treatment plants in the United States use the bluegill fish to monitor the quality of their water. For some time, scientists have been looking for something similar to assess the impact of practices used by logging

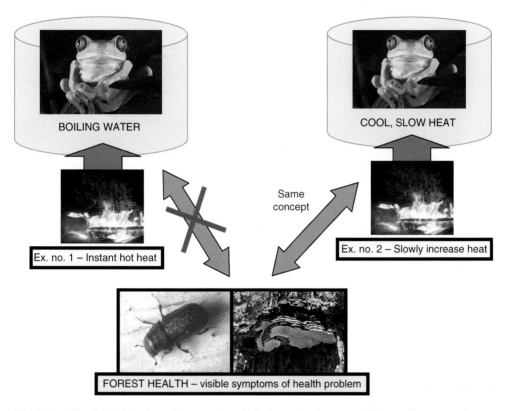

Fig. 7.3. The 'boiled frog' analogy and its similarity to the impacts of forest disease and insect pest problems in a forest. By the time that a forest is attacked by disease or pests (i.e. the first visible signs that a forest has already been weakened by another disturbance), a forest is already less healthy. If no visible symptoms (similar to a frog being in water that is slowly being heated) are apparent, there is no sense of urgency to modify one's behaviour.

companies on biodiversity. The hope is that changes in a specific species of animal or group of species in a particular forest will indicate whether logging is damaging their habitat. So far there does not seem to be any observable common denominator that provides a truly early warning signal to indicate the declining health of a forest.

In 2003, a systematic literature search was conducted to identify animal indicators that would be good candidates to monitor to determine the impact of different logging practises in tropical forests (Azevedo-Ramos *et al.*, 2003). The review concluded that no one animal indicator was effective for all forests, even though logging clearly influenced which animals survived in the forest and also affected their population densities. Each species' response to the different logging practices varied, so that the response of one species to a particular forest practise provided little indication of how practices in different forests might impact the biodiversity of other species. However, the review did find some promising new information, such as that logged-over forests have fewer insect-eating birds and more birds that eat fruits. The difficulty and expense of monitoring animals to assess forest health must also be factored into determining which would be the fruitful paths to pursue. It would be wonderful if something like the miners' canaries was identified for assessing forest sustainability but the complexity of maintaining healthy forests does not provide such simple indicators.

This inability to identify a few simple animal indicators to measure forest sustainability suggests that it would be even more difficult to find any reliable common indicators that are accurate for every situation and every forest. This also highlights the need to better manage global data sets so that causal patterns and trends can be identified early in a project. Mining global data sets should increase our ability to identify indicators to monitor that will provide early warning of impending change to ecosystems and landscapes (see Case 7.2 – Forest Informatics: Need for a Framework for Synthesis, Data Sharing, Development of Information Infrastructure and Standards for Interoperability).

A Challenge: Codification and Consensus in Measuring Sustainable Forestry

Sustainable livelihoods dissociated from sustainable management

Forest certification aims to link sustainable forestry with sustainable development. However, these dual goals have often been difficult to implement simultaneously. Forest certification has been most effective at auditing sustainable forestry in terms of the silvicultural evaluation of timber production, since the science exists to conduct this evaluation. However, sustainable livelihoods for rural communities who reside in forests or are dependent on forests are often harder to implement, especially where it is challenging to build consensus or promote a common vision among multiple stakeholder groups. Furthermore, the science needed to merge the social sciences with the natural sciences is just emerging and it may take some time before a more cohesive and uniform way to implement forest certification becomes a reality.

Forest certification is being used in an increasingly complex social environment (Chapter 3, this volume; see Case 7.3 – Forest Certification, Laws and Other Societal-based Constructs: Tools to Include Social Values in Forest Sustainability). The diversity of views on natural resource use held by society at large, including the many diverse views on sustainable forest management itself, make implementation a challenge. It is imperative that tools are developed to include social science indicators since natural science indicators by themselves will do little to achieve the multiple sustainable development goals that drive certification.

At this stage, social indices are less well developed in the certification process and the emphasis is on economic factors and labour, such as the formation of unions to preserve forest community rights (Vogt *et al.*, 1999). The difficulty of including non-traditional uses of forests in sustainability assessments is discussed in Case 7.4 – Tourism and Sustainability in a Forested Protected Area.

What do communities in the industrialized and developing countries want from certification?

If the goal of certification is to reduce the rate at which tropical forests are converted to non-forest uses, the challenges of illegal harvesting, wars and tenuous political situations, among other things, have made it difficult to implement sustainable management of forests (Table 3.1; Table 7.1; Chapter 5, this volume – 'Influence of wars'). If the goal of certification is to move communities towards sustainable management of their forests, there are many success stories. For example, by having communities begin to document what resources they have in their forests, as a first step, and to develop their own management plans for these resources, certification has become the enabling tool for practising sustainable forest management.

If the goal of certification is empowerment of indigenous communities (e.g. social movement) by giving them rights over resource use from their tropical forests, then certification has been a tremendous success in many cases (see Case 7.5 – The First Certified Community-based Forests in Indonesia: Stories from the Villages of Selopuro and Sumberejo, Wonogiri, Central Java). By acknowledging the customary land-tenure rights of indigenous communities, forest certification has not only been an ideal tool for legalizing and legitimizing rights of forest communities, but has improved socio-economic conditions by providing for sustainable livelihoods. Certification has also created opportunities for communities in tropical countries to actively participate and benefit from carbon trading (see Chapter 6, this volume).

Forest certification is a dynamic tool and is still evolving as it faces challenges to meet increasingly complex situations. The most critical of the challenges that certification faces is whether or not it can improve the livelihoods of local communities by providing higher revenues from sustainably managed forests and if it will at least match revenues that can be obtained from competing land uses. The traditional sources of generating income (e.g. making baskets, hats, bird houses, etc.) require access to global markets, knowledge of marketing strategies with attention to specific product design requirements needed for global economies, and being able to directly deal with retailers to maximize profits (see Case 7.6 – Non-timber Forest Products and Rural Economic Development in the Philippines). These challenges are not insurmountable and many local communities are becoming quite savvy in interfacing with global markets and other societies.

Issues still confronting forest certification today

Most of us do not know where the products we use originated. Consumers are too far removed from the complex process of developing a product to evaluate whether forests are being managed sustainably. This separation makes it important to have verifiable proof of certification. Without it the public is not able to determine whether a product is being created in a sustainable manner. The certification label signifies that the values society expects from forests are being provided and that the health of the forest is not being adversely affected by their purchase. For those consumers that are socially and environmentally conscious, this label allows them to buy products with confidence and without having to spend a considerable amount of time researching a company and its practices. A consumer buys hundreds of products every year; confirming whether sustainability criteria are being met for all of those products is impossible.

The fact that there are so many certification systems available today can be confusing. Multiple systems make it difficult to know which system represents the best sustainability practices and which label adheres to those practices. In the

United States, the development of multiple certification systems from both the industrial sector and the environmental and conservation sectors has resulted in certification protocols that are independent systems and it is often hard to compare protocols. For example, the industry certification system is much broader and less prescriptive, with fewer established indicators (fewer than 100) and relies on professional forest managers to wisely manage their forests. In contrast, the environmental certification system has more stringent guidelines and has over a hundred indicators, including the provision of more environmental services, which gives less management discretion to professional forest managers (Vogt *et al.*, 1999).

A New Challenge to Sustainable Forestry: Managing Urban Forests and Reducing Deforestation at the Urban–Wild-land Interface

The scope of forest management has broadened and shifted considerably from silvicultural manipulations of natural forests or intensively managed forest land (e.g. plantations). Today, forest professionals face a more challenging task in that they need to manage forests in both urban areas and wild-land–urban interfaces. Professionally trained foresters of today must recognize the need to manage all these different types and sizes of forests.

Forests exist along a continuum of different management intensities, as shown in Fig. 7.4. All of these forests need to be managed. These forests provide habitats for endangered species in addition to many other environmental services and they are similar in that they also face problems that jeopardize their health (e.g. exotic and invasive species, wildlife and fisheries conservation issues, responding to climate change and the risk of fire).

The intensity at which trees or forests are managed can define where a forest system exists on the continuum of forest types (Fig. 7.5). For example, an arboretum is an intensively managed landscape that would very quickly change to a very different condition and species composition if not continuously managed. In other words, an arboretum needs continuous intervention by humans to maintain its state or characteristics. Human interventions or disturbances (especially when dealing with natural forests) are in this case primarily in the form of lawn mowing, removing spent flowers or replacing dead plants.

In contrast, a wilderness area is maintained by natural disturbances. Natural disturbances would encompass fires, floods or heavy winds and even the occasional pest outbreak. Today, even wilderness areas are not free of human disturbances (e.g. the effects of air pollution can be felt globally since it is transported atmospherically; fire control has increased the risk of fire in some wilderness areas).

All these forests types (Fig. 7.4) have the following characteristics in common:

- They are dynamic and continuously changing. Disturbances are important factors in maintaining them either in their natural states or in the human-designed state.
- They include flora, fauna and humans. No forest type exists that does not have elements from all of the three components represented in that ecosystem. No ecosystem is 'virgin' or free from human influence.

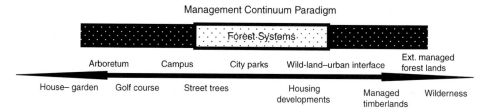

Fig. 7.4. Forest management occurs in a continuum of forest conditions.

Forestry:	**Horticulture:**
The art and science of managing forests to produce various products and benefits including timber, wildlife habitat, clean water, biodiversity and recreation.	The art and science of landscape plant selection, installation and management. The science of cultivating fruits, vegetables, flowers, or ornamental plants and trees.

Urban Forestry:

The art, science and technology of managing trees and forest resources in and around urban community ecosystems for the physiological, sociological, economic and aesthetic benefits trees provide society.

Fig. 7.5. Definitions of forestry, horticulture and urban forestry.

- Lack of appropriate management in any of these forests can impact their health because their structure and function have been modified by past legacies.
- Forests provide solutions to the problem of climate change and are a renewable energy source. Forests in urban and rural environments need to be part of the solution to these problems.

The foremost challenge facing forests today is the increasing threat of urbanization (Stein *et al.*, 2004). Along with the conversion of forests to agriculture land that is occurring in tropical countries, loss of forested land due to urbanization results in massive deforestation. Given these facts, it stands to reason that urban society and urban forestry will play an important role in defining how forests and society interface in the future.

This conversion of forest land to urban areas is a global phenomenon. Stein *et al.* (2004) estimated that a total of 9.1 million ha (22.5 million acres) of urban lands are projected to shift to exurban (lands in the wild-land–urban interface) by 2030 in the United States. From 10 to 20% of this area will be forests that will be converted to other uses needed by urbanites. A few statistics from the United States illustrate this point:

- The population growth in the north-east and the south will result in the most dramatic losses of forest lands in these areas (Stein *et al.*, 2004). During the last

10 years, the state of North Carolina has lost 5% of its forest areas due to the development and expansion of its urban areas (Alig and Butler, 2004).

- The state of Washington lost 2% of its forest areas between 1990 and 2000 (Best and Wayburn, 2001). In the Puget Sound region of Washington, urban growth doubled between 1972 and 1996 and resulted in a loss of a third of the forest cover in this region. Furthermore, since Washington State is expecting more growth and is allowing counties to rezone 25% of their forest land for use in future development, more of the forested area will probably disappear.

The Meyerson model (see Fig. 3.1), which shows a strong link between population density and the amount of deforestation, was developed for the tropics but may have applications worth considering in the temperate regions of the world.

Conversion of forest land to alternative uses is also being driven by the globalization of markets for timber. The shift in global wood markets from the industrialized regions of the world (e.g. Pacific North-west United States) to the less industrialized world has had the unintended consequence of reducing the value of land when it is forested. Since timber prices are low, a landowner can command a higher income when forests are put to other uses. Forested land has even less value when close to urban areas. Since it is more difficult to manage 'working' forests near urban areas, forested land is increasingly being sold for urban development and expansion. With the current global population projections, continuing loss of forest land should be expected unless the value of forest lands can be elevated monetarily or otherwise.

Along those lines, there is greater recognition today that the wild-land–urban interface has to provide habitat for conservation purposes. For example, federally managed forests in the United States are inadequate to provide sufficient forest area for many endangered species (e.g. spotted owls need more forest habitat than exists on public lands). This means that urban planners have to start interacting with foresters and ecologists to determine how to best maintain habitats so that they can continue to be suitable for both conservation efforts and human habitation.

As the wild-land–urban interface or urban growth expands and replaces forests with houses in the western United States, management issues that were only considered relevant for forests in rural areas are now being recognized as problems for urban areas (e.g. fire risk, insect and pest outbreaks) (Andreu et al., 2005). Population growth has resulted in more homes being built in areas with a higher fire risk, where fires are caused by both natural and human factors. In the state of California, more than 100,000 ha (250,000 acres) of forest and rangeland have been burned in wildfires each year since 1950 at a cost of US$900 million to local, state and federal agencies (PIER Collaborative Report, 2005). Not only are there costs associated with the loss of homes near forests due to fires, but the many environmental services that forests provide are also lost (e.g. conservation of endangered species, habitat, water and air quality, carbon sequestration, recreation).

Wildfires in wild-land–urban interfaces result in high insurance costs for replacing homes lost during fires. In fact, the state of California has established that managing forests to reduce fire risk is a priority because of the high insurance costs associated with expensive houses being lost during fires. Investing in managing to prevent fires by thinning trees and shrubs ($235/ha or $580/acre) around expensive homes is comparatively cheaper than the cost of rebuilding million-dollar homes.

In the PIER Collaborative Report (2005), a net benefit of US$600/ha ($1482/acre) was calculated when trees were thinned around homes. In most other parts of the United States, however, thinning costs are prohibitive since there is little economic value to the wood that needs to be removed to reduce fire risk.

Negative environmental impacts and costs have resulted in new approaches to managing these forests, where some economic returns that pay for restoration efforts or reduce the risk of fire are being tested (Andreu *et al.*, 2005). Similar to the new product opportunities that are being considered by rural communities throughout the world, managers of the wild-land–urban interface are exploring the removal of unwanted plant species (e.g. invasive woody species) and converting their biomass to bio-liquids that can be used as a substitute for fossil fuels (Vogt *et al.*, 2005). Most of this woody material, irrespective of size and quality, can be converted to useful products. As the need for managing urban forests increases, novel approaches such as conversion of marginal woody material to effective uses will need to be developed.

Another New Challenge to Sustainable Forestry: Illegal Timber Harvesting

Illegal logging is contributing to escalating deforestation, especially in the tropics, and is negatively impacting the pursuit of sustainable forestry worldwide. In developed countries, such as in the European Union, forest certification has been effective in regulating how forests are managed so that timber harvesting is carried out legally (Table 7.1). The same cannot be said for the less industrialized countries, where much of the illegal timber harvesting occurs and where it is a significant issue. However, obtaining accurate estimates of illegally logged timber is very difficult. The estimated numbers presented in Table 7.1 underestimate the actual magnitude of the loss that is occurring (AFPA, 2004b) and are based on estimates of available industrial roundwood.

Table 7.1. The amount of illegal timber being harvested for the countries that are suppliers of wood (Brazil, Indonesia, Malaysia, West/Central Africa and Russia) and those that are consumers of wood (China, Japan, Europe – EU-15) (from AFPA, 2004b).

Country	Softwood supply suspicious (%)	Hardwood supply suspicious (%)
Russia	17.0	17.0
Indonesia	0.0	58.0
Brazil	0.0	15.0
Malaysia	0.0	11.8
Japan	6.5	5.5
China	31.5	30.6
EU-15	1.2	6.6
Central/West Africa	Not meaningful	30.0
United States	Not meaningful	0.0
Canada	Not meaningful	0.0

Illegal forest activities can include:

harvesting without authority in designated national parks or forest reserves and harvesting without or in excess of concession permit limits; failing to report harvesting activity to avoid royalty payments or taxes; and violating international trade agreements such as the Convention on International Trade of Endangered Species (CITES).

(AFPA, 2004b)

The report written for the American Forest Products Association (AFPA) suggested that from 2 to 4% of the softwood lumber and plywood on the global markets is of suspicious origin while the amount of hardwood lumber and plywood of suspicious origin is much higher (from 23 to 30%; AFPA, 2004b). This report further suggested that this illegal harvest of timber probably depresses global timber prices by 7–16%.

Illegal logging in the tropics and the boreal forests (the two areas of the world with half of the world's forested areas) is very pervasive where the ecological constraints make it difficult to maintain intensive harvesting of timber (see Chapter 2, this volume). Even though forest certification is gaining favour (Chapter 3, this volume), it has not had the desired impact of decreasing illegal harvesting in these environmentally sensitive areas.

Tropical and boreal forests are also found in regions of the world where there is less private ownership of forests or where land tenure is not well established (AFPA, 2004b). Reductions in illegal logging practises appear to be the norm where there is more private ownership of forests. In the European Union and in the United States, where private ownership and clear land tenure exist, illegal logging is less common (Table 7.1). Since one of the goals of certification is to establish ownership rights for rural forest communities, illegal logging should decrease if certification is able to guarantee sustainable livelihoods with an increase in the amount of private ownership of forests.

A Solution for Sustainable Forestry: Manage Biomass Wastes to Produce Energy Sustainably

Biomass wastes: trash in industrialized countries but income in many developing countries

Biomass wastes can be considered a serious management problem in the more industrialized world because of the difficulty of disposing of these materials. Biomass wastes are generated from agriculture, forestry and municipal wastes. According to the PIER Collaborative Report, 60% of the municipal wastes are biomass wastes (paper, cardboard, residual construction wood, waste wood from demolition, stumps, food waste, green wastes from urban environments).

A significant portion of waste materials ends up in landfills and only a small fraction are recycled. In the United States, over 50% of urban wastes are transported to landfills, while approximately 34% of the waste produced by urban dwellers is recycled (the remaining 16% is burned).

Of the many materials that are recycled, paper recycling has been very effective in the United States. It can be recycled four or five times before the fibres are so damaged that they cannot be reused to make more paper products. A 2004 report by AFPA discussed the difficulties of recycling paper products and estimated that only 14 to 16% of the paper products collected as part of curbside recycling efforts could be recycled (AFPA, 2004a).

Wastes in developing countries

In contrast to the developed-world model, organic biomass wastes can be an important source of economic income for scavengers in many developing countries (see Case 7.7 – Importance of Scavenger Communities to the Paper Industry in Mexico). In Mexico, trash collectors who scavenge for waste materials have jobs. This is an informal process of collecting wastes that is carried out by the poorer, unskilled individuals of society (Medina, 2004). Informal scavenger cooperatives are an integral part of waste management in many developing countries.

Recycling wood products (e.g. cardboard) is an important activity in Mexico (see Case 7.7 – Importance of Scavenger Communities to the Paper Industry in Mexico). People cross the border into the wealthier United States, collect cardboard boxes discarded by businesses and transport them back across the border to Mexican towns. This cardboard, collected by the scavengers, provides much of the starting material needed by the wood industry to make paper products. These scavengers have a higher standard of living than other people living in the same community. At the same time, they are providing many environmental benefits because of their activities. Many of the scavenger communities that are involved in recycling wastes are starting to form cooperatives to retain more of the economic return from this activity (Medina, 2004).

Comparison of waste management in developing and developed countries

Waste recycling is dramatically different in the more industrialized countries when compared with the developing countries. The major differences between waste management in the developed and developing countries can be summarized as (Medina, 2004):

Industrialized countries
- Industrialized countries have an abundance of capital; solid-waste management systems have high construction and maintenance costs but low labour costs because fewer people are needed to run the systems.
- The quantity of waste generated in a developing versus developed country differs significantly since the amount of waste produced per person increases as income rises. Cities in industrialized countries have higher waste generation rates than those in developing countries. For example, an average United States resident produces > 1.5 kg of rubbish/day while a person in Cotonou, Benin, generates 125 g of waste/person/day.
- The characteristics of waste generated differ significantly between the industrialized and the developing countries. Wastes from cities in developed countries contain more packaging materials and have a lower density and a higher calorific content than refuse in the developing world. Residents of industrialized

countries consume more processed foods that are packaged in cans, bottles and plastic containers.

Developing countries

- Developing countries have an abundance of unskilled, inexpensive labour and little capital, so the solid-waste management systems do not cost very much and they are more labour-intensive. Here the recycling of municipal wastes is part of the informal economy and is carried out by human scavengers.
- Conventional refuse collection vehicles used in the industrialized countries do not work well in many cities in developing countries because their movement is hampered by narrow and hilly streets that are unpaved.
- Cities in developing countries have a dynamic informal sector, which includes informal refuse collection and scavenging, which provide income for migrants, the unemployed, children, women and handicapped individuals.
- Humans scavenging in developing countries have high health risks (Medina, 2004). Mexico City dump-site scavengers have a life expectancy of 39 years, which contrasts with the 67 years of the general public. In Port Said, Egypt, scavenger communities had an infant mortality of 1/3 (of every three live births, one infant will die before it is 1 year old). Many scavengers live right on the landfills and are exposed to diseases at a higher rate than seen in the general public. For example, more than 35 diseases have been identified in scavenger communities in Manila, Philippines. A few of the diseases are diarrhoea, typhoid fever, cholera, dysentery, tuberculosis, anthrax, poliomyelitis, skin disorders, pneumonia and malaria. The scavenger communities also have a high prevalence of enteric and parasitic diseases, since sanitation, sewage disposal and clean water rarely exist in landfills.

In both the developed and developing world, these respective uses of biomass wastes do not take advantage of the many products that can be produced from them, nor are they converted to alternative products efficiently. Recent developments in energy technology are revolutionizing how biomass can be converted to commercially viable products, such as substitutes for fossil fuel (Vogt *et al.*, 2005). This new technology can decrease the emissions of greenhouse gases into the atmosphere.

Biomass wastes – a disposal problem

Wood wastes are emerging as a significant management problem since this material is mainly being transported to landfills. In the more industrialized world, wood-based material and waste disposal is becoming more problematic as less biomass material is being accepted by landfills. Landfills are increasingly being decommissioned or having a shorter life because biomass wastes are bulky and rapidly fill landfills to capacity. In addition, more communities are refusing to allow new landfills to operate in their neighbourhoods because of previous environmental problems. The magnitude of biomass waste that is generated by society and transported to landfills is unsustainable because we are creating progressively greater amounts of rubbish and we are running out of landfill space.

For example, the state of California is a good example of the significant amount of biomass wastes that are generated annually and are disposed of at landfills. California annually produces 38 million dry tons of biomass that needs to be disposed of (PIER Collaborative Report, 2005). In 1989, a law was passed in California that mandated that by 2000 local municipalities must reduce by 50% the amount of generated wastes going to landfills (PIER Collaborative Report, 2005).

Better approaches to waste management are important issues for industrialized as well as developing countries to consider and adopt. New approaches to waste management can provide environmentally friendly solutions and promote climate-friendly technologies to produce energy for communities. Using new technologies in waste management also has the potential to provide new economic opportunities for rural communities by enabling them to provide products for regional markets.

Is there enough biomass available to produce new products?

Despite the problems of disposing of biomass wastes, questions always arise as to whether there is enough wood available to ensure a sustained supply of biomass as raw materials for new industries. Biomass and biomass wastes are available in large quantities, especially in the industrialized countries. The same cannot be said of many developing countries, where biomass is burned and a scarcity of fuel wood is predicted for the future. The challenge is to collect biomass sustainably in the industrialized countries and to increase the efficiency with which biomass is used in developing countries.

Since biomass is a renewable resource, it has some advantages over fossil fuels in that it is easier to manage biomass to ensure a continuous supply for generating energy or to produce new items. The main advantage in using biomass is that biomass levels from harvested sites can be restored in 1 to 20 years, depending on the species collected, to levels similar to what was originally removed from a site. Fossil fuels, on the other hand, require millions of years to restore.

Current estimates show that the availability of biomass is high in the United States. A recent study by Oak Ridge National Laboratory (ORNL, 2005) showed that the annual availability of biomass and biomass wastes from agricultural and forest lands is 1 billion dry tons. This amount of biomass would sustainably displace 30% of the petroleum consumed in the United States (ORNL, 2005). In the western portion of the United States, there also exists a significant amount of small-diameter wood that poses a high fire risk. This could be used to create products that substitute for fossil fuels in electricity production and for transportation fuels (Vogt *et al.*, 2005).

From a short-term cursory examination of the biomass-to-energy sector, developing alternative products from biomass does not seem to be a viable strategy to pursue in many developing countries when many are faced with a scarcity of fuel wood. However, the consistent scarcity of fuel-wood supply suggests that, as a long-term strategy, many developing countries need to explore how to more efficiently use biomass to generate energy. Technological developments in the energy sector allow biomass to be more efficiently converted to energy. Converting biomass to higher energy sources that can displace some use of fossil fuels seems to be a sustainable alternative to burning it or burying it in landfills. Such an approach would allow a smaller supply of biomass to be used to provide heat for cooking and for heating homes.

Products possible from biomass

How wood is used to make products has changed dramatically within the last decade. Wood or wood fibres are used to make products that are stronger and are capable of being manipulated into many different shapes and sizes (see Case 7.8 – More Efficient Use of Trees to Produce Forest Products). Many of these wood products use small pieces of wood or wood fibres so that each tree is used very efficiently and little wood waste remains.

In addition to the items manufactured by forest-product industries that use small pieces of wood or wood fibres, both agricultural and forest materials and wastes are used to produce an amazing number of products (Paster *et al.*, 2003). Many of them are formed from biomass that has been converted to liquids or gases (Case 7.9 – Energy from Biomass). These products are so completely transformed that they are not recognizable as having originated from a tree.

Plant biomass is used to produce adhesives, fatty acids, surfactants, acetic acid, plasticizers, activated carbon, detergents, pigments, dyes, wall paints, inks and plastics. Some of the products that can be produced from biomass are summarized in Table 7.2.

Biomass can also be converted into many goods traditionally produced from fossil fuels as they are all carbon-based (Case 7.9 – Energy from Biomass):

- Energy.
- Transportation fuels.
- Industrial chemicals/materials.

New products from forest biomass that are climate-friendly technologies

Recent new technological developments in the energy sector are allowing forest biomass to be used in an environmentally friendly manner (Vogt *et al.*, 2005). Since more than half of the world uses wood traditionally as fuel wood for combustion, new technologies need to be further developed and adopted so that wood is used to produce energy more efficiently and in a more environmentally friendly manner.

Table 7.2. Common products from biomass (from Paster *et al.*, 2003).

Biomass resource	Uses
Maize	Solvents, pharmaceuticals, adhesives, starch, resins, binders, polymers, cleaners, ethanol, fabrics
Wood	Paper, building materials, cellulose for fibres and polymers, resins, binders, adhesives, coatings, paints, inks, fatty acids, road and roofing pitch, methanol, ethanol, pharmaceutical precursors Polymers from cellulose: plastics, motion picture film, clear lacquer coating, rayon (fabrics) (www.psrc.usm.edu/macrog/proposal/dreyfus/outcome/plascot/cellace.htm)

One potential energy source that can be produced from wood is methanol. Biomass has been used to produce methanol for over 350 years. The more commonly recognized name for methanol is 'wood alcohol'. However, using biomass as an energy source for production of methanol in the past was more expensive than using natural gas. Today, most methanol is produced from natural gas – a non-renewable resource. However, now that the economics of converting biomass to methanol have changed through the development of new and more efficiently engineered systems that convert biomass to methanol in a matter of minutes (Vogt *et al.*, 2005; Case 7.9 – Energy from Biomass), cheap methanol production is a viable option.

Methanol has many uses that are much more climate-friendly than fossil fuels. Methanol can be used to substitute for or blend with petrol in the transportation sector or it can be used to generate electricity using hydrogen fuel cells (Vogt *et al.*, 2005). The many new emerging products possible from just methanol alone are presented in Box 7.2. In addition to producing many industrial chemical commodities, the production of methanol from wood can help satisfy some of our energy needs (e.g. electricity, transportation fuels) in a climate-friendly way. When wood is used to produce methanol, no fossil fuels are needed (except to initially start the process). Furthermore, since biomass-based energy production using these new technologies has zero net production of CO_2, this technology will be a small, but significant, step towards the reduction of greenhouse gas emissions.

If methanol were blended with petrol or used to charge fuel cells for the production of electricity, carbon emissions avoided by not using these fossil fuels would be 149 to 462 kg carbon for every 1 metric ton of wood converted to methanol (K.A. Vogt, unpublished data). Removing all the small-diameter high-fire-risk wood from a hectare of forest land in Washington State and converting this to methanol would avoid carbon emissions of 2.3 to 7.3 metric tonnes of carbon per hectare of forest land. If wood wastes going to a landfill were converted to methanol to use in electricity production or were blended with petrol, 149 to 462 metric tonnes of carbon emissions would be avoided. Production of electricity using renewable resources will also allow communities to participate in carbon-trading or obtain money from renewable energy certificates.

A Solution for the Future: Integrating Lessons Learned in Conservation and Sustainable Forestry

Conservation literature calls for evaluations and improved accountability

Conservation science in the tropics has up until recently had a biodiversity protection focus and has been driven by conservation of particular species and habitats. As new tools and science have emerged, the importance of including humans as part of the forest landscape has become key (see Fig. 7.6). Factoring in sustainable livelihoods for forest dwellers has gained importance in conservation efforts, as reduction of poverty is one of the primary tenets of sustainability endeavours. Thus, the field of conservation has come to recognize that society cannot be excluded from conservation efforts, a fact illustrated by historical interactions of people with forests (see Chapter 1, this volume; see Cases 2.1–2.4, 3.1–3.5).

Box 7.2. Uses for methanol.

1. Chemical commodities. Methanol is an important alcohol for the chemical industry. Methanol has many markets. It is used in many products found within a typical household, the building materials used in the construction of a home, and even clothing and carpets found in a house. Each year, 12 billion gallons (4.5×10^{10} l) of methanol are consumed worldwide for producing hundreds of essential chemical commodities (from acetic acid and formaldehyde to windshield washer fluid).

2. Transportation fuels and additives:

- Methanol can be used as a petrol blend (e.g. 85% methanol : 15% petrol or diesel), as the primary fuel source for a car, or used to take methyl esters from vegetable oils to produce biodiesel. The cost for converting an existing petroleum storage tank to methanol to produce a new fuelling infrastructure has been estimated by the Methanol Institute to be slightly less than $20,000 per station (E.A. Engineering, 1999).
- If a diesel-like fuel is also desired, syngas can be converted to a non-methanol, hydrocarbon fuel with diesel-like properties (e.g. gas-to-liquid (GTL) fuels). Methanol to dimethyl ether (DME) (Ekbom *et al.*, 2003) can be accomplished in a one-reaction step. As a matter of fact, the same reactor that makes methanol can be tuned to make DME, an ideal diesel fuel with a cetane number of 55+. DME is currently not considered to be a biodiesel because it can also be made from natural gas; however, if made from biomass, it can be included in the definition of biodiesel since it is a diesel fuel and has a biomass origin (R. Upadhye, Lawrence Livermore National Laboratory (LLNL), personal communication).

3. Electricity production. Methanol can be used in turbines to produce electricity (existing technology) or in hydrogen fuel cells to produce electricity (www.idatech.com). Methanol is the preferred fuel to use in fuel cells because it is a one-carbon alcohol and re-forms to hydrogen at lower temperatures. Other alcohols need higher temperatures to re-form them and cost more to use because of having to break carbon bonds, which means that they are more costly to use and increase the price of building a fuel cell.

- If the over-abundant small-diameter wood (2–7 inch diameter or 5–18 cm) that has a high fire risk were totally removed from the forests in Washington, 2.2 gigawatts of electricity could be generated using fuel cells (J. Stewart, LLNL, unpublished report, 2005).
- A house in Seattle (needing 3 kW of power) could obtain 1 month of electricity from 3 metric tonnes of wood using methanol fuel cells (K.A. Vogt, unpublished).
- Methanol is considered a superior fuel for electric power generation using turbines (Methanol Institute, 2005).

4. Reduce pollution in waste treatment plants. Methanol is used by hundreds of waste-water treatment plants to reduce nitrogen pollution (Methanol Institute, 2005).

Box 7.2. *Continued.*

5. Consumer electronics. Methanol fuel cells are becoming the power source for many consumer electronics and fuel-cell vehicles. These consumer electronics are an emerging area of new products that are coming to consumer markets. Methanol produced regionally could be used to supply this demand and keep the distribution system regional or statewide. Today, methanol fuel cells can provide power for 20 h for laptop computers. Similarly, MP3 players can play for 20 h when powered by methanol fuel cells.

6. Jobs, new products from the rural areas for markets. Adopting systems to convert biomass to methanol will result in many jobs being developed in rural areas. These jobs will require professional foresters, engineers and business managers.

7. Economic return to forest landowners. Currently there is little value that can be obtained from forest lands other than selling timber. Forest lands are being lost to urban development because these lands do not have much value when forested. Producing methanol and other products, such as transportation fuels, will provide a new form of economic return to non-industrial forest landowners. Just selling methanol on the market will provide economic value where previously no value existed.

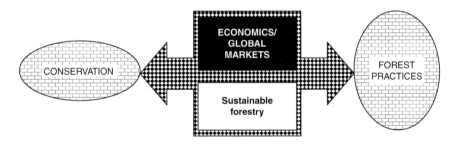

Fig. 7.6. Conceptual diagram showing how conservation and forestry are integral to the production of sustainable forest practices and sustainable livelihoods.

There are a multitude of environmental challenges in the tropics which have made conservation a challenging and yet dynamic field (Vogt *et al.*, 2000). The tropics are experiencing high rates of resource extraction, as well as extraordinary rates of population growth (FAO, 2001). This magnitude of change in population densities frequently occurs with high levels of poverty (Meyerson, 2000), so that conflicts arise over resource uses and their protection (Deacon and Murphy, 1997). When faced with such a challenge, even the best available science will not provide the necessary insight to protect biodiversity, as human demands on natural resources will limit the application of science to conservation (Vogt *et al.*, 2000). Population pressure, combined with weak institutions, poor policies and external debt, creates a multitude of challenges for biodiversity and the sustainable use of natural resources (Deacon and Murphy, 1997; Barrett *et al.*, 2001).

In order to meet the challenges of the future, it is imperative that current conservation mechanisms be analysed and refined so that they can be more effective. Conservation and forest management can learn from each other since they have and continue to face similar challenges when managing a forest at the local level for either resources or conservation (see Case 7.10 – Integrating Conservation and Sustainable Forestry: a Pacific North-west United States Example).

The conservation movement has evolved significantly during the past 20 years and conservation effectiveness has received significant attention in the literature (Christensen, 2003; Murray, 2003). Some of the topics presented include mechanisms to improve conservation efficacy, the accountability of environmental programmes, the need to include the social element into environmental approaches across the landscape and the importance of incorporating the concept of scale into conservation programmes (Moran, 1994; Barrett *et al.*, 2001; Christensen, 2002, 2003; Noss, 2002; Salafsky *et al.*, 2002; Wilshusen *et al.*, 2002; Ellison and Daily, 2003). These are all the same issues facing forests today (see Chapters 2 and 3, this volume).

As stated by Salafsky *et al.* (2002), 'We have yet to fully discover the secrets of effective conservation: forests are still burning, coral reefs continue to be dynamited, wildlife populations are overharvested, urban sprawl worsens, and our global climate is changing with unpredictable consequences.' This does not suggest that the conservation community has not been actively addressing these concerns, but rather that environmental programmes must alter their approaches to improve their effectiveness across the landscape.

Current literature discusses the need to improve conservation practices through the credible auditing of conservation efforts (Christensen, 2003). This is the same role that forest certification is attempting to provide for forest management (see Case 7.3 – Forest Certification, Laws and Other Societal-based Constructs: Tools to Include Social Values in Forest Sustainability). Credible audits will enable the conservation movement to ensure that programmes are achieving their objectives (Christensen, 2003; Ellison and Daily, 2003; Murray, 2003). Salafsky *et al.* (2002) presented three fundamental questions that conservation programmes should be asking if these efforts are to improve:

1. What are the project goals and how has progress been measured?
2. How can project designers and implementers most effectively take action to achieve conservation goals?
3. How can we do conservation better?

These questions are difficult to answer and yet it is critical to do so in order to improve conservation approaches. Forest managers are asking and answering similar questions in the pursuit of sustainable forestry.

The need for programme evaluation is increasingly articulated within the conservation arena (Christensen, 2003; Murray, 2003), although the manner in which successful evaluation should occur remains unclear. Conservation involves both natural and social systems, which makes the work of conservation practitioners extremely complex. The urgency of environmental problems, however, requires that immediate action be taken, regardless of the inherent risks due to the lack of certainty about how best to proceed (Salafsky *et al.*, 2002).

Acknowledge need for conservation accountability

While fiscal accountability has been discussed in the development literature for decades, only recently has it been acknowledged in popular sources (Christensen, 2002; Murray, 2003). According to a recent *New York Times* article, fiscal accountability concerns are now a major factor to consider in conservation (Christensen, 2002). As stated by Nick Salafsky, the co-director of Foundations of Success, 'As a conservation industry, we have to prove we are effective and achieve what we say we are going to achieve. If we can't show that, the attention and resources of society will shift to other problems.'

Donors have begun asking conservation organizations about accountability and long-term effectiveness. In order to improve the efficacy and long-term sustainability of such programmes around the world, it is imperative that social evaluations be conducted of current conservation approaches.

Include scale and context for sustainable conservation

Conservation approaches have begun to reflect the shift towards ecosystem management in which whole systems are considered at larger spatial scales. There are four principles of ecosystem management as presented by Vogt *et al.* (1997). These principles are as follows:

1. Focus on the sustainability of ecosystems, not on the output of products alone.
2. Adopt a holistic understanding of the way all the parts are linked together in an ecosystem and the feedbacks among those linkages.
3. Incorporate a long-term perspective and examine issues at a scale relevant to the functioning of the ecosystem.
4. Recognize that human values shape ecosystem structure and function in myriad ways that can constrain, promote or reduce sustainability.

Designating protected areas in human-dominated landscapes is a difficult task, as the involved parties are confronted with multiple goals promoted by interest groups championing specific and often conflicting agendas (Daily and Walker, 2000). However, management success can be achieved by incorporating the ideology that the integrity of every local area is ultimately dependent on the integrity of the landscape that surrounds it (Noss, 2002).

The long-term survival of forest-dependent communities is determined by the health of forest systems that depend on sustainable management that explicitly considers people to be an integral part of these systems (Sarkar and Ebbs, 1992). Based on this assumption, it is critical that consideration be given to the social, economic and ecological context within which various environmental initiatives are applied, as all of these systems will directly impact the outcome of the proposed initiative. Looking at environmental challenges from a broader spatial scale will intensify the complexity of the issue needing to be monitored; although it will ultimately ensure that the long-term goals of conservation can be achieved across the landscape.

As stated in a speech by John Hudson (2002), senior forestry adviser to Britain's overseas-aid department, 'The opinion to leave things alone is not a reality in this day and age and any long-term solution for tropical forests has to involve better livelihoods for those who live within and around forest systems.' Remote, pristine parks will not survive population and political pressures (Landel-Mills and Porras, 2002) and, while conservation initiatives cannot be expected to 'do it all', it is imperative that the social and political context be considered in the design and implementation of any conservation mechanism (Wilshusen *et al.*, 2002). Actively collaborating with local people can help ensure the long-term viability of any proposed conservation project (Sarkar and Ebbs, 1992; Moran, 1994).

Changes occurring in the conservation field mirror or track many of the issues being faced by those developing sustainable forestry practises (see Case 7.10 – Integrating Conservation and Sustainable Forestry: a Pacific North-west United States Example). Both areas would greatly benefit from linking the two fields since both have learned lessons that could be used to improve practises being implemented within each field. Since conservation and sustainable forestry occur in the same landscape, there is a need for close communication between the groups since they have several similar goals (see Case 7.10 – Integrating Conservation and Sustainable Forestry: a Pacific North-west United States Example). Furthermore, conservation projects in the tropics have already been researching and developing models for integrating sustainable development into conservation. The results and lessons learned from the conservation efforts could aid sustainable forest management since issues of sustainable livelihoods are integral to both and are challenges that both are still trying to solve.

Linking conservation and resource uses not a global trend and sustainable forestry not being practised where needed

The North-west Forest Plan example presented in Case 7.10 (Integrating Conservation and Sustainable Forestry: a Pacific North-west United States Example) illustrates the consequences of failing to link the lessons learned from both conservation and sustainable forest use. Globally, forests are not being sustainably and maximally managed to provide the multitude of services and products (e.g. resource extraction, environmental services and conservation) they can offer. In 2000, 12.4% of the world's forests had protected-area status (FRA, 2000) and were set aside for biodiversity. The remaining 87.6% of the forests are forests where the highest rates of deforestation (14.2% deforestation over a 10-year period) are being recorded.

Clearly there is a need for sustainable management of these resources while at the same time providing sustainable livelihoods for forest dwellers and surrounding communities. Just over half of the global forests are located in the tropical and subtropical ecological zones (FRA, 2000; see Table 7.3). Reducing poverty in these regions is highly dependent on how sustainably these forests can be managed.

Table 7.3. Location of global forests.

	Ecological zones			
	Tropics	Boreal	Temperate	Subtropics
Global forest distribution (%)	47	33	11	9

Forests that are certified and able to provide sustainable extraction of wood products while continuing to provide habitat for biodiversity are mainly located in the industrialized regions of the world. These are the regions of the world where the conservation movement had its beginnings. These are also the regions of the world that used forest materials to power their industrialization efforts in the 1800s. Most of the developed countries, however, have shifted to using fossil fuels instead of forest materials to provide the energy needed for their industries.

Industrialized countries have more certified forest area where sustainable forestry is being practised (FRA, 2000). Industrialized regions of the world are predominantly located in the temperate climatic zone, which has 11% of the global forests (see Table 7.3). This fraction of forest area that is being sustainably managed in these regions is almost equivalent to the amount of land area set aside for protected areas worldwide (12.4%). The future challenge forestry faces is the increase in the amount of forest-land area that can be sustainably managed for the goods and services it provides and for sustainable livelihoods that can generate socio-economic benefits.

CASE 7.1. RESTORATION OF DEGRADED RAINFOREST SITES IN THE WET TROPICS OF NORTH-EAST AUSTRALIA (DAVID LAMB)

Introduction

The wet tropics of north-eastern Australia are occupied by tropical rainforest (around latitude 17°S, longitude 146°E). The high rainfall and good soils have made this area an attractive site for agricultural development. Starting in the late 1800s, farmers began clearing the lowland areas around Innisfail and Cairns for sugarcane. Subsequently, agriculture advanced into the upland areas (above 800 m) and used these lands for dairy farming and various crops. These changes led to much of the flatter land being cleared, leaving most natural rainforest restricted to the steeper areas of the coastal ranges.

Logging has been carried out in many of these forests for over 100 years. This was done using a low-impact polycyclic selection system (see Erskine *et al.*, 2005, for more details on this system). While silviculturalists argued that this allowed ecologically sustainable production, some conservation groups were concerned that the risks to regional biodiversity were too high. In 1988, following a lengthy community debate, logging ceased and most of the remaining rainforest areas (around 1 million ha) were listed on the United Nations Educational, Scientific and Cultural Organization (UNESCO) World Heritage Register. Several tree-planting schemes were then commenced as

part of the government compensation package for timber workers and others adversely affected by the decision. These involved planting trees on underutilized agricultural land in the region that was either degraded or too steep for productive purposes. At the same time a large number of conservation groups and other non-governmental organizations (NGOs) became interested in restoration. The upshot of this has been the development of a number of approaches to restoring forest cover to previously forested sites. The majority of this reforestation has been done on privately owned farms across the region.

It is useful to differentiate the several approaches used. The Queensland state government department responsible for forestry matters (the Department of Primary Industry) had been carrying out plantation establishment on state-owned land in the region for more than 50 years. It tested a variety of species, including many native rainforest species, but subsequently used only one of these (*Araucaria cunninghamii*), together with the exotic species *Pinus caribaea*, in its plantations. The primary objective of these industrial plantations was to provide sawn logs for industry. After 1988, the new reforestation plans adopted entirely different approaches. Current plans have had a much greater focus on providing ecological services (e.g. maintaining biodiversity). These new approaches fall into two broad groups; one group (the 'restoration plantings') has sought to restore as much as possible of the original plant biodiversity to a particular site, while the other (the 'mixed-species plantations') seeks to restore some level of biodiversity (but not necessarily all) to the cleared sites but, at the same time, to also generate some timber production as well.

Restoration plantings

The restoration plantings have been carried out in both lowland and upland situations. A number of approaches have been used but the most common approach involves intensive site preparation (including complete weed control) and dense plantings (often up to 2500 seedlings per ha) carried out at a single time. The amount of fertilization and the types of fertilizer used have varied with sites.

The crucial element in these restoration plantings is the identity and variety of the species used at each site. Good maps of the main types of rainforest present before clearing are used to guide species selection. Species that would have been in these types are then raised in nurseries and planted in randomized mixtures in the field. This is not a trivial task, since many sites would be planted with up to 80 species. A number of these species might only flower and fruit episodically and the task of getting enough seeds or raising large numbers of seedlings of each in the nursery in time to meet some future planting date can be difficult.

Weed control has been carried out regularly until canopy closure. The timing of this depends on location (lowland or upland) and planting density but is usually much less than 4 years.

Most of the areas planted have been comparatively small (less than 10 ha) and most have been expensive to establish. The actual costs were lower because volunteer labour did much of the planting.

The landscape locations for these plantings have varied. Some have been used as buffers or additions to enlarge small rainforest remnants. Others have been used to create corridors between forest remnants while some have been planted as small patches in otherwise extensively cleared areas.

Many sites are now 10 years old and have developed closed canopies and complex structures. Numerous sites are now beginning to acquire species that have colonized from nearby intact natural forest, especially if these are within 1 or 2 km distance. Most have become inhabited by increasing numbers of wildlife.

The mixed-species plantations

These plantations were initially established in a somewhat ad hoc fashion. Most were planted with between two and ten rainforest tree species and each of these was represented by a similar number of individuals. The species were mainly those known to have a high commercial value (based on the pricing structures that evolved during earlier logging in natural forests). Inevitably, some of the species used in these random assemblages were outcompeted and failed. This was because of a difference in growth rates or physiological attributes. Over time more effort was spent on trying to identify complementary species so that more stable mixtures could be assembled.

These mixed-species plantations raise a number of silvicultural dilemmas but perhaps the most difficult is that of making the trade-off between enhanced biodiversity and maximizing productivity. Just how many species should be used and what are the functional consequences of this diversity? Different landowners will make different judgements about this, of course.

Another dilemma that is becoming more evident is the question of just how these multi-species plantations should be managed. Should thinning be done to maximize growth on the best trees (thereby sacrificing some species from the stand) or should it be done in such a way as to retain diversity? Again it is clear that different landowners are adopting different views depending on their original reasons for planting.

Although a lot of effort and resources have gone into this reforestation effort, it is difficult to say just what its ecological consequences have been. The planted areas are small and scattered and have been planted opportunistically rather than as part of a strategic intervention to conserve biodiversity in a fragmented landscape. At the very least, they have provided a new set of stepping stones that will enhance connectivity to some degree. Just how effective this will be remains to be seen.

Additional readings can be found in Erskine *et al.* (2005).

CASE 7.2. FOREST INFORMATICS: NEED FOR A FRAMEWORK FOR SYNTHESIS, DATA SHARING, DEVELOPMENT OF INFORMATION INFRASTRUCTURE AND STANDARDS FOR INTEROPERABILITY (TORAL PATEL-WEYNAND)

One newly emerging area is the cohesive development of informatics for the forestry sector. As can be seen from the number of emerging issues discussed here as well as the topics discussed in the previous chapters, foresters and ecologists today, more than ever before, are faced with a situation where large amounts of data and information are being accumulated from diverse sectors. This need to understand and synthesize large amounts of data is a result of forest science becoming increasingly multidisciplinary in nature. New developments in science and technology have also provided novel opportunities for collecting and organizing data that could greatly expand our modelling and predictive capabilities and provide valuable input for our decision-making processes. While this is an exciting prospect and promises many opportunities, challenges abound and progress will depend on how well that data and information explosion is handled. In addition to the traditional concerns about forests and their sustainable management, there is a growing need for communication at all levels, including presenting forest issues to non-foresters.

Potentially useful and critically important information is plentiful but it is virtually impossible to use in practical ways unless it is easily accessible and presented in a framework that is easy to comprehend and is user-friendly. The sheer quantity and diversity of information requires an organizing framework factoring in information and data at the national, regional and international scale. These regional and national frameworks must also contribute to the global information infrastructure, by making possible the full and open sharing of information and data among nations.

To formulate a strategy that works towards achieving these long-term goals, we need to develop, through public–private partnerships (among government, industry and academia), an objective, accessible database of knowledge that includes:

- what we know about forestry;
- species of all plants and animals, their characteristics and interactions, their habitats and ecosystems;

- how human activities impact these species; and
- what kinds of activities comprise best practices for managing them.

Overall, what is needed is an organized framework for collaboration among regional, federal, state and local organizations and in the public and private sectors. Such a collaborative framework would provide improved programmatic efficiencies and economies of scale through better coordination of efforts.

In the biological and ecological realms, bioinformatics and ecoinformatics are emerging areas where a fair amount of effort is going into setting up distributed information networks at the national, regional and global levels. Well-established information and data exchange standards and networks exist and are being refined for specialized areas such as taxonomy, biodiversity, invasive species and aquatics and for other thematic areas.

In the forestry sector there has been very little movement towards creating an active, integrated global forest informatics network. A forestry network that also allows for access to databases worldwide so that decision-makers, scientists, modellers, managers and the public and private sectors have up-to-date information and data available is critical. A considerable amount of work has gone into the development of tools and metadata standards so that databases can be interchangeable and accessed more easily. The forestry sector has much to gain by actively seeking public and private partnerships in building a global effort that allows for sharing of tools, standards and information that can be modified or even directly used by the forestry sector.

Database accessibility and effective use by a variety of audiences are becoming a common concern worldwide. Duplication of effort, loss of time and loss of collaborative opportunities are often criticisms that are levelled at the lack of accessibility regarding current data and information from collections of ecological and forestry data. While too few data have been the norm, having too many data had never been considered a problem until large amounts of information started being generated, stored and distributed electronically (e.g. data from experiments, analyses, images both real and computer-generated). This availability of electronic data has increased the demand for databases.

Databases have now become the norm as the tools for recording, storing and propagating data. Although data vary in scope and scale, most are specialized and many are freely available on line while others are proprietary. In the forestry sector there have been both public- and private-sector efforts geared to establishing small-scale information networks at the local and national level, but integration is lacking at the broader international level. With current efforts under way in bioinformatics and ecoinformatics to link large-scale terrestrial and oceanic databases to geospatial databases to improve predictive capability at the global level, it is even more important for the forestry sector to be fully engaged in developing data standards and methods for interoperability. Such efforts will increase the utility of information obtained from the inventories and research activities that were collected in the past and where several decades of data exist.

As we raise the threshold for forest informatics, as has been apparent for other informatics approaches, the necessity for capacity building in developing countries will need to be addressed.

Ongoing regional efforts in metadata training are already under way. As networks develop, there will be an increased need to develop the infrastructure that supports these efforts. New approaches, techniques and solutions will also need to be developed in order to translate data from outmoded media into usable formats and to enable the analysis of large forestry and ecosystem databases. Faced with large databases, traditional approaches tend to collapse when doing database management, statistical analysis, pattern recognition and visualization. Software applications that provide more natural interfaces between humans and databases than are currently available will also be increasingly needed.

A global attempt at integrating forestry into the current informatics framework appears to be a necessity. With the development and distribution of tools and standards necessary to facilitate interoperability, the forestry sector is very well positioned to initiate and formulate an effort that is truly global in nature. Engaging in and building on an information network of public and private partnerships allow for meaningful interactions with scientific data and information so that resource use decisions are based on accurate, up-to-date knowledge and sustainable options are available to scientists, decision-makers and the general public.

CASE 7.3. FOREST CERTIFICATION, LAWS AND OTHER SOCIETAL-BASED CONSTRUCTS: TOOLS TO INCLUDE SOCIAL VALUES IN FOREST SUSTAINABILITY (KRISTIINA A. VOGT)

The Basics of Forest Certification

Forest certification needed to be developed because economics, laws and regulations were ineffective in including non-market values (e.g. environmental services desired by society that generated no revenue) when measuring whether sustainable forestry was being practised. These non-market values include more than aesthetic qualities, but placing a monetary value on them is not easy. For example, it is very difficult to place a monetary value on the pure water you drink or the clean air you breathe.

Forest certification is an auditing system that measures whether good forest stewardship is being

practised and the stamp that it provides supposedly creates market-based revenue for non-market values (Vogt *et al.*, 1999). In certification, sustainable forestry or good forest stewardship is assessed by balancing the collection of forest products with the provision of sustainable livelihoods for rural communities or forest-dependent people and by continuing to provide for the suite of other organisms that live in forests. Therefore, forest certification utilizes both social and natural science indicators as an integral part of its audit.

Historically, economics has been the main tool used to determine the costs and benefits of deciding the 'where', 'what' and 'how' of forest resources (Costanza, 1991). Economics, however, does not include:

- externalities associated with a product (e.g. who should pay for environmental disasters if wood impregnated with creosote leaks or decays into water bodies and proves toxic to plants, fish and microbes);
- other costs not directly associated with the products but that might have negative environmental impacts (for example, increasing the growth rate of trees through fertilizer and herbicide use so that a marketable product is available in less than 50 years);
- the social and natural system legacies that change most economic predictions (e.g. natural system legacy = landscape fragmentation that changes how disturbances impact forest edges so that more trees are blown over during a storm, putting the timber supply at risk; social system legacy = the attitudes of a community determining what their landscape should look like, i.e. its aesthetics, such as having either more or less open space, forests, agricultural fields or what activities are acceptable in a forest (hunting, collection of mushrooms, harvesting timber));
- a way to determine a 'real' monetary value for a non-market-based product or ecological service, since these products or services are not traded in global markets. For example, it is difficult to establish a price for the following: global energy balance and climate; space and suitable substrates for human habitation; oxygen, fresh water, food, medicine, fuel, fodder, fertilizer, building materials and industrial inputs; nature's aesthetic, spiritual, historical, cultural, artistic, scientific and educational opportunities.

The legal system, also a societal construct, is a tool that can be used to regulate non-market values. Regulating non-market values using laws can frequently infringe on private and political property rights. In some places, such as the United States, tampering with private property rights is very contentious, since government regulation of what occurs on private lands is not condoned by the majority of the public.

Today, what happens on private lands is an issue for conservation and the practice of sustainable forestry, since public forest lands are inadequate by themselves to provide the habitat needed to sustain species dependent on forests (e.g. spotted owl and marbled murrelet in old-growth forests in the Pacific North-west United States). Private property owners have to be included in the decision-making process if conservation and sustainable forestry are the desired outcomes. However, the polarized nature of resource uses and conservation makes it difficult to build a consensus on what regulations to implement and what would be universally acceptable to all private property owners.

The continued conflicts over regulating and the interpretation of the laws relevant to forests attest to the difficulty of using the legal system to resolve natural resource use conflicts. Today, many of the decisions on resource uses are occurring in the courts. Since the legal system is structured to portray parties as defendants or plaintiffs, it polarizes and creates further conflict among the different groups that are stakeholders in the forests.

There are numerous rules and regulations on how forests should be used and managed. In the United States, laws on forests and forest-dwelling animals exist at federal, state and local levels (e.g. the state of New Jersey has over 100 regulations specifically related to forests). At the federal level in the United States, most of the laws are related to chemicals in the water or air, human health and the protection of endangered species (e.g. 1947 Forest Pest Control Act; 1948 Rodenticide Act; 1970 Clean Air Act; 1972 Federal Environmental Pesticide Control Act; 1972 Federal Water Pollution Control Act; 1973 Endangered Species Act; 1977 Clean Air Act; 1986 Safe Drinking Water Act; 1987 Clean Water Act; 1990 Clean Air Act). These laws and regulations, however, are inadequate to deal with complex environmental problems that have very little economic value placed on forest condition and on maintaining its pristine state. Conflict over environmental issues further polarizes a society divided by multiple attitudes and opinions that are strongly held by different segments of society. Forest certification is one such initiative that was designed to reduce conflict by greening or making resource uses more environmental and providing economic return from the acquisition of environmentally friendly products.

The definition of forest certification clearly states what certification is supposed to accomplish: 'Forest certification is defined as a market based, non-regulatory, voluntary, third-party audit of whether forest ecosystems are being managed in a sustainable manner while providing sustainable livelihoods for rural communities dependent on forests' (Vogt *et al.*, 1999).

Forest certification is different from the economic and regulatory approaches used to control forest uses since it is voluntary. Voluntary approaches to assessing whether forest management is sustainable are typically conducted by a third party or an independent group. These groups are trusted to provide an unbiased assessment of forest uses and do not have a stake in the results of the forest audit. Certification has been used to assess management practices in forests and in agriculture (e.g. coffee production, organic food production). It has been successfully implemented in many tropical countries and appears to be empowering local communities to make decisions in their forests and become part of the global economy (see Case 7.5 – The First Certified Community-based Forests in Indonesia: Stories from the Villages of Selopuro and Sumberejo, Wonogiri, Central Java).

Certification costs money, however, and is not cheap (Taggart, 1999) to implement. Furthermore, being certified also does not guarantee additional financial return to the owners of a forest (Estey, 1999; Taggart, 1999). Except for certain speciality niche products (e.g. wood flooring, cedar shakes used to construct a roof), higher prices in the marketplace have not materialized for products collected from certified forests. Therefore, forest owners or producers of wood products have to include the costs of achieving environmental values deemed important by society as part of the price of their product or decrease their profit margins.

The difficulties of including sustainable development in sustainability assessments have the unintended consequence of minimizing the role of local communities in voicing their views when designing sustainable forest practises. Historical records have repeatedly documented the marginalization of rural communities living near forests. For example, in the 1970s forest-dwelling people were often relocated from their traditional homes in tropical forests because the ideal forest condition was then regarded as one without people (see Chapter 3, this volume). It was not until quite recently that rural communities both in the United States and abroad were being included in development decisions (see Case 7.10 – Integrating

Conservation and Sustainable Forestry: a Pacific North-west United States Example).

Unintended Consequences of Certification: the Globalization of Local Communities

Even though the goals of certification are to pursue sustainable forestry practises while improving the livelihoods of local communities, these two goals alone should not be used to define whether certification has been successful. The impact of forest certification is potentially more significant than had been envisioned by its early developers. Forest certification has provided a vehicle for rural communities to become part of the global economy. For example, many of the certified community forests in the tropics (see Case 7.5 – The First Certified Community-based Forests in Indonesia: Stories from the Villages of Selopuro and Sumberejo, Wonogiri, Central Java) and also in the temperate climatic zones are actively exploring the development of new products from forest materials, especially transportation fuels (see above – 'A Solution for Sustainable Forestry: Manage Biomass Wastes to Produce Energy Sustainably'). Helping communities to develop and produce entirely new products from their forests was never an intended goal of certification.

Forest certification has been less effective in improving the livelihoods of forest-dependent communities when it focused on traditional uses of forest materials. Despite this traditional focus, an unintentional consequence of certification was to change how local communities interface with other global societies and economies.

Today, rural communities extracting resources from forests are playing a leadership role in pursuing economic opportunities in both regional and global markets. In some tropical countries, the international contacts forged when local communities were being certified have opened up new global market opportunities that would not otherwise have existed. These communities made many international contacts and are actively using them to create economic opportunities for their citizens. New opportunities are important for these tropical countries and new forest products need to be developed in an effort to increase the economic options available to them. In addition, these communities need to be introduced to the newer technologies that will improve their energy uses without harming the environment. These communities are working to become part of the global markets.

CASE 7.4. TOURISM AND SUSTAINABILITY IN A FORESTED PROTECTED AREA (STEPHEN F. McCOOL)

Introduction

Sustaining the values that forests provide has become increasingly challenging in an era of complexity, uncertainty, change and contentiousness. Certainly, the very definition of sustainability itself is contested, and no consensus exists on what values forests should sustain (Gale and Cordray, 1991). To a very great extent, sustainability is about divining the public's interest in the future, but that task is made difficult by competing value systems and uncertainty about the preferences of future generations. While this case study is about sustaining values of a wilderness resource, it is also significant from a tourism perspective since the questions and issues are similar. The ambiguities and challenges can be overcome somewhat by viewing sustainability as building resiliency and adaptability into human–natural systems. This construction of sustainability requires, however, not so much scientific expertise but public deliberation, debate and joint fact-finding among scientists, forest managers and affected publics.

This complexity is increased when non-traditional uses of forests, such as recreation and tourism development and promotion, occur in these settings. In many situations, particularly in the United States Pacific North-west and Rocky Mountain regions, the economies of local communities have been restructured over the last two decades. These communities have shifted away from resource commodity processing to a greater dependency on tourism, service industries and occupations, and transfer payments (pensions, social security, dividends, etc.). In these situations, the nature-based amenities that have formed the environmental backdrop for rural communities have now become a source of income to local people. Their stewardship is an issue in sustaining livelihoods, economic opportunity and quality of life.

This case study involves the search for sustainability with the objective of providing lessons for that search and questions for consideration in other similar areas. It is located in a large protected area with a forested environment that is also significant from a tourism perspective.

Planning for Recreation and Tourism in the Bob Marshall Wilderness Complex

The Bob Marshall Wilderness Complex, a 600,000 ha protected area lying astride the continental divide in the state of Montana, is representative of the issues confronting the sustainability of forested ecosystems within protected areas. The Complex, composed of three juxtaposed designated wildernesses (the Bob Marshall, the Great Bear and the Scapegoat), is roadless and has been protected since the 1930s from timber harvesting and other resource commodity activities. The US Department of Agriculture (USDA) Forest Service is charged with its stewardship. It is heavily used by both backpackers and horseback riders engaged in multiple day trips. Visitors either travel through the area on their own or use the services of outfitters to provide tents, food and horses in addition to guiding services. The Complex is often viewed as the 'crown jewel' of the National Wilderness Preservation System, and is an important tourism resource for the state of Montana.

The area is managed under the mandates of the 1964 Wilderness Act, which require it to be 'left unimpaired for future use as wilderness'; [to] 'provide outstanding opportunities for solitude or a primitive and unconfined type of recreation'; and compel the Forest Service to preserve 'the wilderness character of the area'.

In the early 1980s, management of the area was confronted with a variety of issues:

1. Acceptability of visitor-induced biophysical impacts – these impacts occurred at campsites and along trails. Management was confronted with the question of how acceptable these impacts were in a wilderness setting and how to manage them.

2.   Conflicts between horseback riders and hikers – hikers often had to walk along trails frequently covered in manure and stay at campsites that had been extensively impacted by horses and mules.

3.   Amounts of solitude that could be offered – how much opportunity to experience solitude should the agency provide and what do visitors expect? Management needed to determine if the outstanding opportunities for solitude were being provided and how to manage visitors to sustain those opportunities.

4.   Other issues, such as big-game hunting (primarily for elk, deer and bear) and the role of outfitters in the wilderness, were also generating conflict over how the Complex should be managed.

An attempt to address these issues in the early 1980s with a plan drew significant public opposition; the plan could not be implemented because there was little public engagement and it confused technical prescription with social values and judgements. The plan had been written in the typical fashion: agency experts applying the best technical and scientific knowledge to solve a set of problems and then releasing the plan to an unprepared public, who responded in a negative fashion. This type of planning, termed 'rational-comprehensive planning', grew out of progressive era approaches to resolving public problems: apply the best technical knowledge to a social problem using experts (to eliminate political influences). Science and efficiency took preference over values; public preferences and responses were marginalized. It was thought by the agencies searching for solutions to the problem that involving only experts and scientific knowledge would result in 'objective' decisions.

However, the 'objectivity' of plans often mirrored the frequently hidden value systems and perspectives of the experts writing the plans. And such approaches to planning implicitly assumed consensus on goals and scientific agreement on cause–effect relationships (Thompson and Tuden, 1987). The status of contemporary forest management in North America, however, is contentious, complex and filled with scientific and technical uncertainty.

Some Lessons Learned

In the case of the Bob Marshall Wilderness Complex, the reasons for the first planning attempt

failure were many. Certainly its failure included the contentious environment in which the planning occurred. In addition, the conflict over how to resolve the issues identified above and the lack of knowledge about visitor use patterns, location and amounts of impacts, and the effectiveness, efficiency and equability of management actions designed to address these problems contributed to this failure. An administrative decision was made to start the planning process over again, but to change the paradigm guiding this process.

The fundamental lesson learned was to involve the public throughout the planning process, from problem definition and framing, through the development of alternative strategies and actions, to the selection of a preferred alternative. This process was designed to develop consensus (McCool *et al.*, 2000) about objectives of the plan, to build trust by discussing issues and making decisions in public, and to create ownership in the plan to ensure its implementation and success. This type of planning had been proposed by Friedmann (1973). Termed 'transactive planning', it involves establishing dialogue among all those affected and involved in a planning issue – managers, scientists, publics – which would lead to mutual learning (about each other, the planning process and the subject matter), which in turn would result in societal guidance – actions to sustain the values of the Bob Marshall Complex in this case.

A second lesson learned from this search for sustaining values was to combine the technical, rational planning process with public engagement (see Fig. 7.7). Transactive planning was combined with a new wilderness planning process being developed at the time by Forest Service scientists termed the 'limits of acceptable change' (LAC) (Stankey *et al.*, 1984, 1985; McCool and Cole, 1998). The LAC

Technical planning process   +   Public involvement (learning, consensus building)

Successful planning

Fig. 7.7.   Schematic of the process that guided development of a plan leading to the sustainability of values provided by the Bob Marshall Complex.

process is oriented towards identifying the acceptability or appropriateness of visitor-induced impacts in protected areas and determining the types of management actions that would be both effective and acceptable. While effectiveness is a technical issue, acceptability is a social or political one.

The public engagement was implemented through the creation of a task force consisting of about 35 individuals representing managers, scientists (both biophysical and social) and differing public stakeholders. The task force met several times per year (day-long meetings) and followed the nine-step LAC process (Stankey *et al.*, 1984). The plan was completed after about 5 years of deliberation. Monitoring the implementation of the plan occurs every year in an annual meeting with members of the public, many of whom were active in the mid-1980s task force.

A third lesson was that each type of task-force member brought different expertise and knowledge to the planning process: scientists provided the technical background about recreational use patterns, expectations, motivations, campsite and trail impacts, wildlife habitat requirements and impacts of grazing on native vegetation; managers provided knowledge about what actions and strategies were administratively feasible and what were consistent with agency policy and law; and the public provided experiential and local knowledge about specific places in the area and indicated what dimensions of the plan might not be politically or socially acceptable. All three types of participants were necessary to the construction of a plan that would be effective, that would be scientifically sound and that could be implemented in an environment characterized by declining fiscal resources and social conflict.

Being specific and explicit where possible was a fourth lesson. Ways to deal with campsite impacts, expectations of solitude and grazing, specific and quantifiable standards of how much impact was acceptable were developed. These standards varied by specific zones within the wilderness. For example, in the most pristine regions of the wilderness, campsite impacts were limited to 100 square feet of barren soil. In the least pristine, up to 1000 square feet was allowed. By establishing these standards, all three groups had the same understanding of the maximum impact that could be allowed in order to sustain important wilderness and tourism values.

Discussion

Planning in the 'fishbowl' (Sargent, 1972) – where all decisions and information are public and visible – is not necessarily the easiest and most efficient process (indeed, public engagement is often criticized for the 'extra' costs it entails) but often results in more explicit consideration of value issues such as sustainability (e.g. what should be sustained and how), identification of the distributional costs of proposed management actions, development of trust among participants and creation of ownership (Lachapelle and McCool, 2005) in the resulting plan.

While the process in the Bob Marshall Complex is still ongoing, in the sense that the public is still engaged in implementation, it exemplifies critical questions in the search for sustaining the values provided by forested ecosystems. For example, how do we know that a plan or planning process is successful? What do we mean by success? Planning literature often refers to plans that have been written but sit on a shelf gathering dust. Should plans be a living document that are not only constantly referred to but are always in a state of evolution? In many respects, a plan is a set of actions developed under certain assumptions about the future. Those assumptions often prove faulty, and thus the plan must be modified accordingly. But what about concepts such as sustainability? Should values to be sustained be relatively permanent? What about changing public preferences? How should managers deal with these? In one sense, such changes are no more problematic than political philosophies, which change with each new election or transition of power.

Sustaining forest values and building resiliency require coordination, deliberation and the input of a variety of value systems. In the Bob Marshall Complex, this was done by lengthy, intimate and continuous engagement of the public, managers and scientists, each on an equal footing, each 'at the table' and each contributing constructively to the planning process. The process pointed out the need to redefine what planners do (facilitation vs technical operations), understand that many sustainability issues are value-based rather than technical and commit resources to developing opportunities for learning and consensus building in the planning process.

CASE 7.5. THE FIRST CERTIFIED COMMUNITY-BASED FORESTS IN INDONESIA: STORIES FROM THE VILLAGES OF SELOPURO AND SUMBEREJO, WONOGIRI, CENTRAL JAVA (ASEP S. SUNTANA)

We plant trees and develop our forest far beyond any expectation to gain rewards and acknowledgement. We do it due to our expectation to provide sustainable forestry to our future generations. Moreover, we need to respect the nature, so the nature can respect us.

(Mulyono at Wijaya, 2005)

In a break with the past, forest stakeholders now seem more concerned about certification than with the latest forestry regulation or with initiatives from the United Nations.

(Bass, 2003)

Introduction: Background, Problems and Development Initiation

Why the two communities were prime candidates for certification

Wonogiri district, one of the regions in Central Java province in Indonesia, has been known as a dry and barren area. Selopuro and Sumberejo are two of the villages in Wonogiri that frequently experience clean-water shortages during the dry season and flooding during the rainy season. Agricultural lands in these two villages, as well as in some neighbouring villages, have low-quality topsoil, which barely supports the growth of agricultural crops. The landscapes in the villages of Selopuro and Sumberejo are dominated by rocky lands.

In 1960, Parmo and Siman (Jarak hamlet), Tayat and Misman (Pagersengon hamlet) and Warimin (Sidowayah hamlet) – who, like most Indonesians, have only one name – were concerned about the situation described above and started introducing new ways of thinking regarding how to counter the underprivileged conditions in Selopuro village (Wijaya, 2005). At the same time, Mbah Sularjo – a hamlet chairman in Sumberejo village – also started considering finding out how to alleviate poverty in his hamlet (Wijaya, 2005). Many activities were pursued in order to provide economic benefits, to improve the ecosystem condition and to provide significant steps to alleviate poverty. Among the activities pursued was the planting of

acacia (*Acacia auriculiformis*), teak (*Tectona grandis*) and mahogany (*Swietenia mahogani*). In 1972, the World Food Programme (WFP) provided financial support and advised forest farmers to terrace their lands and to cultivate eucalyptus (*Eucalyptus alba*), acacia (*A. auriculiformis*) and calliandra (*Calliandra calothyrsus*). In 1985, forest farmers in Sumberejo established a number of forest farmers groups, known as 'Kelompok Tani Hutan Rakyat' (KTHR).

Land preparation, tree planting and related forest management activities became more organized with the establishment of KTHR. At that time, KTHR members began to develop tree nurseries as well as to prepare the lands for planting tree seedlings by breaking and clearing big rocks. The groups also decided to focus on planting mainly teak and mahogany trees, mostly because of economic considerations; the two species provide higher prices than other commercial species of trees.

Nowadays, after more than 30 years of developing community-based forestry, the two villages are among the greenest areas in the Wonogiri district. Teak, mahogany and other species and varieties of trees have been planted on every possible piece of land, from maize fields to herbal medicine plantations, from housing compounds to public buildings. These tree plantings have provided a better environment for their livelihoods and also better ecosystems for wildlife.

According to Persepsi (2005), in 2004 the forest farmers of the two villages shared about 17.5% forested land as well as sharing 70,000 m³ raw timber production in the Wonogiri district. Economic benefits from managing community-based forestry provide at least 20% of forest farmers' incomes.

Like any other forest farmer groups in Indonesia, forest farmers of Selopuro and Sumberejo consider income from forests as seasonal. Forest farmers would sell wood only if they are forced to do so, particularly 'when things are too difficult to handle, financially', explained Sutanto, one of the forest farmers (*Jakarta Post*, 2005).

Strict discipline and a code of conduct for how forests were used were an integral part of the forest farmer organization. For example, whoever fells a tree has to plant ten to 25 new trees (*Jakarta Post*, 2005).

The practices established by the forest farmer organizations started gaining recognition from other farmers in neighbouring villages, local government officers, academia, funding agencies, research institutions, and so on and so forth. Their achievements demonstrated their willingness to provide better solutions for the community. These two communities on their own found successful solutions to the following problems that they were facing (Persepsi, 2005): (i) regarding their ability to produce food when faced with an inability to control rainfall patterns (i.e. lack of spring water during the dry season and flooding in the raining season) and the poor conditions of their agricultural lands; and (ii) poverty alleviation and no longer practising the unsustainable approaches to obtaining rural livelihood practices that had occurred in the past. The manner in which these communities pursued solutions to their problems made them ideal candidates to participate in forest certification.

The certification process of the villages of Selopuro and Sumberejo

With significant support from a local non-governmental organization, namely Persepsi (Perhimpunan Untuk Studi Pengembangan Ekonomi dan Sosial or Society for Social and Economic Development Studies), forest farmer groups agreed to join the 'Community-based Forest Management (CBFM) certification programme'. CBFM was under the aegis of the Indonesian Ecolabelling Institute (LEI).

Forest farmer groups in both Selopuro and Sumberejo villages together with Persepsi established the 'Forest Farmers Group for Forest Certification (KPS)'. This organization was to coordinate the management of the forest areas and obtain certification status for forests in both villages. The forest areas being certified were then considered as a forest management unit (FMU) in the CBFM certification programme of LEI.

In Selopuro village, there are eight KPSs (Pagersengon, Jarak, Sudan, Selorejo, Watugeni, Sidowayah, Tulakan and Pendhem) with 682 members (households). The total forest area included in the KPS is 262.77 ha. In Sumberejo

village, there are eight KPSs (Kalinekuk, Semawur, Ngandong, Wates, Rembun, Rowo, Puthuk, Gembuk) that include 958 households. The total area in Sumberejo is 526.19 ha.

After thorough administrative and field assessments were conducted by Penusahaan Terbatas Mutu Agung Lestari (one of the LEI-accredited certification bodies), forest certification was granted to the two KPSs or farm forestry organizations on 22 October 2004. The certificates were officially granted at the first LEI General Assembly held in Jakarta. Dr Emil Salim, former Minister of Environment and one of the founding fathers of LEI, presented the certificates.

There are many organizations that have been involved with the process of certification and the continuing management of the forest lands in these two villages. Most of them worked with KPS on the institutional development as well as providing technical assistance regarding forest management techniques and business development (Persepsi, 2005). The names of the organizations involved in this certification process are: Persepsi, World Wide Fund for Nature (WWF) Indonesia, LEI, Gesellschaft für Technische Zusammen arbeit (GTZ), *pemerintahan desa* (village government), Dinas Lingkungan Hidup (district environmental office), Kehutanan dan Pertambangan (LHKP) Wonogiri, Balai Pengelola DAS Solo, forest products industries, Trade and Industrial Department, Koperasi Agro Niaga Jaya (ANJ) Wonogiri, Pusat Kajian Hutan Rakyat/Community Forestry Research Centre (PKHR) – Gadjah Mada University, Wonogiri district government, Paguyuban Pedagang Kayu (Association of Wonogiri Timber Traders) Wonogiri, and Bengawan Solo Watershed Forum (consists of 14 districts in Central Java, Jogjakarta and East Java provinces). This list demonstrates how the success of the certification process needed the partnering of many organizations and groups, who each had something to contribute to the programme.

Persepsi, ARuPA (ARuPA is an Indonesian name for Volunteers Alliance for Saving Nature) – also a local NGO – and LEI agreed to start providing further assistance to expand the community-based forest certification programmes to other regions. There was a special interest to include other villages in Wonogiri to become certified because of the encouraging results obtained in the villages of Selopuro and Sumberejo. Accordingly, Persepsi is developing a plan to extend the programme to cover up to 20,000 ha of community forests in Java (TNC–WWF ALLIANCE@News, 2005). There are at least four villages interested in pursuing forest certification because of this success. These villages are Girikikis, Sejati, Tirtosworo and Guwotirto.

Community-based Forest Management and Certification

There are lots of concepts and programmes that explain and promote community-based forest management. The Center for International Environment Law (CIEL, 2002) claimed that most of the existing community-based natural resources management (CBNRM) policies and programmes are still limited in scope and are predicated on an assumption of state ownership of land and other natural resources. If this statement is correct, it has implications for community-based forest management in Indonesia. The government of Indonesia, for instance, retains primary management authority and merely grants legal rights to local communities to use and benefit from certain natural resources in a defined area in return for local communities agreeing to assume certain duties. The government of Indonesia also tends to provide 'community-based' forestry programmes with a 'prescriptive' approach and does not take into account the full range of services that the forest provides for the local community in Indonesia (Yuniati, 2005). On top of that – as happened in industrial-based forestry – the Ministry of Forestry highly guides the technicalities of forest management towards timber production.

Forest farmers in Selopuro and Sumberejo are practising community-based forest management and not a CBNRM or co-management. They work extremely hard in avoiding having their forest practices come under the influence of either the local forestry office or the state-owned company in the neighbouring area. According to CIEL (2002), if it is community-based forest management, then the community should exercise and have the primary decision-making authority.

A number of changes have been occurring recently in Indonesia that allow the villagers themselves stronger control and management over community-based forests. For a number of years, national and local NGO activists, academia and even several forestry-related government officers have been demanding a critical shift in the forest management regime, i.e. from state-based forest management to community-based forest management. The fall of the authoritarian regime in 1998 opened up opportunities for civil society to have a better dialogue with both executive agencies and legislative bodies concerning the issue. As a result, the Forestry Act was approved by the Indonesian House of Representatives on 24 September 1999. This Act recognizes community-based forest management systems (Undang-Undang Kehutanan/ UUK No. 41/1999). Moreover, the People's Consultative Assembly (MPR) enacted Decree no. IX/MPRI/2001 on Agrarian and Natural Resources Management reforms. The decree is to some extent influenced by the intensive lobbying of the civil society movement (mostly academia and NGO activists), which aimed to ensure that community-based property rights as well as community-based forest management are legal under Indonesian law.

Regardless of the 'multi-stakeholders' process of bringing these two important laws into effect, several people have criticized them. Some stakeholders do not like how the acts were formulated. Some have objected to the content of the UUK No. 41/1999 and how the act was being interpreted and implemented by central government (Yuniati, 2005).

Since the decree on Agrarian and Natural Resources Management provides a mandate and accountability both to the President of Indonesia and to the Parliament to provide any necessary acts to reform all regulations regarding agrarian and natural resources management, the government has considerable influence on how these laws are being interpreted and implemented. The decree also provides a broader opportunity for using customary law as the guiding principle for determining that communities would make decisions in forests under the auspices of the community-based property rights or community-based forest management. Although MPR RI decree No. IX/MPRI/2001 has been widely accepted by the public (due to the content of the decree), both the Parliament and the President of Indonesia are reluctant to implement it.

The lack of implementing this decree has made the general public assume that any effort – such as rigorous lobbying with either government institutions or Parliament – to gain acknowledgment from the state concerning the importance of CBFM practices will have little chance of succeeding. The situation has also triggered stakeholders to seek a better instrument to assess forest management practices so that they can be environmentally and socially acceptable by the Indonesian society and the international community. Forest certification and the fact that it is a credible tool and globally accepted may provide a mechanism that activists can use to provide significant social, economic and political recognition for community-based forest management. Certification is recognized and accepted by the public through its transparency and multi-stakeholder consultation process.

Since the beginning of the development process (1993–1998), the LEI has considered recognizing community-based forest management as the tool to achieve sustainable forest management. LEI decided to develop a particular forest certification scheme to evaluate the performance of community-based forestry practises. This certification scheme is completely different from any of the other systems that exist to evaluate forests – both natural production forests and plantation forests – managed by companies.

Benefits, Challenges and Opportunities

The Selopuro and Sumberejo success stories demonstrate that community-based forestry can fulfil certification standards. The challenge now is to help the communities secure commercial and other benefits as a result of the certification.

There are apparently potential buyers from local (Bali and Surabaya) and international (France and the Netherlands) markets that have been considering buying wood from certified forest managed by forest farmers in Seloputro and Sumberejo at a premium price. WWF Indonesia and its partners are developing market links between the communities and a Dutch buyer (which has expressed strong interest in a long-term partnership with the two communities) as well as Indonesian companies in Bali and Jogjakarta (WWF, 2005). 'With the premium price facility, they can sell their wood products for 15 percent to 30 percent more than the usual price,' said Taryanto Wijaya of Persepsi (*Jakarta Post*, 2005).

Certification also works if it changes how communities can access credit to develop their industries. These communities have improved their opportunities to get significant credit from commercial banks and co-ops. It has also increased the opportunity for the forest farmer groups to have financial and technical support from international communities (see 'Unintended Consequences of Certification: the Globalization of Local Communities' in Case 7.3).

Broadened (eco-sensitive) markets have begun to emerge. Fortune 100 companies, although most of them still initially fail to recognize the widespread use of wood and paper in their day-to-day business operations, believe that purchasing with an eye to environmental responsibility can actually improve financial performance (Metafore, 2005). Forest certification could become a significant tool to improve the image of the companies in the Fortune 100.

At the village level, there have been many improvements in the environmental conditions of the surrounding landscape due to unintended consequences of certification. For example, many headwaters are found along the villages' Nekuk River providing water for people not only in Selopuro and Sumberejo but also in neighbouring villages. Social interaction among forest farmers has improved, particularly through KPS.

Community-based forest management systems provide another potential benefit in terms of bio-energy. In an era when the price of fossil fuel has become too high for forest farmers and when the availability of fossil fuel sometimes becomes rare, utilizing certified woody biomass – from thinning and pruning activities – as raw materials for bio-energy and bio-fuel could become a tremendous opportunity for forest farmers and for the entire community members in fulfilling their energy needs (see above – 'A Solution for Sustainable Forestry: Manage Biomass Wastes to Produce Energy Sustainably'). To take advantage of this opportunity, appropriate technology transferring biomass into bio-energy needs to develop. Since the two villages are located in rural areas, the technology should be simple, cheap and easy to apply.

Besides a number of potential benefits, there are some existing and potential problems that need to be addressed. Some of them are, but are not limited to: (i) financial and managerial problems in dealing with the immediate need for short-term cash among KPS members; (ii) weak marketing and bargaining position in (certified) forest product trading; (iii) difficulties in diversifying (certified) forest products; and (iv) lack of information regarding how to get more benefits from forest certification programmes.

Community-based forestry is a champion model for achieving sustainable forest management in Indonesia. Since forest certification provides public recognition and benefits from eco-sensitive markets, it is important that such markets are available for CBFM practitioners.

Seloputro and Sumberejo villages have shown that rural people are capable of managing and preserving their forests while still enjoying the many advantages they provide, not only for themselves but also for neighbouring villages.

CASE 7.6. NON-TIMBER FOREST PRODUCTS AND RURAL ECONOMIC DEVELOPMENT IN THE PHILIPPINES (IVAN L. EASTIN AND SHELLEY L. GARDNER)

Background

Maling Cruz graduated from the University of the Philippines with a baccalaureate degree in agricultural economics. She was recently hired as a field marketing specialist with the Community Fibre Farmers Association (CFFA), a non-governmental organization working to improve the welfare of handicraft workers and the small farmers who produce the raw materials for the handicraft industries in the Philippines. After a short period of training in the metropolitan Manila headquarters of CFFA, Maling was sent to her project site, Kalibo, located in the western Visayas region of the country (Fig. 7.8). In Kalibo, Maling's project involved working to help the abaca farmers increase the revenue they earned from processing their abaca plants into abaca fibres.

Although she has been working in Kalibo for almost a year now, Maling is becoming increasingly frustrated with her project. She has received little cooperation from the local agents and traders involved in the abaca business. In addition, the remote location of her project site has made it difficult to communicate with and receive support from the CFFA staff members in Manila or from the government agencies tasked with promoting the development of the fibre-based industries, such as the Fibre Development Authority. The Executive Director of CFFA had asked her to return to Manila to present her plan for helping the local abaca farmers improve the revenue stream from their abaca crop. Recognizing her need for expert guidance to help resolve the abaca farmers problems, Maling has contacted your market research group. She has asked you to provide a set of strategic marketing recommendations that address the primary challenges confronting the abaca farmers: lack of bargaining power on the part of rural abaca farmers, unavailability of credit-granting institutions at the local level and farmers' lack of market and pricing information.

Introduction

Abaca/abakà/Manila hemp (*Musa textiles* Nèe), which is in the *Musaceae* family (banana family), is probably one of the most important cultivated plants endemic to the Philippines. In appearance it is almost identical to the banana plant, to which it is closely related, although its fruits are smaller, have more seeds and are not edible (Fig. 7.9). Rather, the economic importance of this plant can be attributed to the fibres located in the stem of the abaca plant. The fibre was used by Filipinos long before the Spanish occupation. When Magellan arrived at Cebu, the weaving of abaca fibre was widespread in the islands, and the plant is reported to have been growing wild in much the same places as those in which it is now cultivated.

It was probably the use of abaca in traditional swidden agricultural practices (shifting agriculture) that led to its wide use in cultivation today. Cultivation is carried on now to such an extent that it is questionable whether any wild plants remain. Abaca has also been introduced into other tropical countries, but most fibre still comes from the Philippines. Abaca fibres have a very high tensile strength and were historically used as a raw material in the manufacture of ropes within the marine cordage industry. However, the recent development of lower-cost synthetic fibres has reduced the importance of abaca fibre in the production of ropes and cordage.

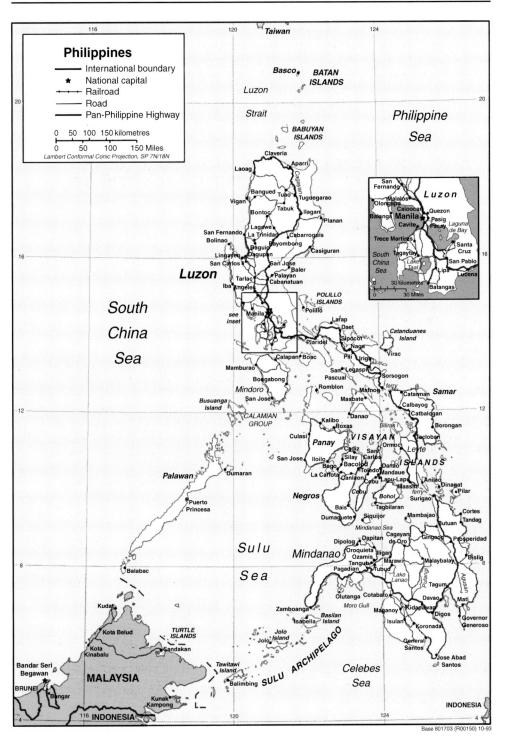

Fig. 7.8. The Philippines.

In recent years, abaca fibres have been increasingly used in the production of speciality papers (including currency and hand-made papers) and handicrafts for the export market (including place mats, handbags, rugs and doormats). Locally, abaca is still used for a wide variety of products, including textiles, baskets, hats, trays, bags, laces, lampshades, belts, matting and furniture. In addition, there is still a strong international demand for the raw fibre, which is used to produce ropes and high-quality paper products.

Abaca primarily grows in the Bicol Region, the Visayas and the island of Mindanao, where the climate, soil and abundant rainfall all favour its growth. Abaca is a shade-tolerant species that generally grows best in secondary and depleted forests where it is sheltered from direct sunlight (Fig. 7.9).

The abaca plant is mature and ready to be harvested when it flowers/fruits. If the plant is harvested too early, the strength of the fibres will be weak and the yield of fibres will be low. Following harvest, the abaca plant typically produces suckers from the stalk, which grow into new plants. The tendency for vegetative reproduction of the abaca plant reduces the necessity of replanting a field following harvest, although a small amount of replanting is always necessary. The average size of an abaca farm on the Philippines is less than 1 ha in area, suggesting that the majority of abaca farms are owned by small farmers.

The Production of Abaca Fibres

The commercial fibres are the fibrovascular strands located on the outer surface of the leaf sheath that surrounds the stem of the plant (Fig. 7.10). After the abaca plant flowers, it is cut down and the stems are collected at a central production location. The fibre strands are separated from the leaf sheath using a long, sharp knife (Fig. 7.11). The peeled leaf sheaths

Fig. 7.9. Abaca plant and the abaca production process in rural Philippines. Abaca plants closely resemble banana plants but produce an entirely different fruit. Photo by Ivan Eastin.

Fig. 7.10. Abaca plant and the abaca production process in rural Philippines. A cross-sectional view of an abaca stem. Photo by Ivan Eastin.

Fig. 7.11. Abaca plant and the abaca production process in rural Philippines. A razor-sharp knife is used to separate the leaf sheath from the abaca stem. Photo by Ivan Eastin.

Fig. 7.12. Abaca plant and the abaca production process in rural Philippines. Once the leaf sheath has been separated from the stem, it is peeled off. Photo by Ivan Eastin.

are then removed from the stem and discarded (Fig. 7.12). Since the fibres are only found in the thin outer layer of the leaf sheath, the bulk of the leaf sheath is discarded (Fig. 7.13). The thin fibre strands are then passed through a comb to separate the abaca fibres (Fig. 7.14). Fibres can be produced using either a hand comb (called hand stripping) as seen in Fig. 7.14 or a machine spindle (called spindle stripping). The abaca fibres are then collected and laid in the sun to dry for approximately 4 h (Fig. 7.15). After drying, the fibres are collected and they can either be graded or bundled together ungraded into 40 kg bundles ready for sale to the abaca agents, who travel from Libacao to purchase the fibres.

The higher-grade abaca fibres are derived from the leaf sheaths located closest to the centre of the abaca stem. While the Bureau of Fibre and Inspection Service defines a total of 15 grades of abaca fibre, this system is only practised at the Grading and Bailing Establishment and by the

abaca exporters. In practise, there are only three grades that are employed by the farmers and agents. The fibres located close to the centre of the stem usually produce the highest grade of abaca fibres (A+/A grade or *primera* grade). Fibres located in the middle region of the stem usually produce a medium grade of abaca fibres (B grade or *segunda* grade). Finally, fibres located in the outer portion of the stem usually produce the lowest-grade fibre (C grade or *tersera* grade).

Given the relationship between fibre grade and the location of the leaf sheath within the stem, abaca fibres can usually be graded as they are produced. While the location of the fibre within the stem is usually a good indication of the fibre grade, the final determination of fibre grade is based upon fibre colour and texture. *Primera*-grade fibres are generally white to ivory white in colour and are soft to the touch. *Segunda*-grade fibres are a light brown colour while *tersera* fibres are a

Fig. 7.13. Abaca plant and the abaca production process in rural Philippines. The outer layers of each leaf sheath contain the fibres. Photo by Ivan Eastin.

Fig. 7.14. Abaca plant and the abaca production process in rural Philippines. A hand comb is used to separate the fibres in rural areas. Photo by Ivan Eastin.

darker reddish brown, coarse-textured and rough to the touch. While it is not difficult to grade the abaca fibres into the three basic grades, many farmers do not understand the fibre grading system and the agents and traders have not been proactive in providing them with this type of training.

Higher grades of abaca fibres are used in the production of paper, yarn and thread. Medium-grade abaca fibres are used in the production of fibre crafts (e.g. rugs, place mats, doormats, handbags, hats, coasters and hot pads). Lower-grade abaca fibres are usually sold as raw fibre or cordage (marine rope).

Grade A+ fibres are usually produced for a specific order, and this grade of fibres can be sold on a kilogram basis, rather than in the typical 40 kg bundle. However, many farmers prefer to sell abaca fibres in ungraded bundles for a lower aggregate price rather than take the time and effort to segregate the fibres based on grade as they are produced.

Fig. 7.15. Abaca plant and the abaca production process in rural Philippines. The abaca fibres are left to dry in the sun for about 4 h before they are collected and graded. Photo by Ivan Eastin.

This is due to the fact that the agents and traders only buy fibres in 40 kg bundles and will not generally purchase partial bundles from farmers.

Pricing

There is a different price structure for hand-combed fibres and spindle-stripped fibres. Customers in the export markets prefer the higher-quality machine spindle-stripped fibres and they are willing to pay a 15–20% premium for these fibres. Despite the two-tiered pricing structure for abaca fibres, most farmers continue to produce hand-stripped fibres because they lack the capital required to buy spindle stripping machines and they have no access to credit.

The prices paid to the farmers for abaca fibre depend on the grade of the fibre and the location where they are sold to the agents (Table 7.4). The margin that the local agent receives from the trader is usually about ₱3 to ₱4 per kilogram. Interestingly, farmers are discouraged from bypassing the local agents and trying to sell their fibres directly to one of the traders in Kalibo through a discriminatory pricing policy. This discriminatory pricing policy is set up so that the farmer actually receives a lower price from the trader than from the local agent.

The traders discourage direct sales by the farmer for several reasons: (i) these individual sales are usually small; (ii) individual farmers are not willing to provide a regular, reliable supply of fibres to the trader; and (iii) bypassing the traditional channels weakens the trader–agent relationship. In the past, farmers and development workers have found that, while agents are generally willing to release price information upon

Table 7.4. Prices of abaca fibre paid by fibre agents.

Fibre grade	Julita/Libacao (₱/kg)	Kalibo (₱/kg)
A+	48.0	47.0
A	22.4	21.9
B	20.2	19.3
C	13.0	11.0
Ungraded	15.0	14.0

Note: Exchange rate: ₱26/US$1.

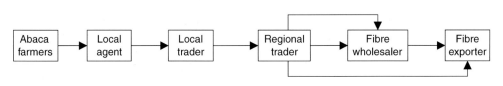

Fig. 7.16.   Distribution channel for abaca fibre in the Philippines.

request, the traders are not usually willing to divulge price information, particularly the prices paid for abaca fibres by exporters as well as the price of abaca fibre in the export market. However, it has been estimated that farmers receive less than 25% of the export price for their fibres.

Distribution

The farmers that Maling was working with lived in Barangay Julita, about a 1 h drive from the town of Libacao on the island of Panay in western Visayas. The local buying agent collects abaca fibre in the town of Libacao. He then sells the fibre to one of the three traders of abaca fibre located in the town of Kalibo. The traders then ship the abaca fibres to other traders on Luzon or Cebu, or they might sell their fibres directly to fibre exporters located in Manila, Legaspi and Cebu City (Fig. 7.8).

   While some farmers have the technical skills to segregate their fibres into the three basic fibre grades, most of them don't usually bother grading their fibres. The reason for this is that it is difficult for a farmer to produce 40 kg of graded fibres. Therefore, they prefer to sell their fibres ungraded, a practice referred to in the trade as 'all-in'. In large part, this practice is due to the fact that these farmers are under-capitalized, have no access to credit and usually need money to pay for living expenses. In general, they cannot afford to wait until they produce a full bundle of graded fibres, despite the fact that they would receive a much better price for the graded fibres.

   Another factor influencing the farmers' decision to produce ungraded fibres is the traders' refusal to purchase fibres directly from farmers. Finding themselves with no ability to bypass the local agents and sell directly to the large traders leaves them with little option but to accept the lower prices being offered by agents. Recently, however, some traders have begun to hint that they might be willing to purchase fibres directly from farmers if they

were willing to work together and provide larger volumes of fibres on a regular basis.

   Agents and traders generally grade the fibre into three to six grades, depending on their arrangements with the traders in Kalibo and their technical grading skills. However, the traders in Kalibo primarily perform an accumulation function within the distribution channel. Abaca fibres can pass through as many as six agents before reaching the final exporter, although a channel length of four agents and traders is more typical (Fig. 7.16). Exporters are generally the ones who separate the abaca fibres into the fifteen different fibre grades and sell them into specific domestic or export markets.

Maling's Assessment

During her meeting with her advisory group, Maling discussed a variety of strategic options. At the end of the meeting, the advisory group identified the following factors as the primary marketing challenges restricting local abaca farmers' ability to increase their sales revenues:

- Farmers have traditionally worked independently to produce and sell abaca fibres, despite the fact that it's difficult to produce enough fibres to meet the 40 kg bundle specification.
- The 40 kg fibre bundle requirement discourages farmers from grading their fibres, forcing them to settle for the lower composite fibre price set for ungraded fibres.
- Traders discourage farmers from selling direct and bypassing the local agents.
- Individual farmers lack the bargaining power to negotiate favourable sales terms for their products.
- Traders are willing to accredit farmer associations as agents but the traditional independence of farmers limits the use of this strategy by local farmers.

- Individual farmers lack access to credit, which restricts their ability to purchase the mechanical fibre stripping machines that would allow them to sell their fibres for a higher price.
- Lack of access to credit often forces farmers to sell their fibres ungraded in order to obtain cash to cover their immediate living expenses.
- Farmers lack access to reliable market price information.
- Farmers lack technical training on how to grade abaca fibres.
- Many farmers lack technical training on the fibre stripping process (e.g. appropriate tooth size for the different fibre grades).

Working with her advisory group, Maling noticed that this large group of factors could be categorized into three underlying problems: lack of bargaining power on the part of rural abaca farmers, unavailability of credit-granting institutions at the local level and farmers' lack of market and pricing information. She knew that whatever strategy they devised would have to address each of these underlying problems and would need to be appropriate for the culture and technology of the local farmers. Maling is now looking towards your market research group to help develop a strategy to resolve these problems and improve the economic benefits for abaca farmers.

Additional sources of information: Brown (1920–1921), Clay (1996), Conklin (1954, 1975), Fair Trade Organization, http://www.fairtrade.org.uk; Freese (1997, 1998), FAO (1989), Josiah (1999), Philippine Fair Trade Organization, http://www.philippinefairtrade.com; Rutten (1990).

CASE 7.7. IMPORTANCE OF SCAVENGER COMMUNITIES TO THE PAPER INDUSTRY IN MEXICO (MARTIN MEDINA)

This case presents a brief analysis of the role scavengers have played in supplying raw materials to the Mexican paper industry.

The Historical Origins of Paper

The Chinese invented paper in the 1st century (Munsell, 1980). Samples of paper made from linen rags have been found dating from this century. Eventually, knowledge of paper-making spread to Korea, Japan, Central Asia, Tibet and India. When the Arabs, in the course of their eastern expansion, neared Samarkand, they too became acquainted with the production of paper and paper mills were subsequently set up in Baghdad, Damascus and Cairo and later in Morocco, Spain and Sicily (Munsell, 1980).

During most of the history of paper-making, paper was made of post-consumer rags; that is, worn, torn and old textile materials discarded by individuals (Anon., 1941). Linen was the most desirable raw material sought by paper-makers. The process involved macerating the rags into pulp, drying the pulp and then pressing it to form sheets. Paper is composed of fibres interlaced into a compact web.

In Samarkand, Baghdad, Cairo and Damascus, paper was made from rags in the 9th century (Anon., 1941). Rag collectors salvaged hemp ropes, fishing nets, footwear soles and robes to be sold to paper-makers. During the 12th and 13th centuries, Bedouins and Arabs looted Egyptian graves in order to recover burial clothes, to be reused or sold for recycling to paper-makers.

Paper-making from rags was introduced to Europe – via Spain – by the Arabs in the 10th century and it extended throughout Europe and North Africa between the 11th and 13th centuries (Anon., 1941). During that period, rag collectors in Spain were known as *pannorum collectores* or as *pannicolorum collectores* (collectors of rags and 'little rags', respectively). And post-consumer rags used in paper-making were called *resuris veterum pannorum* ('shaved old rags') in the 12th century. Therefore, ragpickers have played a crucial role in supplying raw materials to the paper industry throughout its history. Ragpickers are often believed to operate in the margins of society and the economy. Historical evidence demonstrates that ragpickers have never been marginal to the development of the paper industry (Anon., 1941; Valls, 1980).

The Development of the Mexican Paper Industry

Using rags to make paper

The Mexican paper industry was created in 1590, when the first paper mill opened in the vicinity of Mexico City (Lenz, 1990; Sanchez, 1993). This was also the first paper mill to operate in the Americas. Throughout the 17th, 18th and 19th centuries, the Mexican paper industry made paper from old rags, which were usually scarce. The rags collected in colonial Mexico were in short supply due to the fact that inhabitants used their clothes as long as possible and discarded them infrequently and in small quantities. During this period, rag collectors were known as *traperos* (Lenz, 1990; Sanchez, 1993).

The recovery of discarded rags for papermaking during the 17th and 18th centuries was of such economic importance that it commanded royal attention (Lenz, 1990). Felipe III of Spain, for instance, signed the *Reglamento de Libre Comercio de Indias* (free trade law between the Spanish Crown and its territories in the Americas) in 1778, which exempted rags collected in the Spanish possessions in the Americas from the payment of import tariffs. This *Reglamento* attempted to encourage Mexican *traperos* to increase their gathering of rags, which would be exported to Spain, transformed into paper and part of the paper sent back to New Spain (Lenz, 1990).

Why recycled paper is used to make paper

Obstacles to using wood to make paper

It was in the early 20th century that the Mexican paper industry began making paper from wood pulp, but the switch from rags to pulp did not alleviate the shortage of raw materials (Medina, 2003). Mexican Indians own most of the forested areas in the country, but many lack deeds or their ancestral rights to the land have not been recognized. The lack of definition of property rights has led to the plundering of forestry resources by outsiders, as well as reluctance from investors to put their money into commercial timber plantations.

Moreover, the remaining woodlands in the country are located in remote and inaccessible areas. Since the Mexican government does not subsidize the construction of access roads, the cost of road construction must be considered in each logging project. Road building can account for about 50% of a logging project's total costs.

Finally, the small scale of logging operations and the use of outdated technology drive up the cost of the timber obtained to such a degree that the prices of domestic forest products often exceed international prices. The previous factors translate into an insufficient domestic supply of pulp: Mexican logging operations provide only 40% of the country's consumption of fibre (Medina, 2003).

Recycled fibre important for the Mexican paper industry

Due to the impracticality of achieving backward vertical integration with the forestry sector, the Mexican paper industry has undertaken vigorous efforts to increase the use of recycled fibre. In 1984, the Mexican paper industry used 58.3% waste paper as a fibre source, while in 1994 it had increased to 73.8%. Correspondingly, primary fibre (wood pulp and sugarcane bagasse) utilization decreased from 41.7% in 1984 to 26.2% in 1994.

The industrial consumption of the cardboard collected by scavengers illustrates these efforts. Mexico lacks recycling programmes, so the paper industry does not have alternative sources of raw materials. Contemporary scavenging fills this gap and plays a critical role in the supply of raw materials to the Mexican paper industry (Cortez, 1993; Medina, 1997).

The Mexican paper industry actively seeks and consumes discarded paper collected by scavengers (Fig. 7.17). The cardboard collected by scavengers cost the industry 300 Mexican pesos a ton in June 1994, while a ton of US market pulp cost the equivalent of 2250 pesos plus transportation costs (Medina, 1997). Thus, imported market pulp is seven times more expensive than domestic sources of discarded cardboard. By engaging in recycling, the paper industry not only saves raw material costs but also provides wages for scavengers. The construction and operating costs of a paper mill consuming waste paper are a fraction of those of a plant using wood pulp. Faced with such a large difference in costs, the Mexican paper industry has integrated vertically with scavengers via middlemen.

Fig. 7.17. (a) Mexican scavengers collecting cardboard wastes in a United States border town. (b–c) Vehicles used by Mexican scavengers to bring cardboard wastes back from a United States border town. Photos by Martin Medina.

As a result of the North American Free Trade Agreement (NAFTA), market barriers to trade in most paper and paperboard were phased out in the year 2003. The Mexican paper industry is currently trying to survive by upgrading its processes and by lowering its costs, which means maximizing the use of waste paper and cardboard collected by scavengers (Medina, 2003).

Why scavengers scavenge

Scavenging in Mexico and other developing countries exists due to economic factors, widespread poverty, high unemployment, the lack of a safety net for the poor, as well as industrial demand for inexpensive materials. Scavenging provides an income for migrants who have recently moved to cities, the unemployed, widows, children, the uneducated and the elderly.

Reliable data on scavenging activities are hard to obtain, but the World Bank has estimated that scavenging worldwide supports 1–2% of the urban population in developing countries. A recent estimate put the number of people in Mexico relying on scavenging for their livelihoods at around 1 million. A study conducted by the author in formal refuse collection and scavenging activities in three Mexican cities calculated an economic impact of over US$20 million.

Further, recycling of cardboard and paper can render environmental benefits, such as energy savings, diminished use of water and reduction in water and air pollution, compared with the manufacture of paper from virgin resources. Scavenging reduces the amount of solid wastes that need to be collected, transported and disposed of by municipalities. In this way, it reduces municipal expenses in collection vehicles, facilities and personnel, and extends the life of local dumps/landfills. Thus, scavenging in Mexico has a significant social, economic and environmental impact (Medina, 2005).

Conclusion

In conclusion, scavengers are an important component in the operation of the recycling system and have played a critical role in supplying raw materials to the Mexican paper industry. Scavenging activities improve the competitiveness of the Mexican paper industry because of the cost savings realized. For its entire existence, the Mexican paper industry has had backward vertical integration with scavengers. Rag and cardboard collectors, therefore, have never operated in the margins of the Mexican economy.

Scavenging can represent a perfect illustration of sustainable development in a developing country: jobs are created, poverty can be alleviated, raw material costs to industry are lowered – improving competitiveness – resources are conserved, pollution is reduced and the environment is protected.

CASE 7.8. MORE EFFICIENT USE OF TREES TO PRODUCE FOREST PRODUCTS (DAVE G. BRIGGS)

Introduction

Someone building a house in 1910 would discover that essentially every wooden component, structural framing, floor underlayment and floor, wall and roof sheathing, interior trim, moulding and cabinets, was made of lumber. Shelving and furniture were also made from solid wood parts manufactured from lumber. Someone building a house in 1960 would discover that plywood had replaced lumber sheathing and underlayment and was also commonly used in shelving and cabinets. Particle board was also becoming common as an underlay and in shelving, countertops and furniture.

Today, a home builder is confronted with a much greater diversity of wood products to select from. While traditional lumber, plywood and particle board are still available, there are many new 'engineered composites'. These composites consist of wood elements, ranging in size from large and obvious to tiny and imperceptible, that are held together by adhesives and possibly combined with non-wood materials such as plastics, cement or metal. Finally, some non-wood products are designed to mimic the appearance of wood. These changes in the diversity of wood products are the topic of this case, which focuses on the following question: What are these new wood products and why are they becoming more common?

What are the New Wood Products?

The new wood products appearing on the market can be classified in many different ways. Here they will be placed into three broad groupings: (i) engineered wood products; (ii) wood/non-wood products; and (iii) hybrid products that combine other products. New wood products are described below:

- Engineered wood products. Engineered wood products are defined here as pieces of wood that are glued together to form products with specific definable strength properties. The pieces of wood used in these products range from lumber to pulped fibres. Some examples are:

    - Glued laminated (glulam) beams. Pieces of lumber, usually non-destructively tested to determine their mechanical properties, are joined end to end and glued in a stack to form long beams of large depth and width. Knowledge of the properties of the lumber pieces allows the manufacturer to design beams with very exact and uniform properties to meet any architectural specification.
    - Laminated veneer lumber (LVL). Pieces of veneer, similar to those used in making plywood and non-destructively tested to determine their mechanical properties, are joined end to end and glued in a stack to form a long, wide billet with a thickness equivalent to a standard lumber thickness. The billet is cut off at the desired length and ripped to produce pieces with a width similar to standard lumber. Use of pretested veneers allows the design and production of a lumber substitute with exact and uniform properties.
    - Oriented strand board (OSB), laminated strand lumber (LSL) and parallel strand lumber (PSL). Logs are flaked into short (7–10 cm) or long (20–25 cm) strands that are dried, coated with an adhesive, laid out in a thick mat, and pressed to form a large panel with a specific density that largely defines the strength properties of the product. In OSB the mat consists of layers where the long axis of the strands in one layer is perpendicular to the long axis of strands in adjacent layers. The mat thickness is controlled so that the pressed panel will have a thickness equivalent to common plywood thicknesses. The large panel from the press is then manufactured into lengths and widths that are used in many of the same applications as plywood. The mat is controlled so that the final panel thickness will be equivalent to the standard thickness of lumber. The large panel from the press is then ripped to produce pieces with lengths and widths similar to standard lumber.

    - Particle- and fibre-based products. These are based on small wood particles (particle board) or pulped wood fibres (fibreboard) that are mixed with an adhesive, formed into a mat and pressed into panels. Fibreboards can also be moulded or extruded into a variety of shapes and hence are made into doors, mouldings and trims. The source of raw material for these products is often byproducts from other wood manufacturing industries.

- Wood/non-wood products. These are products in which pieces of wood are combined with a different material, such as a metal, fibreglass, cement or plastic. Examples are:

    - Wood and cement composites. Typically small wood particles or wood fibres are mixed with sand and cement and formed into building blocks, roofing tiles and siding products. Some of these products are textured, moulded and coloured to resemble natural wood grain and provide colour choices.
    - Wood and plastic composites. Small wood particles or, more recently, powdered wood (wood flour) is mixed with one of several plastic resins to produce moulded or extruded 'wood–plastic' lumber, window components, siding, etc.

- Hybrid products. These products are made by combining two or more other products into a new application. For example:

    - I-beams and I-joists. The flanges of the 'I' are commonly either non-destructively tested lumber or LVL, whereas the web, which provides depth to the 'I', may be plywood, OSB or metal tubing.

Why are these New Wood Products becoming More Common?

Adoption of these products often leads to replacement of older wood products. Thus, the market for sheathing and floor underlayment in housing has seen plywood replace lumber and now OSB replace plywood. Solid sawn lumber (especially in larger dimensions) is losing markets to glulam beams, LVL, oriented strand lumber (OSL) and PSL and I-beams and I-joists. Particle board is losing certain markets to new fibreboard products. When these substitutions occur, the displaced product usually does not disappear; instead it often retains a presence in speciality niche markets.

There are a number of factors that are driving trends towards greater use of these and other new wood products. Three categories (recovery economics, limitations imposed by tree size, wood variability and dimensional stability) are discussed next.

Recovery economics

Round-wood logs consumed to make lumber and plywood often make up 50–70% of the product cost. These logs must be of sufficiently large diameter, length and quality and yet the yield of high-value lumber or veneer is only about 50%. The remaining 50% consists of low-value chips and other by-products sold to pulp and paper and other industries. In contrast, many of the new products can use smaller, lower-quality, cheaper logs and obtain yields between 80 and 95%. The ability to use lower-cost raw material and obtain higher yields gives these new products a significant cost advantage over traditional lumber and plywood products.

Limitation of tree size

Many construction applications must span large room openings in houses, garages, office buildings, libraries, etc. Traditionally such spans were met by using solid sawn lumber beams but this is becoming more difficult because of the opposing trends towards larger room spans but smaller-diameter trees. The diameter and length of a log and hence the dimensions of lumber that can be obtained are dependent on tree size. The average tree diameter is becoming smaller and furthermore each tree has

progressively smaller-diameter logs at greater height along the stem. For example, in the Pacific North-west United States, the diameter of the average tree harvested changed from 70 cm in 1976 to 41 cm in 1997 and today the average log diameter is 29 cm. This makes it difficult to obtain and accumulate long lumber pieces of the width, thickness and quality needed for many applications. Products based on small wood elements overcome these size and quality constraints.

Wood variability

The wood of different species is obviously different; hence there are traditional preferences for how certain species are used. For example, Douglas fir and southern pine are used for strong, stiff construction members while white pine and ponderosa pine are used for mill-work items, etc. However, as almost any consumer knows, there is a great deal of variation among the pieces of a given species at the local lumber supplier. There are differences between trees of the same species due to genetics, local growing conditions, etc. Furthermore, during its lifetime the growth processes of a tree produce great differences in wood properties even within a single tree.

The following paragraphs provide four examples of sources of within-tree wood variability:

- Anisotropic nature of wood. A tree grows by seasonally adding new height growth to its top, making it taller. Simultaneously it covers itself with a new growth ring of wood, increasing the stem diameter. Thus, the new height increment and new diameter increment of wood are at the top of the tree and just underneath the bark. The stem diameter is smaller higher in a tree because the height increment is younger and hence has fewer growth rings surrounding it. A cross-section through the stem at a given height reveals a set of concentric growth rings reflecting the time since the tree reached that height; the most recent ring is the outermost. Most wood cells are long, small-diameter tubes with an open centre called the 'lumen' for transporting water and nutrients from the soil to the foliage, where photosynthesis occurs. The cell wall is composed of cellulose, lignin and hemicellulose, which provide strength and stiffness. Cells produced early in the growing season ('early wood') are generally larger in diameter, with thin walls and a large lumen, while those produced later in the growing season ('late wood') are smaller in diameter, with thick walls and a small lumen; late-wood cells are much denser, stronger and stiffer than early-wood cells. The majority of wood cells have their long axis aligned parallel with the length axis of the stem, branch or root in which they are found. As a result of the geometric arrangement, wood properties such as stiffness and strength, shrinkage and swelling, etc., differ in three mutually perpendicular directions; along the length axis, along the radius from centre to bark, and along the tangent to the concentric rings; hence wood is referred to as an 'anisotropic' material. Anyone who tries to split a log into firewood quickly learns which of these three directions is easiest. Anisotropic behaviour is one major source of wood variability.

- Juvenile and mature wood. For a number of years, the growth rings that form around a height increment develop wood with different properties from those of wood-formed growth rings that surround the same height increment in later years. The terms 'juvenile wood' and 'mature wood' are often used to refer to these regions of 'inner' and 'outer' growth rings in any cross-section. Depending on species and the property of interest, the transition from juvenile to mature wood may be five to 20 or more rings from the centre of the cross-section. Although the change in properties is gradual, juvenile wood tends to have lower strength and stiffness and greater longitudinal shrinkage than mature wood. The higher longitudinal shrinkage of juvenile wood compared with mature wood is a major cause of warp in lumber, since sawing lumber from a log yields many pieces with wood from both regions. The patterns of changing wood properties associated with juvenile and mature wood are another major source of wood variation.

- Knots and grain direction. Tree branches, which become knots in products, are necessary for a tree to orient foliage to gather sunlight for photosynthesis. However, wood cells within a branch are oriented along the branch length; hence the anisotropic directions of branch wood are tilted with respect to the directions in stem wood. The direction of stem wood in close proximity to a branch is also distorted. These 'grain deviations' due to knots and other factors are another source of variation; a 30° deviation of grain from the length axis of a piece of lumber reduces its compression, bending and tensile strength by 50 to 75%.

- Dimensional stability: Wood is also known as a 'hygroscopic' material, since the cellulose in wood attracts and holds water. Any wood product is always losing or gaining water in an effort to achieve equilibrium with the humidity and temperature of its environment. This gain or loss of moisture is accompanied by swelling or shrinking. Small changes lead to annoying 'dimensional stability' problems, such as windows and doors that swell tight during one part of the year and become loose at another time. These changes may also be accompanied by warping. Issues associated with moisture change, warp and dimensional stability are another source of variation.

Engineered, composite and hybrid products overcome issues of tree size limitations and wood variation by the following:

- Non-destructive testing and sorting of lumber and veneer into narrow, precise strength and stiffness classes. Product designers can then select and arrange components to design a finished product with specific properties. This can be done repeatedly to produce numerous finished product items that are nearly identical in performance.

- Technology to make flakes, particles and fibres breaks down the original wood structure into numerous small pieces, which can be thoroughly mixed, including mixing with material from other trees, to create a very homogeneous raw material.

- Adhesives used to glue small wood components together plus other additives can restrict reaction to moisture, reducing issues associated with warp and dimensional stability.

- Combining wood with metals, cement, plastics, etc. provides designers with greater flexibility to create innovative products with improved performance capabilities.

CASE 7.9. ENERGY FROM BIOMASS (RAVI UPADHYE)

Approaches to Obtaining Energy from Biomass

There are several ways of obtaining energy from plant sources:

1. The biomass can be combusted in a boiler to raise steam, which can be used to drive a generator to produce electricity. Despite its simplicity, however, this method has serious drawbacks. Due to the incomplete combustion of combustible volatiles, the overall thermal efficiency of the process tends to be rather low. In addition, uncombusted volatiles have a potential to contribute significantly to environmental problems if biomass combustion is carried out on a large scale.

2. Biodiesels can be made from crops grown specifically for that purpose. Examples are soybeans and maize, from which oil can be extracted and processed to make biodiesels. Biodiesels can be used directly in diesel engines.

3. Ethanol can be made by fermentation from starch or sugar from certain plants. Examples are maize, barley and beets, which contain either starch or sugar in significant proportions. Ethanol can be either used as an additive in conventional petrol-powered engines or fed directly into internal combustion engines specifically tuned to accept ethanol as the fuel. It can also be reformed to generate hydrogen, which can be fed to a fuel cell.

4. Both the pathways listed above generate residues that are depleted in oils, starches or sugars. The main components of these agricultural residues are cellulose and hemicellulose, with small fractions of lignin. Some examples are bagasse from sugarcane and maize husks and stems. Forest wastes (mainly derived from trees) are comprised of cellulose, hemicellulose and lignin as their major components (see Chapter 6). Unlike starch and sugar, at the present time cellulose and hemicellulose are difficult to ferment. The yields are rather low, and chemical processing to convert cellulose to sugars may be needed before fermentation can occur. Ethanol from cellulose (EFC) holds great potential due to the widespread availability, abundance, and relatively low cost of cellulosic materials. However, although several EFC processes are technically feasible, cost-effective processes have been difficult to achieve. Only recently have cost-effective EFC technologies

begun to emerge. There is currently no technology to obtain ethanol from lignin by direct fermentation. Thus fermentation cannot be used to obtain ethanol from a major portion of forest wastes.

5. For biomass containing significant portions of cellulose, hemicellulose or lignin, an alternative method of extracting energy is gasification. In this process, the biomass is mixed with air and steam at high temperature (~500°C) to obtain a mixture (called producer gas) of nitrogen, hydrogen, carbon monoxide, carbon dioxide, steam, particulates and tars. If oxygen is used as the oxidizing medium, the resulting gas (called syngas) has no significant amounts of nitrogen.

Both producer gas and syngas can be combusted directly in a boiler, producing steam to generate electricity. They tend to burn more cleanly, without causing environmental problems. They can also be combusted in internal combustion engines designed to accept them as fuels.

The syngas can be converted into a large variety of products (Fig. 7.18). Properly balanced syngas can be directly converted into methanol over a commercially available catalyst. Methanol can be used as an additive in the current petrol engines or fed directly into specially tuned engines. For example, Indy race cars run on methanol. In addition, it can be fed directly into direct methanol fuel cells (DMFCs) to obtain electricity. Finally, it can be re-formed to produce hydrogen, which can be fed to a number of fuel cells to obtain electrical power.

Additionally, dimethyl ether (DME) can be made from methanol in one step, or can be obtained directly from syngas (bypassing the methanol step). DME is an excellent diesel fuel, with a cetane number of 55+. It can be used directly in diesel engines. It is non-toxic and routinely used as a propellant in hairsprays.

Technologies to Produce Methanol from Biomass

There are four technologies that constitute a system to produce methanol (or DME) from biomass:

1. Biogasification: gasification of waste biomass to yield syngas (mixture of hydrogen and carbon monoxide, with a small amount of carbon dioxide).

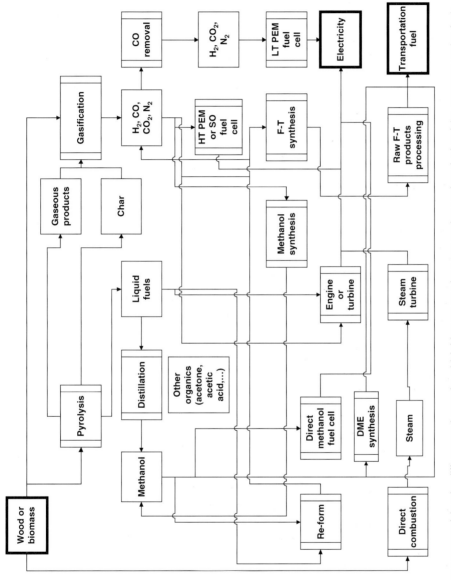

Fig. 7.18. Diagram summarizing the different pathways by which biomass can be converted to methanol, electricity or transportation fuels.

2. Gas clean-up: removal of particulates and other impurities from the syngas.

3. Balancing the syngas: chemical adjustment of the syngas to obtain the optimal ratio of hydrogen to carbon monoxide.

4. Methanol or DME synthesis: synthesis of methanol or DME from the syngas.

The last three technologies are commercially available, and no research and development will be needed to implement these. Biogasification has been demonstrated on a small scale, using air (instead of oxygen) to make producer gas. Replacing air with pure oxygen to produce syngas and scale-up to a commercially viable unit are the two developments needed in this technology.

The major drive in the United States today is to use wood biomass to produce ethanol. Wood biomass should be used to produce methanol since wood chemical composition makes this an ideal liquid to produce. Agricultural crops are easier to convert to ethanol since ethanol is mainly produced from the conversion of sugar compounds

and ethanol cannot be made from cellulose and lignin at present (see Chapter 6, this volume).

We believe that, in order to make complete and effective use of the world's biomass resources, both ethanol and methanol are necessary. Products that can be easily converted to ethanol by fermentation should be used for this purpose. The waste products from agriculture and forestry operations that cannot easily be converted to ethanol should be used to produce methanol or DME, using known technologies as outlined above. Needless to say, these two pathways are not in competition but are, rather, complementary. Wood biomass has been used to produce methanol for over 350 years, as agricultural crops have been used to make ethanol. The efficiency with which both agricultural crops and forest biomass are being converted to liquid products has dramatically increased. The number of different products and how they can be used to substitute for fossil fuels is discussed earlier in this chapter – 'New products from forest biomass that are climate-friendly technologies'.

CASE 7.10. INTEGRATING CONSERVATION AND SUSTAINABLE FORESTRY: A PACIFIC NORTH-WEST UNITED STATES EXAMPLE (KRISTIINA A. VOGT)

Certification is an important tool that needs to be adapted and used to reduce the risks of negative environmental impacts, despite the fact that it has many challenging constructs that are difficult to implement. The complexity of issues that need to be considered in an audit of forest sustainability is highlighted by a retrospective examination of the North-west Forest Plan. The North-west Forest Plan (Chapter 3, this volume) was implemented more than a decade ago as a conservation plan for federal lands in California, Oregon and Washington. The goals of the Plan were to make these forests ecologically, economically and socially sustainable (Stokstad, 2005). These are the same goals held by certification programmes. So the Plan serves as an excellent example of the challenges facing forest certification. The evaluation of the Plan, a form of audit, showed that some goals were achieved while others were not.

Goals reached:
- Old-growth forests – even with increased fires occurring in forests, more old-growth forests were preserved.

- Marbled murrelet – populations of this endangered seabird have been stable, even though the reasons why are not clear to scientists.
- Riparian habitats – watersheds are improving.

Goals not reached:
- Retaining land in forests has been difficult because of the high incidence of catastrophic forest fires.
- Spotted owl – the population of spotted owls is declining, primarily due to competition from an invasive species, the barred owl, and the loss of habitat on private lands.
- Timber and economics – timber sales never reached the projected levels, so jobs have been lost in rural forest-dependent communities, there has been a loss of income for timber industries and the United States Forest Service, which manages federal forest lands.
- Adaptive management areas – research to inform management on the best approach to manage forests was to occur on research areas set aside for this purpose. These reserves did not play the role expected because of lawsuits,

bureaucracy and the lack of adequate funding to support this research.

Little forest management and harvesting have been allowed to occur in the federal forests of the Pacific North-west because of an inability to develop consensus among the resource use and conservation stakeholders. The need to 'survey and manage' before any timber could be harvested meant a significant amount of site data needed to be collected before any timber cutting. The requirement for data was used by some groups to prevent the harvesting of timber (Stokstad, 2005). The 'survey and manage' requirement meant that large amounts of data needed to be collected or sites monitored before decisions could be made on how forests were to be used. Since a broad range of species needed to be surveyed, several years of data collection would be required. Sufficient funds to pay for this data collection were not available, which also slowed the process of satisfying the 'survey and manage' requirements.

The requirement of the 'survey and manage' programme made it very difficult to cut any timber (Stokstad, 2005). Today, the lack of cutting timber has had severe repercussions. Not only did the timber-dependent communities lose their livelihoods but this has contributed to the high risk of catastrophic fires today in the drier inland forests.

This retrospective analysis on how successfully the North-west Forest Plan was implemented shows that unexpected consequences resulted from the Plan:

- Timber-dependent communities were not given new job opportunities.
- An invasive species, the barred owl, caused further endangerment of an already listed species, the spotted owl.
- Increased fire risk to some forests because they contained too much biomass.

The issues addressed by the Plan continue to haunt people making decisions regarding forests in the western United States. These unexpected consequences associated with decisions made in previous years have contributed to the undervaluing of forest lands. Maintaining forests in the drier inland landscapes will be difficult due to their high fire risk, and more forest lands at the increasingly large urban–wild-land interface will probably be converted to urban developments because they have a higher monetary value in today's housing markets (see earlier in this chapter – 'A New Challenge to Sustainable Forestry: Managing Urban Forests and Reducing Deforestation at the Urban–Wild-land Interface').

8 It's a Small World After All

DANIEL J. VOGT AND PATRICIA A. ROADS

> Never doubt that a small group of thoughtful, committed citizens can change the world.
> Indeed, it is the only thing that ever has.
>
> Margaret Mead

Emerging Trends: Forest Decisions are Democratic at Local Levels but Globally Influenced

Throughout this book we have discussed the interconnectivity of societies and forests of the world. Often this association has been from the aspect of either forest neglect or over-exploitation by society, but there are many examples of forest conservation that can be traced back to 2000 years ago. In Chapter 3 we discussed the current trends of greater participation by local communities in forest uses and conservation, which increases the democratization of the decision-making process. Another trend that emerged during the last three decades is the globalization of forest uses and conservation. Even though globalization suggests the need to manage resources at larger spatial scales, globalization is causing us to realize that it is a small world after all.

Two cases will be briefly mentioned in this chapter, i.e. Miami Township in Montgomery County, Ohio, and Papua New Guinea. Despite the fact that global markets are driving forest uses, these cases demonstrate how local communities can be an important part of the decision-making process in forest conservation. The two stories are from totally different areas of the world; one is from an industrialized country while the other is from a developing one. Both have a form of democracy that works from the top down as well as from the bottom up. They also have in common a desire and the will to maintain the liveability of their communities by protecting the quality of their environments.

The fable below summarizes very effectively the need for greater participation on the part of the general public in helping to develop consensus on resource use and conservation decisions about forests. Resolving conflicts and eliminating polarized decision-making regarding forests and natural environments will occur when a greater proportion of the lay public becomes knowledgeable and engaged in the decision-making process. Special interest groups today orchestrate much of the decision-making process regarding forest uses. This results in multiple agendas that do not allow us to implement the goals of integrated conservation and sustainable development projects.

The Fable: The Mouse and the Mousetrap (unknown author)

A mouse looked through the crack in the wall to see the farmer and his wife opening a package.

The mouse wondered, 'What kind of food could the package contain?'

He was devastated to find that the package contained a mousetrap. Retreating to the farmyard, the mouse approached the barnyard animals, where he proclaimed the warning. 'There is a mousetrap in the house; there is a mousetrap in the house!'

The chicken clucked and scratched, raised her head and said, 'Mr Mouse, I can see that this is a grave concern to you, but it is of no consequence to me. I cannot be bothered by it.'

The mouse turned to the pig and told him, 'There is a mousetrap in the house.' The pig sympathized but said, 'I am so very sorry, Mr Mouse, but there is nothing that I can do about it but pray. Be assured you are in my prayers.'

The mouse turned to the cow. She said, 'Wow, Mr Mouse, I'm sorry for you. But it's no skin off my nose.' So the mouse returned to the house, shoulders slumped, head lowered with worry and depression dragging him further down. It saddened him to realize that he would have to confront the farmer's mousetrap without the help of any friends. That very night a sound was heard throughout the house. The sound was an indication that the mousetrap had caught its prey.

The farmer's wife rushed to see what had been caught. In the darkness she did not see that it was a venomous snake whose tail had got caught in the trap. The snake bit the farmer's wife. The farmer rushed her to the hospital and she returned home with a fever.

Now everyone knows you treat a fever with fresh chicken soup, so the farmer took his hatchet to the farmyard for the soup's main ingredient. Unfortunately the wife's sickness continued, so family, friends and neighbours came to sit with her around the clock. To feed them, the farmer had to butcher the pig.

Still, the farmer's wife did not get better. In a few weeks she died; and so many people came to her funeral that the farmer had to slaughter the cow to provide meat for all that attended.

The only animal left was the mouse.

The Moral

The next time you hear or read about the loss of tropical forests being burned for grazing cattle or growing marijuana, think of the mousetrap having snapped shut. Do not think that the problem is in another part of the world and has no consequences for you. At the same time realize that whatever you and your neighbours do can affect other people as well. We all live on, and share, the same world and a problem located somewhere else in the world could be a problem for us all.

Forest sustainability facing new challenges

Superimposed on these emerging trends of society and forests is the fact that those factors affecting forests historically are still impacting them today, but we also have new and different activities and disturbances occurring today. For example, if climate change occurs and changes the frequency and/or quantity of rainfall or even the temperature, the size, composition, health and/or location of the forest biomes will change. Since the success of human civilizations has historically been enhanced by the presence of forests (see Chapter 2, this volume), changes in global climate patterns, and hence forests, could determine where future civilizations will be found. As previously mentioned, more than half of the globe is still dependent on fuel wood for energy needs (mainly by combusting wood in a very non-environmentally friendly way), so our dependence on forests has not decreased through time. In fact, we should be exploring very different uses for forests while considering the impact on the environment from the beginning. One solution to the excessive emissions of CO_2 into the atmosphere and possible climate change is to use biomass in an environmentally friendly manner to help with our energy needs. Biomass conversion to energy products must use recent technology developed in the energy sector if it is going to be environmentally friendly (see Chapter 7, this volume).

In addition to potentially shifting the locations of forests around the world and influencing sites of civilization development, climate change is also expected to directly impact human health. Disease vectors and infectious diseases are found throughout the world and are expected to alter with changing climates (see Chapter 5, this volume, and Case 4.4 – Malaria and Land Modifications in the Kenyan Highlands). Since temperatures have such a strong effect on disease vectors and their ability to proliferate, areas of the world where mosquito-transmitted diseases such as malaria are prevalent will move in conjunction with the movement of these new forests. In fact, malaria mostly originates from forests and is even called a 'forest disease'. It was prevalent even in the eastern lowlands of North America when the early European colonialists arrived (see Chapter 5, this volume). This change in regional locations of diseases can potentially be disastrous, as shown by what happened to the indigenous populations in North America, who were not immune to the diseases brought by European colonialists in the 1500s (Mann, 2005). Up to 95% of the indigenous population in the United States was estimated to have died from diseases carried by the colonists prior to the arrival of the Pilgrims in 1620 (Mann, 2005).

How we managed forests in the past had little effect on human health. But, today, the rapid spread of human populations, coupled with land-use changes, is increasing the spread of diseases that emerge at local levels and, because of our global economy, quickly spread globally. People can travel in even greater numbers to any part of the globe within a day or two for either business or vacations. Unfortunately, we can carry germs and viruses from one area of the globe to another and not even know it. Germs or viruses do not need a ticket to move to different parts of the world or through airport security. As yet, we cannot detect diseases being carried by a traveller.

It is at the local level in developed countries that the impact on human health receives media coverage, even though it may have originated in a less developed country and then spread throughout the world. Acquired immune deficiency syndrome/human immunodeficiency virus (AIDS/HIV), which was first transmitted to humans by monkeys in a remote area of Africa, is an example of this. In addition, this year the concern over avian flu is topping the list of possible medical emergencies. This flu is said to be as deadly as or more deadly than the pandemic of 1918.

As methods of human transportation become more reliable, faster and available in previously inaccessible areas, the spread of dangerous pathogens will increase and they will spread faster. We may see disasters such as the ones created when Native Americans (in both North America and South America) were exposed to smallpox after Europeans and Spaniards began to colonize the Americas (Mann, 2005).

In addition to the potential impact of climate change, there are several other new activities and disturbances occurring in forests that did not occur in the past. For example, national and state forests in the United States are being used to produce illegal methamphetamine and to grow marijuana (National Drug Intelligence Center, 2003; Robinson, 2005). The following quotation from the *Seattle Times* article written by Robinson (2005) describes the impacts of using forests to grow marijuana:

> The marijuana 'growers poach wildlife, spill pesticides, divert water from streams and dump tons of trash'. In the past year, 100,000 marijuana plants have been removed from California national parks, including 44,000 from Sequoia. 'We've found AR-15s, shotguns, rifles, knives strapped to poles, and crude crossbows' says the chief ranger at Sequoia. 'Growers bushwhack a couple of miles into the woods, carrying 25-pound tanks of propane, 50-pound sacks of fertilizer, pesticides, and hoes. Periodic food drops supplement poached animals. The farmers clear the understorey of foliage, leaving a canopy for camouflage; they cut terraces in the slopes, run irrigation hoses from creeks and rivers for miles and carve out a sprawling camp.'

The Department of Natural Resources in the state of Washington, United States, spends a considerable amount of funds trying to clean up illegal methamphetamine labs that are being constructed on their lands (National Drug Intelligence Center, 2003). The pesticides and caustic and flammable chemicals from these labs leave a legacy of chemical contamination in these forests. In the state of Washington, the number of discovered labs that needed to be removed and the chemicals cleaned up increased from 308 in 1998 to 939 in 2001 (National Drug Intelligence Center, 2003).

Producing drugs on the premises of public forest lands has similar repercussions to those wars have had, and continue to have, on restricting the ability of the people to collect resources from the forest or to even pursue recreational activities there (see Chapter 5, this volume). People do fear going into forests because of safety concerns, since those making the drugs defend these illegal drug operations. 'Methamphetamine production is causing serious safety and environmental concerns in Washington . . . Further, the production, distribution, and abuse of methamphetamine are more commonly associated with violent crime than any other drug' (National Drug Intelligence Center, 2003). Today, the major difference between wars and these illegal drug-producing operations is that drug operations are found in forests located in the industrialized world while wars have been most common in the developing world.

At this point in history, we should realize how important forests are to our survival. We need to recognize that science and society have to work together to guarantee that all our resource uses are sustainable. Knowledgeable science without informed societies willing to listen to its advice and act on its recommendations is like trying to ride a bicycle that has no wheels. You can't get very far with it and the journey will be very uncomfortable, perhaps even painful.

Another important take-home message that the next discussion highlights is the development of a common vision at the local to regional levels for how forests are used.

Many suggest that developing a common vision among different stakeholders at the local and regional scales for forest uses will be pivotal in allowing us to manage forests sustainably for rural livelihoods and to provide the many products and values from forests sought by society. Globalization and democratization of decision-making will continue to influence the future uses of forests and thus determine forest sustainability for society in a global landscape.

Grass-roots Movements and Common Visions at the Local Level for Sustainable Management of Human-dominated Landscapes

The issues faced by the first case in this chapter (i.e. the county in Ohio) are playing out right now in many other parts of the world. It is crucial that we recognize that communities can develop consensus on local issues. A common vision that is shared and developed by local communities can make a difference in how decisions for growth management and conservation of forests are made. If a common vision is not developed, the continued loss of forests will probably continue unabated, since urban expansion is occurring at an unprecedented rate (see Chapter 7, this volume).

Since we recognize that protected lands alone are insufficient to provide the complexity of habitat needed by animals that are threatened or endangered by human land-use changes, conservation goals need to become integrated into the local community decision-making process. Achieving such consensus is integral to the implementation of integrated conservation and development projects. Some of the cases discussed in this volume highlight how decision-making at the local level has helped to develop consensus on resource uses and has encouraged these communities to implement sustainable forest practises (see Case 2.2 – The Impact of Indigenous People on Oak–Pine Forests of the Central Himalaya; Case 3.4 – Indian Forest: Land in Trust; Case 7.5 – The First Certified Community-based Forests in Indonesia: Stories from the Villages of Selopuro and Sumberejo, Wonogiri, Central Java). Other cases in this volume show how the exclusion of the local people (or segments of the local population) in the decision-making process resulted in the polarizing of their decisions and made it impossible to successfully implement sustainable practices and sustainable livelihoods (see Case 2.1 – Nepal, Community Forests and Rural Sustainability; Case 2.3 – Dead-wood Politics: Fuel Wood, Forests and Society in the Machu Picchu Historic Sanctuary; Case 3.1 – Debt-for-nature Swaps, Forest Conservation and the Bolivian Landscape; Case 3.2 – Cattle, Wildlife and Fences: Natural Disaster and Man-made Conflict in Northern Botswana; Case 3.3 – Forest Communities in China and Thailand).

There have been many barriers to achieving the integration of projects that result in sustainable livelihoods for local people while also pursuing conservation goals (Sayer and Campbell, 2004) but success stories are occurring at many local levels. These local-level successes need to be examined for the elements that made them successful and then used only when considered within the context of the larger landscapes. It is at this broader level that other constraints may arise. For example, two adjacent towns, with a common river travelling through them, have each converted most of their forests into asphalt surfaces. Each town has independently determined that the additional increase in their own impermeable surfaces will not significantly increase risks of the river flooding but neither has considered the results of both towns increasing their impermeable surfaces.

Small places have big ideas – Montgomery County, Ohio, United States

Quite often it takes the success of a grass-roots movement to make people realize that they have the ability to change the situation. Everyone can make a difference when they have concerns and then take responsible action. One community in the state of Ohio, United States, provides an excellent example. Miami Township, Montgomery County, Ohio, contains approximately 26,000 people and has over 34 km^2 (21 square miles) of land within its boundaries. Miami Township is a southern suburb of the city of Dayton. In fact, Dayton is well known for its rich heritage of inventions and innovations (it has more patents per capita than any other city in the United States), is the home of the Wright Brothers and the birthplace of aviation and is also the home of the Dayton Peace Accords.

Both the Great Miami River and a major interstate run through the centre of Miami Township. It contains a variety of business districts, several light-industrial parks and housing in an array of economic ranges and densities. In the Township's approximate centre there is one of the largest malls in the area, Dayton Mall. The south-western portion, approximately 25% of the land in the township, consists of agricultural property or large wooded lots with single-family dwellings. In addition to the farms this portion of the Township contains areas of deciduous forests.

Travelling through some of the back roads you can almost believe that you have gone back in time 100 years and are alone in the middle of a forest where there is little evidence of civilization. Winding in and out of the trees are small streams, whose water tumbles over rocks and boulders on its way to the river. Who would not be refreshed by the sound of these streams and the sight of water sparkling in the sun? Any season you can become mesmerized gazing at the trees; in the spring buds slowly swell and finally emerge to show their many shades of green. Fully leafed they provide an umbrella of shade for hot summer days. In the autumn their foliage changes colours, displaying burgundies, reds, yellows and browns. Winter finds them wrapped in layers of white snow until spring comes again. The birds also flit about in these backwoods areas and don't seem to mind the few people who are fortunate enough to be aware of the magic to be found in this area. Along the river you find bike paths that allow you to travel from one end of the county to roughly the middle of the state. On warm days you can find families fishing at the river's edge or picnicking in any one of a number of parks that are in placed throughout the township. Most people that live in the south-western area of the township appreciate the opportunity to enjoy the natural landscapes that exist here and would be repulsed by even thinking about its loss.

Like most suburbs of large cities, Miami Township has had developmental pressures and expects them to continue into the future. Citizens in the south-western portion of the township saw large areas of either deciduous forests or farmlands being turned into housing divisions. These divisions put the maximum number of houses on the property that existing zoning laws allowed. This was done without regard to the possible adverse effects on surrounding homeowners or the consequences for the area's rural ambience. Environmental concerns played no part in the development plans. Some developers and speculators were so narrowly focused that they saw nothing wrong with removing every tree and scraping the land bare to make the construction cost as low as possible.

The rural atmosphere was being replaced by sprawl. Concerned citizens in this area regarded this change as unacceptable and were unwilling to allow the uncontrolled loss of forests and farmlands to continue. Numerous legal entanglements initiated by local residents and a successful vote, when the issue was placed on the ballot, prevented large sections of rural land from being turned into more sprawl. At the urging of its citizens and under the direction of its elected officials, members of the zoning staff gathered information on feasible ways to protect the rural nature of the township while maintaining viable development.

After several years of work and numerous discussions among elected officials, citizens and developers alike, new zoning laws were passed. These laws enabled citizens to provide more input when decisions regarding development are concerned. In a continuing effort to not only maintain the high standard of life in the township, but to further improve it, the zoning staff is working to implement changes that will make Miami Township more environmentally friendly. Some of these improvements include more extensive landscaping criteria for businesses or developers wanting to build in the township and retaining or planting native trees are an integral part of the plan. The zoning staff, working with local business owners, are consulting with nationally recognized city planners to make Miami Township a more 'walkable community'.

This small community and a large portion of its business owners, both locally based and national chains, are beginning to realize the importance of eliminating the 'concrete jungle' that was considered progress until the 1990s. The mentality that designed layouts of malls with acres of asphalt-covered parking lots is being replaced with designs of more environmentally friendly malls with park-like settings. More walking access to shops and bike paths that are incorporated into already existing roadways will start this project in that direction. Implementing these plans and other features that may be required will not be an easy job, nor will it be inexpensive. However, the importance of these projects and their completion cannot be overemphasized. Even though these plans are in the early stages, the people of this community have the vision and ability to achieve their objectives. These changes are long-ranging and are being done not only to improve the aesthetic appeal of the area but also to safeguard the sustainability of the land for future generations.

Other communities throughout the United States and other countries around the world have come to the same conclusion and have made, or are making, the changes required to maintain their environments. Even business communities are trying to adjust their perspectives to consider including environmental criteria as part of normal business operations. Making the changes required to provide a healthy environment for future generations may seem expensive, but not making the changes now will be even more expensive and changes may be impossible to accomplish if left too long.

Papua New Guinea and what we may learn from its past

Papua New Guinea is a good case for examining how a civilization has sustainably managed its environment in the past. The story that follows is important for several reasons:

- First – people have been living in Papua New Guinea for approximately 46,000 years without totally destroying their environment! How did they do it?
- Second – the four major food crops used by the inhabitants were first domesticated here and later spread throughout the region. This sophisticated agriculture has been going on for approximately 7000 years – which makes it the most successful and longest-lasting example of sustainable food production yet.
- Third – global intrusion in this area is increasing and the local population is being exposed to attitudes regarding land and wealth that are totally alien to their culture. Will this country become another possession-oriented society and will it sacrifice its forests to obtain those goods, or will it help teach the rest of the world how to use its own resources in a sustainable way?

On the warm, humid breeze there is a heavy scent of tropical flowers. Blended with this is a hint of salt from the ocean waves rippling on the shore. With so much area covered by rainforest, you can almost taste the colour green. The chatter of thousands of rainbow-hued birds mixes with the babble of animals concealed in the trees. The sound creates a less than musical chant, which grates on the nerves at first but then seems to melt into the background and becomes barely noticeable. At night, the brilliance of the stars in the Southern Cross shines on the land and sprinkles diamonds on the beach.

Welcome to Papua New Guinea!

Geography

The information provided below has been summarized from PNG Embassy (2005) and Wikipedia (2005). The Independent State of Papua New Guinea, commonly referred to as Papua New Guinea or PNG, occupies the eastern half of the tropical island of New Guinea and includes numerous smaller islands and atolls. The smaller islands that belong to Papua New Guinea include the Bismark Archipelago, New Britain, New Ireland and the North Solomons. Some of these islands are volcanic, with dramatic mountain ranges, and all are relatively undeveloped. This chain of islands is located between the Coral Sea and the Pacific Ocean, at the north-east of northern Australia and east of Indonesia.

Approximately 130–65 million years ago, both Australia and New Guinea were part of the ancient supercontinent, Gondwanaland. During former ice ages when sea levels where lower, this area was exposed, forming a land bridge. This earlier land connection between Australia and New Guinea explains the close genetic link among species. Several species of marsupials, including kangaroos and possums, are located in Australia and New Guinea, but nowhere else in the world.

Papua New Guinea has approximately 462,800 km^2 (178,688 square miles) of land area, which does not include its smaller islands and atolls. The central part of the island features a rugged, wide ridge of mountains known as the Highlands, which is characterized by sheer slopes, sharp ridges, fast-running rivers and the scars of numerous landslides. Dense rainforests are found in the lowlands and along the coastal areas. The mainland ranges from beaches to coastal swamps and rough fjords to dry savannahs, with deep valleys and steep mountains in the interior. Nearly 85% of the main island is carpeted with tropical rainforest, containing vegetation that is a combination of Asian and Australian species. The country is also home to an impressive variety of exotic birds, including almost all of the known species of 'birds of paradise'. It is also home to more species of orchids than any other country in the world.

New Guinea has the largest remaining intact block of tropical forests in the Asia–Pacific region and the largest tropical rainforest after the Amazon and the Congo. The island's coastal systems contain some of the most pristine and largest tracts of mangroves in the world, while the lowlands and mountain ranges contain as much as 50 million ha of tropical forests, which are abundant with plant and animal life, some of which exists nowhere else on earth (Worldwildlife, 2005). Papua New Guinea includes a number of ecoregions: Admiralty Islands lowland rainforests, Northern New Guinea montane rainforests, Central Range montane rainforests, South-eastern Papuan rainforests, Huon Peninsula montane rainforests, Southern New Guinea freshwater swamp forests, Louisiade Archipelago rainforests, Southern New Guinea lowland rainforests, New Britain–New Ireland lowland rainforests, Trobriand Island rainforests, New Britain–New Ireland montane rainforests, Trans Fly savannah and grasslands, Northern New Guinea lowland rainforests and freshwater swamp forests and Central Range subalpine grasslands.

History

The information provided below has been summarized from Denham *et al.* (2004). Archaeological evidence indicates that humans arrived on New Guinea from 45,000 to 60,000 years ago. They travelled by sea from South-east Asia during an ice-age period when sea levels were lower and the distances between land masses were shorter. These migrants arrived in several waves and because of the terrain of the island each group developed culturally in different ways. This territory is so densely forested and its topography so forbidding that a large portion of the island's inhabitants remained isolated from each other for millennia.

Although the first arrivals were hunter-gatherers, they managed the forest to provide food. There are indications of sophisticated forms of agriculture having been practised in New Guinea at approximately the same time that agriculture was developing in the prosperous days of Mesopotamia and Egypt. New findings on plant exploitation at several areas in the Highlands of New Guinea by archaeological, archaeobotanical and palaeoecological research endorse previous interpretations that New Guinea was a centre of independent agricultural origins by at least 6950–6440 BCE and probably much earlier. This agricultural activity was sophisticated enough, despite the limited availability of tools (made from wood or bone), for the inhabitants to be able to manage the formidable terrain, provide enough food to flourish and also domesticate numerous plants without total deforestation or major soil erosion. Even by today's standards, this is an accomplishment.

Other studies across the Highlands also show evidence of mid-Holocene forest clearance. Human clearance of primary forest at Kelela Swamp in the Baliem Valley occurred prior to 7800 BCE, with continued disturbance and establishment of extensive grasslands by 3000 BCE (Denham *et al.*, 2004).

The palaeoecological evidence indicates that fire was used throughout the Upper Wahgi Valley from at least 7000–6500 BCE for clearing forested areas to establish and maintain an agricultural landscape. Previous research suggests that mounded cultivation enabled the multi-cropping of plants with different edaphic (soil) requirements. Water-tolerant plants such as taro (*Colocasia*) were planted within the wetter areas between the mounds while water-intolerant plants such as sugarcane (*Saccharum*), bananas (*Musa*), yams (*Dioscorea* spp.), edible pitpit (*Setaria palmifolia*) and mixed vegetables were planted or staked on these 'island beds'. Based on phytogeographic and genetic evidence, taro and bananas, along with breadfruit (*Artocarpus altilis*), sago (*Metroxylon sagu*), sugarcane and some yam species, are all considered to be indigenous to and domesticated in the New Guinea region.

Contrary to previous ideas, the first farming in this area was not derived from South-east Asia but emerged independently in the Highlands. Plants such as bananas were probably first domesticated in New Guinea and later diffused into the Asian continent.

To maintain soil fertility in their gardens and farmlands, indigenous farmers added weeds, grass, old vegetation and organic matter as compost. They used their rubbish, ash from cooking fires, rotten wood and animal manure as mulches and fertilizers to improve the soil. With heavy rainfall in some areas of up to 816 cm (400 inches) per year, ditches had to be dug around their fields to lower the water table and prevent waterlogging of their crops. This is in addition to the positioning of their crops on and around mounds to best match the specific crop's edaphic requirements. Crops that fix atmospheric nitrogen were rotated with other crops to help maintain the soil's nitrogen levels. To use all available land, terraces were constructed on steep slopes, soil retention barriers were erected and excess water was removed by using vertical drains. All this was done thousands of years ago with little more than wooden tools.

Originally, the Highlands of Papua New Guinea were covered with oak (*Quercus*) and beech (*Fagus*) forests, but after thousands of years of continuous activity the heavily populated areas were completely converted to agricultural use. The Highland lifestyle relies on wood for building homes, fences, tools, cooking utensils, weapons and fuel for both cooking and heating their homes.

The inhabitants discovered the best tree to grow to provide their wood need was casuarina (*Casuarina oligodon*), which is native to the Papua New Guinea Highlands. It belongs to a group of several dozen tree species with leaves that resemble segmented pine needles now found throughout the Pacific Islands, Australia, South-east Asia and parts of East Africa. One among its many attributes is the fact that it is a very hard wood (commonly called ironwood) that is easy to split. Additional benefits to planting casuarinas are (Denham *et al.*, 2004):

- they are fast-growing;
- its wood is excellent to use for both building and as a fuel wood;
- it can enhance soil fertility by increasing soil nitrogen levels because its roots have nodules that fix atmospheric nitrogen;
- planted in gardens it increases the soil's fertility and it shortens the amount of fallow time the land requires before crops can be planted;

- when used in areas with steep slopes its roots help to hold the soil and reduce erosion; and
- farmers also claim that the casuarina tree reduces garden infestations of the taro beetle (Diamond, 2005). This alone is of major importance, if it is true, because taro (similar to the potato) is a food staple of the Papua New Guineans.

The Highlanders found natural sproutings of casuarina seedlings along stream beds and transplanted them in groves near their villages. This transplanting guarantees that they will have a continuing supply of trees for all the varied uses that they have found for the casuarinas. From pollen evidence, approximately 1190–970 BP there was a notably large surge in the amount of casuarina pollen found in two locations about 800 km (500 miles) apart. These data imply that the inhabitants of Papua New Guinea have been practising agroforestry or a form of silviculture at least since that time (Denham *et al.*, 2004).

European discovery

In 1511 the Portuguese Antonio de Abreu made a voyage from the Aru Islands to the Moluccas and probably sighted the coast of New Guinea. However, it was Don Jorge de Meneses, also a Portuguese explorer, who is credited with its European discovery in 1526 or 1527. Meneses named the island Ilhas dos Papuas, Island of the Fuzzy Hairs (Papua is a Malay word). The Spaniard Inigo Ortiz de Retes later called it New Guinea because he thought the people looked similar to those of Guinea in Africa. Further exploration followed, including landings by British explorers, e.g. Bougainville, Cook, Stanley and John Moresby (PNG Embassy, 2005; Wikipedia, 2005).

Numerous European traders, adventurers and gold hunters visited in the 16th and 17th centuries, but no claims were made on New Guinea until the Dutch asserted authority over the western part of the island in 1824 (PNG Embassy, 2005; Wikipedia, 2005). Germany took possession of the northern part of the island in 1884. Three days later the British declared a protectorate over the southern region. In 1902 the British possession was placed under the authority of the Commonwealth of Australia. With the outbreak of the First World War, Australian troops took the northern part of the island from the Germans. After the Second World War the eastern half of New Guinea reverted to Australia while Indonesia took Dutch New Guinea in 1963. Papua New Guinea was granted self-government and full independence in 1975.

Culture

Most Papua New Guineans are Melanesians, even though their physical and cultural characteristics vary. It is in fact the most heterogeneous country in the world. This diversity, reflected in a folk saying, 'For each village, a different culture,' is perhaps best illustrated in the number of local languages (Wikipedia, 2005). About 650 Papuan languages have been identified; of these only 350–450 are related. The remainder appear to be totally unrelated to each other or to any other major language group. In a population of fewer than 5 million people, there are between 600 and 700 indigenous societies with a distinct language and unique culture for each. There is no such thing as a typical Papua New Guinean.

The Papua New Guinea society is based on centuries-old traditions from a Melanesian heritage (PNG Embassy, 2005; Wikipedia, 2005). Much of the inherited social structure and values, from matters affecting day-to-day life and gardening to marriage and death, have remained unchanged for centuries. The responsibility for the daily management of the garden, animals and children lies with the women. Men and older boys primarily hunt or in some cases work outside the village. Social alliances are based on family, clan and tribe. Ownership of material wealth is vested in the household and controlled by the eldest male. Fundamental notions of reciprocity and family obligations still hold true in today's society. Wealth is not traditionally accumulated for its own sake, but to bring prestige by giving it away in elaborate ceremonies, which places an obligation on the receiver.

The traditional Papua New Guinea social structure includes the following characteristics:

- The practice of subsistence economy.
- Recognition of bonds of kinship with obligations extending beyond the immediate family group.
- Generally democratic relationships with an emphasis on acquired, rather than inherited, status.
- A strong attachment to the land.

As one citizen of Waghi River Valley stated, 'My village is like a tree. The leaves change, but the roots of culture stay the same' (MSNBC, 2005).

Most Papua New Guineans still strongly adhere to traditional social structure, even though there are complaints from the village elders that the modern world is corrupting the younger generation (PNG Forum, 2005).

Long before Europe envisioned the concept of democracy Papua New Guinea communities were reaching decisions by consensus, not by the most powerful individual of the community.

Attitudes regarding the land are also extremely different from those in many other countries (PNG Embassy, 2005; Wikipedia, 2005). Land is not 'owned' by any one individual, it is not considered a commodity to sell, and it is a permanent and integral part of the community. Land is owned by generations of clans. Individuals may sell their land-usage rights, but the land itself is not sold. The clans, as a group, may lease the land to the government. If outside investors want to acquire land, they must negotiate with the government for leases. The land is considered the life of the people of Papua New Guinea and they believe that its proper care is their duty. This basic difference in attitudes regarding the 'value of land' has created great difficulties between the indigenous peoples and foreign-based companies.

The first conflict between these ideological differences occurred in 1987 (all information below on the Bougainville Crisis is from McIntosh, 1990). After 20 years of land disputes and rainforest devastation, villagers of the Solomon Islands forced a Unilever subsidiary, Levers Pacific Timbers, to cease operations by using civil disobedience. Unfortunately this clash was quickly forgotten and any lessons learned by the government and foreign companies were ignored. Nothing better illustrates this than the consequences of the ecocide of the Bougainville Copper Ltd (BCL) mine at

Panguna. The original trouble started in 1963 when the parent company, Conzinc Rio-Tinto Australia (CRA), was granted a prospecting licence by the Australian colonial government. Local people objected to the presence of geologists in their area. No consultation between the elderly women, who held land on behalf of the matrilineal clans, and any official, either of the government or CRA, had taken place. The rightful landowners argued that the copper would not rot in the ground so they wanted additional time for their people to educate themselves regarding the facts so that they could better decide what was in their best interest. Colonial administrators, trying to build a base for future economic development, ignored their requests. Small disturbances and confrontations between the two groups continued and resulted in prison sentences for five villagers in 1966.

After increasing compensation, obtaining police protection and establishing a Village Relations Office to handle claims, the copper mine at Panguna was opened for commercial production in 1972 (McIntosh, 1990). Bougainville Copper Ltd had been granted a 'tailings lease' over the whole Jaba River Valley. This allowed them to discharge the mine's tailings as cheaply as possible by dumping them directly into the Jaba River and its tributary, the Kawerong. Toxic wastes such as cyanides and heavy metals from the mine's copper and gold processing were discharged into the river system. The Nasioi, local landowners, who farm the surrounding land and fish in the river for their livelihood, found most marine life in the estuary destroyed and their farmlands contaminated by the polluted groundwater.

Basil Peutalo of the Papua New Guinea Catholic Commission for Justice, Peace and Development commented to the Tripartite Consultation of the Melanesian, Indonesian and Australian Council of Churches in Papua New Guinea on 18 August 1989:

> This ecocide was done without warning, without permission having been asked or granted, and in areas where the inhabitants had thought that they would not be touched by the mining activities. Here is a people who fear that they are no longer in control of their destiny and land. They are losing control of the patrimony of their children. For thousands of years, our ancestors lived out their interconnectedness with the natural world. However, this view of nature and the relationship of the human person with it is challenged today by a spirit of utility which views the earth as property to be used. The huge amount of money that goes with such destructive activities has become an attractive wrapping around the negotiations with local peoples.
>
> (McIntosh, 1990)

After a 9-year revolt on the island of Bougainville, attributed to traditional land rights, economic issues and this ecocide, active hostilities ended with a truce in 1997 and a permanent ceasefire was signed in 1998. A peace agreement was signed in 2001 between the government and ex-combatants. This civil war quickly grew into a war for independence, displaced roughly 40,000 people and cost the lives of approximately 20,000.

Economy

Information on the economy is summarized from PNG Embassy (2005) and Wikipedia (2005). Papua New Guinea is richly endowed with natural resources but, due to the rugged terrain, its exploitation has been hampered. These resources have been attributed to Papua New Guinea's position in the 'Ring of Fire' and the shift of continental tectonic plates. Mineral deposits including oil, copper and gold can be found throughout the islands. There are extensive reserves of natural gas and other known fields are yet to be developed. Today these items account for approximately 72% of export earnings.

The economy is dominated by mineral and petroleum exports. However, the agriculture, forestry, fishing and manufacturing sectors account for a significant portion of the nation's gross domestic product. Subsistence and semi-subsistence farming supports 85% of the population. Most villages are self-sufficient and some small surpluses are available for trading in local markets.

Palm oil, coffee, cocoa and copra (coconuts) are the cornerstones of agricultural exports. Additional cash crops include tea, rubber, cardamom, chillies and sugar. Today, local markets consume a large portion of the agricultural production, but the government is considering the potential for expanding agricultural production. Major reforms in fiscal and trade policies are being considered to induce innovation in agriculture and livestock projects to attract investors.

Fish exports are primarily confined to shrimp. To increase economic growth in this sector, fishing boats of other nations catch tuna in Papua New Guinea's waters under government licences.

Due to low global prices for timber, the forestry industry declined in 1998 but has been slowly recovering in recent years. Today timber is one of the nation's major exports. The government recognizes the value of its rare and magnificent natural forests and has implemented major policy changes aimed at creating an environment for sustainable forestry. In 1993 it formed the Papua New Guinea Forest Authority to reform the forestry processes and devise a programme to manage the tropical rainforest. This group's goal is to achieve a balance between the use of forests for conservation purposes and for the production of wood products. Many groups from around the globe are working with Papua New Guinea citizens along with both national and local governments to provide the assistance and education needed to make certain the nation's resources are sustainably managed.

Unfortunately, after this book went to press, reports started coming out of Jakarta from an environmental group that illegal logging was taking place in Papua New Guinea and that an estimated 250,000 ha (625,000 ac) of virgin forest are being harvested each year. The reports claim that PNG laws regarding timber production are being ignored by Malaysian-owned interests and the government is not enforcing its own laws. These illegal timber shipments are destined for China, Japan, South Korea, Europe and North America.

Until recently, tourism in Papua New Guinea hasn't provided much in the way of economic return. Several major obstacles have caused this. First, the terrain has made it difficult for the country to develop the necessary infrastructure to promote tourism fully. Transportation into the Highlands required an aeroplane until a limited road network was created for four-wheel-drive vehicles. Secondly, the government continues to allocate priority to the development of the infrastructure, but the country is vast and providing the resources is a constant challenge for the government. Despite these problems, the development of tourism is being actively encouraged.

What is the Future of Societies Enveloped by Global Markets?

European involvement in Papua New Guinea has disrupted this human value-centred society and tried to replace it with a system where economic value is the gauge by which to measure one's worth. Unbridled resource consumption carries a cost, not only for individuals but for societies as well. If any time or any place is able to teach us anything about sustainable living, it may be now and in this remote corner of the globe. We need to learn from the past and not have developing nations become industrialized, yet repeating the mistakes of the past.

It is useful to look at what aspects of Papua New Guinea have allowed these people to survive on a small land base. The Papua New Guinea case highlighted how the cultural norm and lack of ownership of resources allowed these communities to build consensus on resource uses. A similar philosophy has been the core of the American Indian Nations or tribes and how they approach the management of their resources. They now have rights to their forests (see Case 3.4 – Indian Forest: Land in Trust, and Case 3.5 – Forest Management and Indigenous Peoples in Western Canada) and are making decisions on how to sustainably manage their lands. These indigenous groups in North America have to consider the future capacity of the forests to provide resources, i.e. sustainable management, because the tribes cannot move elsewhere if they over-exploit and degrade their resources and lands. How well they manage their resources today will determine the ability of future generations to survive from forests.

In these two examples (PNG and Miami Township), local people built a consensus on what they found to be acceptable practises related to forests or resources. The power of many local communities building consensus when resource uses and conservation conflict occur will create changes in the landscape. The process of democratization of decisions regarding forests started during the 1980s and is still evolving. The right model for how this democratization should unfold will probably determine the future look and quality of forests. It will not be enough to have local communities making decisions but what is required is that we learn to use the knowledge of people indigenous for any given area (e.g. PNG). Miami Township in Montgomery County, Ohio, consists of informed citizens who are changing how forests are being managed.

Local communities working at providing solutions are crucial for implementing integrated conservation and development projects. Science also needs to be an important part of the process when it is used to help define what future conditions are possible, given local changes that occur in the landscape. This is relevant since we are in a global economy, but local values drive decisions. The local values and decisions that might impact the landscape need to be supported by facts from the sciences of ecology, conservation and disturbances.

We also need to recognize that science does not have all the answers today, since we are frequently asking new questions that were not asked in the past. We need to use science in achieving the goals of sustainable livelihoods and sustainable forests and conservation efforts. Our use of science has to be 'adaptive' and incorporate new information as it becomes available. Science should not be used as though we have the answer today. Science also has to include human values and attitudes and be able to demonstrate when pursuing these values will or will not be sustainable in human landscapes.

The uncertainty in our scientific understanding should not be a barrier to using science in decision-making. Humans have faced many challenges in using technology and the sciences to modify our environment or to restore it after it has been altered. The key to being sustainable is to be able to learn from the mistakes after they have occurred. A couple of examples will be briefly mentioned because they demonstrate how humans learn and are very adaptable. The Apollo 13 is a simpler example of humans learning to deal with the malfunctioning of technology. The other example is an 'experiment' that was conducted in Arizona with the creation of the Biosphere 2. The Biosphere 2 demonstrates the difficulty of making decisions when the science was not advanced enough.

Apollo 13 is an example of a relatively simple environment that contained structures (e.g. equipment, humans, etc.) and processes (e.g. humans respiring CO_2 into the atmosphere, CO_2 being scrubbed out of the atmosphere, etc.) that humans have built using technology. On 11 April 1970, the National Aeronautics and Space Administration (NASA) launched the Apollo 13 space capsule with a lunar module attached and a three-man crew aboard. About 6 minutes after lift-off a little vibration was felt. The centre engine of the S-II stage had shut down 2 minutes earlier. This malfunction forced the remaining four engines to burn 34 s longer than planned. In addition this required that the S-IVB third stage burn 9 s longer so that Apollo 13 could get into orbit. There were only a few minor surprises over the first 2 days, but for the most part Apollo 13 was looking like the smoothest flight of the programme. Things seemed to be going fine; as a matter of fact, the ground crew was commenting on how bored they were. At 55 h, 55 min into the flight, oxygen tank number 2 blew up, causing tank number 1 to fail. About 200,000 miles (320,000 km) from Earth the Apollo 13 command module lost most of its supply of oxygen, electricity, light and water. The command module was powered down and the crew moved into the lunar module for the long trip home. The attitude at NASA was that no American had been lost in space and now was not the time to start. As the world watched, NASA personnel worked around the clock to bring the crew and spacecraft safely back to Earth. In this case, technology failed but the crew of Apollo 13 survived. NASA also had to adapt to a risk that they had not predicted had a high probability of occurring.

Much technology had gone into the building of the Apollo space capsule. Our technology, however, can only predict the probability of some events occurring but does not include unforeseen events or the repercussions of unknown events. This shows the need to have backup or buffering capacity in any ecosystem so that an ecosystem can continue to function sustainably. The next example highlights how considerable science and the latest technology went into building the Biosphere 2 – a completely enclosed system – but how its management needed to become adaptive because of unforeseen disturbances that made it unliveable for humans.

Biosphere 2 is another example of a human-built environment but is one that is more complex than the Apollo capsules (Marino and Odum, 1999). This large, airtight, greenhouse-looking complex was built just north of Tucson, Arizona, United States, by Edward Bass for about $150 million dollars and covered an area of about 1.27 ha (3.15 acres). Inside this simulation of Biosphere 1 (our Earth), they integrated a variety of biomes (e.g. human habitat, agricultural plots, rainforest, thorn scrub/grassland, desert, marsh/mangroves, and even an ocean with corals). A biospherian team of four men and four women was enclosed in the Biosphere 2 in September 1991 for 2 years. The idea was to see if the humans could survive in a closed system with these biomes without any additional inputs (e.g. air, food, water, etc., exclusive of electricity). It turns out that the structures and functions within the Biosphere 2 were much more complex than anticipated. For example, the agricultural plots did not produce as much food as expected (pest problems), plants did not produce enough oxygen to replace that consumed by humans and other organisms especially the decomposer microbes, some plants and ocean algae grew too aggressively and had to be constantly managed by the overworked biospherians.

The Biosphere 2 example is extremely useful for humans to think about when they restore or rebuild biomes that have been previously altered by human activities. It shows that our scientific knowledge is evolving with time and we need to consider how we include new knowledge into our sustainable management frameworks. Some might have considered the Biosphere 2 a failure but in fact it has taught us about the importance of including uncertainty in our scientific understanding when we design systems to be sustainable in the future. The Biosphere 2 also contributed a tremendous amount of information on restoration ecology because it was entirely comprised of many human-constructed biomes.

Not only do we need to include uncertainty in our sustainable management designs but we need to learn from the past. The discussion of the history of forest fear, use and conservation will inform us of why particular human behaviours or activities occurred and should allow us to not repeat the same mistakes again. At the same time we need to try to anticipate problems before they occur.

People cannot assume that someone else will take care of whatever problem needs fixing. We are a globalized society – information is available on a global level but the decisions that impact human landscapes are often started at the local or regional levels. People need to be making decisions that balance the uses of the world's resources while still conserving them for the future. 'Think global, act local' are four powerful words that provide a solution to how forest uses can be kept sustainable.

Forests have mainly been used for extraction of resources and some suggest (Robinson and Redford, 2004) that linking livelihoods with conservation has not worked because we have to move beyond just the extraction of natural resources. We have made a case in this book that this does not have to be the case. Technology can allow us to use natural resources more efficiently and in a more environmentally friendly manner. Technology will be a crucial element of future forests because it allows economic values to be developed for forests so that forests do not only have higher economic return from being converted to urban homes. Forests have intrinsic value to society so they need to not just be converted to non-forest conditions such as development (Montgomery County is an example of balancing resource uses with the local values for what is wanted from the landscape). In other locations, forests have to take advantage of the new ways that forests can contribute to energy production that is environmentally friendly. This is especially relevant in developing countries that are still mainly dependent on forests for their energy use in very environmentally unfriendly manners.

References

Aber, J., McDowell, W., Nadelhoffer, K., Magill, A., Bernston, G., Kamakea, M., McNulty, S., Currie, W., Rustad, L. and Fernandez, I. (1998) Nitrogen saturation in temperate forest ecosystems – hypotheses revisited. *BioScience* 48, 921–934.

Adams, H.R., Sleeman, J.M., Rwego, I. and New, J.C. (2001) Self-reported medical history survey of humans as a measure of health risk to the chimpanzees (*Pan troglodytes schweinfurthii*) of Kibale National Park, Uganda. *Oryx* 35, 308–312.

AFPA (2000) Pulp and Paper Factbook. Available at: www.afandpa.org (accessed 30 September 2000).

AFPA (2004a) Paper recycling: quality is key to long-term success. A report compiled by the American Forest and Paper Association. March 2004. Available at: www.afandpa.org (accessed 18 March 2004).

AFPA (2004b) Illegal logging and global wood markets: the competitive impacts on the US wood products industry, prepared by Seneca Creek Associates LLC and Wood Resources International LLC. Available at: www.illegal-logging.info/news.php?newsId=717 (accessed 9 September 2005).

Agee, J.K. (1993) *Fire Ecology of Pacific Northwest Forests*. Island Press, Washington, DC.

Agee, J.K. (1998) The landscape ecology of western forest fire regimes. *Northwest Science* 72, 24–34.

Agrawal, A. (2005) Community, intimate government, and the making of environmental subjects in Kumaon, India. *Current Anthropology* 46, 161–190.

Agrios, G.N. (1997) *Plant Pathology*, 4th edn. Academic Press, New York.

Aiken, G.R., McKnight, D.M., Wershaw, R.L. and MacCarthy, P. (1985) An introduction to humic substances in soil, sediment, and water. In: Aiken, G.R., McKnight, D.M., Wershaw, R.L. and MacCarthy, P. (eds) *Humic Substances in Soil, Sediment, and Water: Geochemistry, Isolation, and Characterization*. John Wiley, New York, pp. 1–9.

Alexander, E.B., Kissinger, E. and Cullen, P. (1989) Soils of southeast Alaska as sinks for organic carbon fixed from atmospheric carbon dioxide. In: Alexander, E.B. (ed.) *Proceedings of Watershed '89: A Conference on the Stewardship of Soil, Air, and Water Resources*. R10-MB-77, USDA Forest Service, Alaska Region, Juneau, Alaska, pp. 203–210.

Alcxopolous, C.J., Minns, C.W. and Blackwell, M. (1996) *Introductory Mycology*. John Wiley and Sons, New York.

Alig, R. and Butler, B. (2004) *Area Changes for Forest Cover Types in the United States 1952 to 1997, with projections to 2050.* General Technical Report PNW-GTR-613. USDA Forest Service, PNW Research Station, Portland, Oregon.

Allan, B.F., Keesing, F. and Ostefeld, R.S. (2003) Effect of forest fragmentation on Lyme disease risk. *Conservation Biology* 17, 267–272.

Amorosi, T., Buckland, P., Dugmore, A., Ingimundarson, J.H. and McGovern, T.H. (1997) Raiding the landscape: human impact in the Scandinavian North Atlantic. *Human Ecology* 25, 491–518.

Anderson, K. (1993) Native Californians as ancient and contemporary cultivators. In: Blackburn, T.C. and Anderson, K. (eds) *Before the Wilderness: Environmental Management by Native Californians.* Ballena Press, Menlo Park, California, pp. 151–174.

Andreu, M.G., Vogt, K.A., Vogt, D.J., Edmonds, R.L., Oliver, C. and Gara, R. (2005) Emerging bio-energy technology solutions to reduce fire risk along the wildland urban interface (WUI). In: Land, D.N. (ed.) *Emerging Issues Along Urban/Rural Interfaces: Linking Science and Society.* The Center for Forest Stability, Auburn University, Auburn, Alabama, pp. 189–193.

Anon. (1941) *The Story of Paper-Making.* Butler Paper Company, Chicago, Illinois.

Appelstrand, M. (2002) Participation and societal values: the challenge for lawmakers and policy practitioners. *Forest Policy and Economics* 4, 281–290.

Aradóttir, Á.L. and Eysteinsson, Þ. (2005) Restoration of birch woodlands in Iceland. In: Stanturf, J.A. and Madsen, P. (eds) *Restoring Temperate and Boreal Forested Landscapes.* CRC Press, Boca Raton, Florida, pp. 195–209.

Aradóttir, Á.L., Arnalds, Ó. and Archer, S. (1992) Hnignun gróðurs og jarðvegs (The degradation of vegetation and soil). *Græðum Ísland* IV, 73–82.

Arnalds, A. (1987) Ecosystem disturbance in Iceland. *Arctic and Alpine Research* 19, 508–513.

Asner, G.P., Palace, M., Keller, M., Pereira, R., Jr, Silva, J.N.M. and Zweede, J.C. (2002) Estimating canopy structure in an Amazon forest from laser range finder and IKONOS satellite observations. *Biotropica* 34, 483–492.

Assmann, E. (1970) *The Principles of Forest Yield Study.* Pergamon Press, Cambridge, Massachusetts.

Auclair, A.N.D., Worrest, R.C., LaChance, D. and Martin, H.C. (1992) Climatic perturbation as a general mechanism of forest dieback. In: Manion, P.D. and LaChance, D. (eds) *Forest Decline Concepts.* American Phytopathological Society Press, St Paul, Minnesota, pp. 38–58.

Azar, C. and Rodhe, H. (1997) Targets for stabilization of atmospheric CO_2. *Science* 276, 1818–1819.

Azar, C., Lindgren, K. and Andersson, B.A. (2003) *Hydrogen or Methanol in the Transportation Sector?* Report to the Swedish Transport and Communication Research Board, Department of Physical Resource Theory, Chalmers University of Technology, Göteborg, Sweden.

Azevedo-Ramos, C., De Carvalho, O. Jr and Nasi, R. (2003) Animal Indicators, a Tool to Assess Biotic Integrity after Logging Tropical Forests. Available at: www.cifor.cgiar.org/publications/pdf_files/polex/Pazevedo-ramos0301.pdf (accessed 15 April 2004).

Barbour, A.G. and Fish, D. (1993) The biological and social phenomenon of Lyme disease. *Science* 260, 1610–1616.

Barrett, C., Brandon, K., Gibson, C. and Gjertsen, H. (2001) Conserving tropical biodiversity amid weak institutions. *BioScience* 51, 497–502.

Bass, S. (2003) Certification – A new vehicle for securing public benefits from forests. In: Meidinger, M., Elliot, C. and Oesten, G (eds) *Social and Political Dimensions of Forest Certification.* Forstbuch Verlag, Remagen-Oberwinter, Germany, pp. 27–50.

Beard, K.H. and Pitt, W.C. (2005) Potential consequences of the coqui frog invasion in Hawaii. *Diversity and Distributions* 11, 427–433.

Beard, K.H., Eschtruth, A.K., Vogt, K.A., Vogt, D.J. and Scatena, F.N. (2003) The effects of the frog *Eleutherodactylus coqui* on invertebrates and ecosystem processes at two scales in the Luquillo Experimental Forest, Puerto Rico. *Journal of Tropical Ecology* 19, 607–617.

Beard, K.H., Vogt, K.A., Vogt, D.J., Scatena, F.N., Covich, A., Sigurdardottir, R., Siccama, T.C. and Crowl, T.A. (2005) Structural and functional responses of a subtropical forest to 10 years of hurricanes and droughts. *Ecological Monographs* 75, 345–361.

Becker, E. (2002) Two acres of farm lost to sprawl each minute, new study says. *New York Times*, 4 October 2002, p. A19.

Bertrand, M.R. and Wilson, M.L. (1997) Microhabitat-independent regional differences in survival of unfed *Ixodes scapularis* nymphs (Acari: Ixodidae) in Connecticut. *Journal of Medical Entomology* 34, 167–172.

Best, C. and Wayburn, L.A. (2001) *America's Private Forests. Status and Stewardship.* Island Press, Washington, DC.

Bethel, J.S., Turnbull, K.J., Briggs, D.G. and Flores, J.G. (1975) Timber losses from military use of herbicides on the inland forests of South Vietnam. *Journal of Forestry* 73, 228–233.

Bilby, R.E., Fransen, B.R. and Bisson, P.A. (1996) Incorporation of nitrogen and carbon from spawning coho salmon into the trophic system of small streams: evidence from stable isotopes. *Canadian Journal of Fisheries and Aquatic Sciences* 53, 164–173.

Binns, W.O. and Redfern, D.B. (1983) *Acid Rain and Forest Decline in W. Germany.* Forestry Commission Research and Development Paper 131, Forestry Commission, Edinburgh, UK.

Birrell, A.M. (1999) *Chinese Mythology: An Introduction.* Johns Hopkins University Press, Baltimore, Maryland.

Black, H.C. (ed.) (1994) *Animal Damage Management Handbook.* Gen. Tech. Rep. PNW-GTR-332, USDA Forest Service, Pacific Northwest Station, Portland, Oregon.

Blaustein, R.J. (2001) Kudzu's invasion into southern United States life and culture. In: McNeeley, J. (ed.) *The Great Reshuffling: Human Dimensions of Invasive Species.* IUCN, World Conservation Union, Gland, Switzerland, and Cambridge, UK, pp. 55–62.

Blöndal, S. (1995) *Innfluttar trjategundir I Hallormsstadaskogi* (Exotic tree species in Hallormsstadur Forest). Iceland Forest Service, Reykjavik, Iceland.

Bo tree (2005) Encyclopædia Britannica. Encyclopædia Britannica Premium Service. Available at: www.britannica.com/eb/article-9015801 (accessed 13 October 2005).

Boyce, J.S. (1929) *Deterioration of Windthrown Timber on the Olympic Peninsula.* Technical Bulletin 104, USDA, Washington, DC.

Brechin, S.R., Wilshusen, P.R., Fortwangler, C.L. and West, P.C. (2002) Beyond the square wheel: toward a more comprehensive understanding of biodiversity conservation as a social and political process. *Society and Natural Resources* 15, 41–64.

Brisson, D. and Dykhuizen, D.E. (2004) OspC diversity in *Borrelia burgdorferi*: different hosts are different niches. *Genetics* 168, 713–722.

Browder, J.O. (1992) Social and economic constraints on the development of market-oriented extractive reserves in Amazon rainforests. *Advances in Economic Botany* 9, 33–41.

Brown, S. (2002) Measuring, monitoring, and verification of carbon benefits for forest-based projects. *Philosophical Transactions of the Royal Society of London* A 360, 1669–1683.

Brown, S. and Pearson, T. (2005) *Cost Comparison of the M3DADI System and Conventional Field Methods for Monitoring Carbon Stocks in Forests.* Report to the Nature Conservancy, Conservation Partnership Agreement, Winrock International, Arlington, Virginia.

Brown, S., Gillespie, A.J.R. and Lugo, A.E. (1989) Biomass estimation methods for tropical forests with applications to forest inventory data. *Forest Science* 35, 881–902.

Brown, S., Pearson, T., Slaymaker, D., Ambagis, S., Moore, N., Novelo, D. and Sabido, W. (2005) Creating a virtual tropical forest from three-dimensional aerial imagery: application for estimating carbon stocks. *Ecological Applications* 15, 1083–1095.

Brown, W.H. (ed.) (1920–1921) *Minor Products of Philippine Forests*, Vol. I (1920), Vol. II (1921) Vol. III (1921). Bureau of Printing, Manila.

Bunnell, F.L. and Johnson, J.F. (eds) (1998) *The Living Dance: Policy and Practices for Biodiversity in Managed Forests*. University of British Columbia Press, Vancouver, Canada.

Bunnell, F.L. and Kremsater, L.L. (1990) Sustaining wildlife in managed forests. *Northwest Environmental Journal* 6, 243–69.

Buol, S.W. (1991) Pedogenesis of carbon in soils. In: Johnson, M.G. and Kern, J.S. (eds) *Sequestering Carbon in Soils: A Workshop to Explore the Potential for Mitigating Global Climate Change.* EPA/600/3-91, USEPA, Environmental Research Laboratory, Corvallis, Oregon, pp. 51–55.

Burger, A.E. (2005) Dispersal and germination of seeds of *Pisonia grandis*, an Indo-Pacific tropical tree associated with insular seabird colonies. *Journal of Tropical Ecology* 21, 263–271.

Bushley, B. (2003) Impacts of community forestry on community development among user groups in Nepal's Western Terai. MSc thesis, University of Washington, Seattle, Washington.

Calvin, W.H. (2004) *A Brief History of the Mind: From Apes to Intellect and Beyond.* Oxford University Press, Oxford, UK.

Campbell, S. and Liegel, L. (Technical Coordinators) (1996) *Disturbance and Forest Health in Oregon and Washington.* Gen. Tech. Rep. PNW-GTR-381, USDA Forest Service, Pacific Northwest Station, Portland, Oregon.

Carey, A.B., Horton, S.P., Biswell, B. and Dominguez de Toledo, L. (1999) Ecological scale and forest development: squirrels, dietary fungi, and vascular plants in managed and unmanaged forests. *Wildlands Monograph* 142, 4–71.

Carpenter, D. and Cappuccino, N. (2005) Herbivory, time since introduction and the invasiveness of exotic plants. *Journal of Ecology* 93, 315–321.

Castellano, M.A. (1996) Outplanting performance of mycorrhizal inoculated seedlings. In: Mukerji, K.G. (ed.) *Concepts in Mycorrhizal Research.* Kluwer Academic Publishers, Dordrecht, the Netherlands, pp. 223–301.

Cederholm, C.J., Johnson, D.H., Bilby, R.E., Dominguez, L.G., Garrett, A.M., Graeber, W.H., Greda, E.L., Kunze, M.D., Marcot, B.G., Palmisano, J.F., Plotnikoff, R.W., Pearcy, W.G., Simenstad, C.A. and Trotter, P.C. (2000) *Pacific Salmon and Wildlife: Ecological Contexts, Relationships, and Implications for Management. Wildlife–Habitat Relationships in Oregon and Washington.* Washington Department of Fish and Wildlife, Olympia, Washington.

Centers for Disease Control and Prevention (2003) *Morbidity and Mortality Weekly Report* 51, 1169.

Chaloner, D.T. and Wipfli, M.S. (2002) Influence of decomposing Pacific salmon carcasses on macroinvertebrate growth and standing stock in southeastern Alaska streams. *Journal of the North American Benthological Society* 21, 430–442.

Chapin, M. (1992) The coexistence of indigenous peoples and environments in Central America. *The National Geographic Society, Washington, DC, Research and Exploration* 8(2), 2–10.

Chernov, Y.I. (1985) (translated by D. Love) *The Living Tundra.* Cambridge University Press, Cambridge, UK.

Christensen, J. (2002) Fiscal accountability concerns come to conservation. *New York Times*, 5 November 2002, p. D2.

Christensen, J. (2003) Auditing conservation in an age of accountability. *Conservation in Practice* 4, 12–19.

CIEL (Center for International Environment Law) (2002) *Whose Resources? Whose Common Good? Towards a New Paradigm of Environmental Justice and the National Interest in Indonesia.* Washington, DC, Copy of the book can be found at: www.ciel.org/Publications/Whose_Resources_3-27-02.pdf

Ciesla, W.M. and Donaubauer, E. (1994) *Decline and Dieback of Trees and Forests. A Global Overview.* FAO Forestry Paper 120, Food and Agricultural Organization of the United Nations, Rome, Italy.

Clark, D.B., Read, J.M., Clark, M.L., Cruz, A.M., Dotti, M.F. and Clark, D.A. (2004) Application of 1-m and 4-m resolution satellite data to ecological studies of tropical rainforests. *Ecological Applications* 14, 61–74.

Clay, J.W. (1996) *Generating Income and Conserving Resources: 20 Lessons from the Field.* World Wildlife Fund, Baltimore, Maryland.

Columbia Encyclopedia (1995) 6th Edition. Available at: www.encyclopedia.com/html/c1/caravel.asp (accessed 9 August 2005).

Compton, J.E., Church, M.R., Larned, S.T. and Hogsett, W.E. (2003) Nitrogen export from forested watersheds in the Oregon coast range: the role of N_2-fixing red alder. *Ecosystems* 6, 773–785.

Conklin, H.C. (1954) The relation of Hanunóo culture to the plant world. Thesis, Yale University, New Haven, Connecticut.

Conklin, H.C. (1975) *Hanunóo Agriculture: a Report on an Integral System of Shifting Cultivation in the Philippines.* Forestry Department Development Paper 12, FAO, Rome, 1957, reprinted Elliot's Books, Northford, Connecticut.

Connell, J.H. and Orias, E. (1964) The ecological regulation of species diversity. *The American Naturalist* 98, 399–414.

Corcoran, C.M. (1999) Rehabilitation of former US military lands bordering the Panama Canal. *Journal of Sustainable Forestry* 8, 67–79.

Cortez, C. (1993) El sector forestal Mexicano: entre la economia y la ecologia? In: *Comercio Exterior.* University of Mexico, Mexico City, Mexico, pp. 370–377.

Costanza, R. (1991) Assuring sustainability of ecological economic systems. In: Costanza, R. (ed.) *Ecological Economics: the Science and Management of Sustainability.* Columbia University Press, New York, pp. 331–343.

Cottrell, A. (1986) *A Dictionary of World Mythology.* Oxford University Press, Oxford, UK.

Daily, G.C. and Walker, B.H. (2000) Seeking the great transition. *Nature* 403, 243–245.

Daszak, P., Cunningham, A.A. and Hyatt, A.D. (2000) Emerging infectious diseases of wildlife – threats to biodiversity and human health. *Science* 287, 443–449.

Davidson, H.R.E. (1964) *Gods and Myths of Northern Europe.* Penguin Books, London, UK.

Dawkins, H.C. and Philip, M.S. (1998) Tropical Moist Forest Silviculture and Management. A History of Success and Failure. CAB International, Wallingford, UK.

Deacon, R.T. and Murphy, P. (1997) The Structure of an environmental transaction: the DNS. *Land Economics* 73, 1–24.

Denham, T.P., Haberle, S.G. and Lentfer, C. (2004) New Evidence and Revised Interpretations of Early Agriculture in Highland New Guinea. Available at: http://palaeoworks.anu.edu.au/publications.html (accessed 15 September 2005).

Diamond, J. (2005) *Collapse: How Societies Choose to Fail or Survive.* Penguin Group, New York.

Dighton, J., White, J.F. and Oudemans, P. (2005) *The Fungal Community – Its Organization and Role in the Ecosystem.* Taylor and Francis, Boca Raton, Florida.

DOE (2003) Energy Information Administration/Monthly Energy Review August 2005. Available at: www.eia.doe.gov/emeu/mer/pdf/pages/sec1_5.pdf (accessed 20 August 2005).

DOE (2004) Feedstock composition glossary for researchers in the Information Resources section of the Office of Biomass Program (OBP) of the US Department of Energy's Office of Energy Efficiency and Renewable Energy (EERE). Available at: www.eere.energy.gov/biomass/feedstock_glossary.html (accessed 15 September 2005).

Drake, J.B., Knox, R.G., Dubayah, R.O., Clark, D.B., Condit, R., Blair, J.B. and Hofton, M. (2003) Aboveground biomass estimation in closed canopy neotropical forests using lidar remote sensing: factors affecting the generality of relationships. *Global Ecology and Biogeography* 12, 147–159.

Draper, G. and Meyer, M.A. (1995) The contribution of the Emperor Asoka to the development of the humanitarian ideal in warfare. International Committee of the Red Cross, International Review of the Red Cross no. 305, pp. 192–206. Available at: www.icrc.org/web/eng/siteeng0.nsf/html/57JMF2 (accessed 9 August 2005).

Drengson, A. and Taylor, D.M. (eds) (1997) *Ecoforestry: the Art and Science of Sustainable Forest Use.* New Society Publishers, Gabriola Island, British Columbia, Canada.

Dunlap, R.E., Gallup, G.H., Jr and Gallup, A.M. (1993) Of global concern. Results of the health of the planet survey. *Environment* 35, 7–39.

Durant, W. (1954) *Our Oriental Heritage.* Simon and Schuster, New York.

Durant, W. (1957) *The Reformation. A History of European Civilization from Wyclif to Calvin: 1300–1564.* Simon and Schuster, New York.

Duryee, T. (2004) Power shift. *The Seattle Times,* 26 September 2004. Available at: www.seattletimes.com (accessed 26 September 2004).

E.A. Engineering (1999) *Report to the Methanol Institute.* Available at: www.methanol.org (accessed 15 August 2005).

Edmonds, R.L., Agee, J.G. and Gara, R.I. (2005) *Forest Health and Protection.* Waveland Press, Prospect Heights, Illinois.

Ekbom, T., Berglin, N. and Lindblom, M. (2003) High efficiency methanol/dme production from biomass via black liquor gasification as renewable motor fuels. In: Proceedings of *Bioenergy 2003. International Nordic Bioenergy Conference,* Jyväskylä, Finland, pp. 440–442.

Elliott, C. (1996) Paradigms of forest conservation. *Unasylva* 47, 3–6.

Ellis, P.B. (2003) *A Brief History of the Druids.* Carroll and Graf, New York.

Ellison, K. and Daily, G. (2003) Making conservation profitable. *Conservation in Practice* 4, 12–19.

EPA (2001) Draft Inventory of US Greenhouse Gas Emissions and Sinks: 1990–1999. Available at: www.epa.gov/globalwarming/publications/emissions.htm (accessed 14 December 2004).

EPA (2004) US Greenhouse Gas Emissions and Sinks: 1990–2002. EPA 430-R-04-003 (April 15, 2004). Available at: http//yosemite.epa.gov/oar/globalwarming.nsf/content/ResourceCenterPublicationsGHGEmissionsUSEmissionsInventory2004.html (accessed 3 February 2005).

Epstein, P.R., Diaz, H.F., Elias, S., Grabherr, G., Graham, N.E., Martens W.J.M., Mosley-Thompson, E. and Susskind, J. (1998) Biological and physical signs of climate change. Focus on mosquito-borne diseases. *Bulletin of the American Meteorological Society* 79, 409–417.

Erdman, L.W. (1953) *Legume Inoculation: What it Is. What it Does.* Publication no. 2003, USDA, Washington, DC.

Erskine, P., Lamb, D. and Bristow, M. (2005) *Reforestation in the Tropics and Subtropics of Australia Using Rainforest Tree Species.* Rural Industries Research and Development Corporation, Canberra, Australia.

Estey, J.S. (1999) Chain of custody as an impediment to certification. In: Vogt, K.A., Larson, B., Gordon, J., Vogt, D.J. and Fanzeres, A. (eds) *Forest Certification: Roots, Issues, Challenges, and Benefits.* CRC Press, Boca Raton, Florida, pp. 285–291.

Eswaran, H., Van den Berg, E. and Reich, P. (1993) Organic carbon in soils of the world. *Soil Science Society of America Journal* 57, 192–194.

Everest, J.W., Miller, J.H., Ball, D.M. and Patterson, M.G. (1991) *Kudzu in Alabama, History, Uses, and Control.* No. ANR 65, Alabama Cooperative Extension System, Auburn, Alabama.

FAO (1989) *Impact of Changing Technological and Economic Factors on Markets for Natural Industrial Fibres: Case Studies on Jute, Kenaf, Sisal, and Abaca.* Economic and Social Development Paper 77, FAO, Rome.

FAO (1998) FAO Yearbook of Forest Products 1996. Table 4. Production, trade and consumption of forest products, 1996. Available at.: www.fao.org/docrep/W9950E/w9950e24.htm (accessed 6 June 2005).

FAO (1999) Global trends in Forest Products. Available at: www.fao.org//docrep/W9950E/w9950e05.htm (accessed 27 August 2005).

FAO (2001) Global Forest Resources Assessment 2000. Summary Report Food and Agriculture Organization website. Available at: www.fao.org/forestry/fo/fra/index.jsp (accessed 1 March 2003).

Farmers Association (2000) *Icelandic Agricultural Statistics 2000*. The Farmers Association of Iceland, Reykjavik, Iceland.

Finnbogadóttir. (1992) Prologue: Thingvallavatn. *Oikos* 64, 1.

Finney, B. (2005) Voyaging and isolation in Rapa Nui prehistory. Available at: http://pvs.kcc.hawaii.edu/rapanui/finney.html (accessed 7 October 2005).

Fisher, R.F. (1995) Soil organic matter: clue or conundrum? In: McFee, W.W. and Kelly, J.M. (eds) *Carbon Forms and Functions in Forest Soils. Eighth North American Forest Soils Conference, Gainesville, Florida*, May 1993. Soil Science Society of America, Madison, Wisconsin, pp. 1–11.

Flenley, J.R. and King, S. (1984) Late quaternary pollen records from Easter Island. *Nature* 307, 47–50.

Forest Survey of India (1999) State of Forest Report, Dehradun. Available at: http://envfor.nic.in/fsi/sfr99/sfr.html (accessed 1 July 2005).

Forseth, I.N. and Innis, A.F. (2004) Kudzu (*Pueraria montana*): history, physiology, and ecology combine to make a major ecosystem threat. *Critical Reviews in Plant Sciences* 23, 401–413.

Forseth, I.N. and Teramura, A.H. (1987) Field photosynthesis, microclimate, and water relations of an exotic temperate liana, *Pueraria lobata*, kudzu. *Oecologica* 71, 262–267.

Foster, D.R. (1988) Species and stand response to catastrophic wind in central New England. *Journal of Ecology* 76, 135–151.

FRA (2000) Global Forest Resources Assessment 2000, FAO Forestry Paper 140. Available at: www.fao.org/forestry/site/fra/en (accessed 15 September 2004).

Franklin, J.F. (1989) Towards a new forestry. *Journal of American Foresters* 95, 37–44.

Franklin, J.F. (1992) Scientific basis for new perspectives in forests and streams. In: Naiman, R.J. (ed.) *Watershed Management. Balancing Sustainability and Environmental Change*. Springer-Verlag, New York, pp. 25–72.

Franklin, J.F. (1993) Preserving biodiversity: species, ecosystems, or landscapes. *Ecological Applications* 3, 202–205.

Franklin, J.F. and Forman, R.T.T. (1987) Creating landscape patterns by forest cutting: ecological consequences and principles. *Landscape Ecology* 1, 5–18.

Franklin, J.F., Cromack, K., Jr, Denison, W., McKee, A., Maser, C., Sedell, J., Swanson, F. and Juday, G. (1981) *Ecological Characteristics of Old-growth Douglas-fir Forests*. Forest Service Pacific Northwest Forest Range Experimental Station General Technical Report PNW-118, USDA, Portland, Oregon.

Franklin, J.F., Shugart, H.H. and Harmon, M.E. (1987) Tree death as an ecological process. *BioScience* 37, 550–556.

Freese, C.H. (ed.) (1997) *Harvesting Wild Species: Implications for Biodiversity Conservation*. John Hopkins University Press, Baltimore, Maryland.

Freese, C.H. (1998) *Wild Species as Commodities*. Island Press, Washington, DC.

Friedmann, J. (1973) *Retracking America*. Anchor Press/Doubleday, Garden City, New York.

Fujimori, T., Kawanabe, S., Grier, C.C. and Shidei, T. (1976) Biomass and primary production in forests of three major vegetation zones of the northwestern United States. *Journal of the Japanese Forestry Society* 58, 360–373.

Furniss, R.L. and Carolin, V.M. (1977) *Western Forest Insects*. Forest Service, Miscellaneous Publication 1339, USDA, Washington, DC.

Gale, R.P. and Cordray, S.M. (1991) What should forests sustain? Eight answers. *Journal of Forestry* 89, 31–36.

Gash, J.H.C., Nobre, C.A., Roberts, J.M. and Victoria, R.L. (eds) (1997) *Amazonian Deforestation and Climate*. John Wiley & Sons, Chichester, UK and New York.

Gende, S.M., Quinn, T.P., Willson, M.F., Heintz, R. and Scott, T.M. (2004) Magnitude and fate of salmon-derived nutrients and energy in a coastal stream ecosystem. *Journal of Freshwater Ecology* 19, 149–160.

Giurescu, C.C. (1980) *A History of the Romanian Forest*. Romanian Academy Publishing House, Bucharest, Romania.

Goldammer, J.G. and Stocks, B.J. (2000) Eurasian perspective of fire: dimension, management, policies, and scientific requirements. In: Kasischke, E.S. and Stocks, B.J. (eds) *Fire, Climate Change, and Carbon Cycling in the Boreal Forest*. Ecological Studies, Vol. 138, Springer Verlag, New York. pp. 49–65.

Golley, F. (1993) *A History of the Ecosystem Concept in Ecology: More than the Sum of the Parts*. Yale University Press, New Haven, Connecticut.

Goodland, R.J.A., Asibey, E.O.A., Post, J.C. and Dyson, M.B. (1991). Tropical moist forest management: the urgency of transition to sustainability. In: Costanza, R. (ed.) *Ecological Economics. The Science and Management of Sustainability*. University of Columbia Press, New York, pp. 487–515.

Gower S.T., Vogt, K.A. and Grier, C.C. (1992) Above- and belowground carbon dynamics of Rocky Mountain Douglas-fir: influence of water and nutrient availability. *Ecological Monographs* 62, 43–65.

Greenland, D.J., Wild, A. and Adams, D. (1992) Organic matter dynamics in the soils of the tropics – from myth to complex reality. In: Lal, R. and Sanchez, P.A. (eds) *Myths and Science of Soils of the Tropics*. Publication Number 29, Soil Science Society of America (SSSA), Madison, Wisconsin, pp. 17–34.

Gresham, C.A., Williams, T.M. and Lipscomb, D.J. (1991) Hurricane Hugo wind damage to south-eastern US coastal forest tree species. *Biotropica* 23, 420–426.

Grier, C.C. and Logan, R.S. (1977) Old-growth *Pseudotsuga menziesii* communities of western Oregon watershed: biomass distribution and production budgets. *Ecological Monograph* 47, 373–400.

Grier, C.C., Vogt, K.A., Keyes, M.R. and Edmonds, R.L. (1981) Biomass distribution and above- and belowground production in young and mature *Abies amabilis* zone ecosystem of the Washington Cascades. *Canadian Journal Forest Research* 11, 155–167.

Grier, C.C., Lee, K.M. and Archibald, R.M. (1984) Effect of urea fertilization on allometric relations in young Douglas-fir trees. *Canadian Journal Forest Research* 14, 900–904.

Grier, C.C., Lee, K.M., Nadkarni, N.M. and Klock, G.O. (1987) *Productivity of Forests of the United States and Its Relation to Soil and Site Factors and Management Practices*. Forest Service Technical Report, USDA, Washington, DC.

Grove, R.H. (1992) Origins of Western environmentalism. *Scientific American*, July, 22–27.

Grove, R. (1997) The island and the history of environmentalism: the case of St Vincent. In: Teich, M., Porter, R. and Gustafsson, B. (eds) *Nature and Society in Historical Context*. Cambridge University Press, Cambridge, UK, pp. 148–162.

Guha, R. (1989) *The Unquiet Woods: Ecological Change and Peasant Resistance in the Himalaya*. Oxford University Press, Oxford, UK.

Haberle, S.G. and Chepstow-Lusty, A. (2000) Can climate influence cultural development?: a view through time. *Environment and History* 6, 349–369.

Halldórsson, G. Sigurðsson, O. and Ólafsson, E. (2002) *Dulin veröld: smádýr á Íslandi*. Mál og Menning, Reykjavik, Iceland.

Hannah, L., Carr, J.L. and Landerani, A. (1995) Human disturbance and natural habitat: a biome level analysis of a global data set. *Biodiversity and Conservation* 4, 128–155.

Hansen, E.M. and Lewis, K.J. (eds) (1997) *Compendium of Forest Diseases*. American Phytopathological Society Press, St Paul, Minnesota.

Hardin, G. (1968) The tragedy of the commons. *Science* 162, 1243–1248.

Harmon, M.E., Ferrell, W.K. and Franklin, J.F. (1990) Effects on carbon storage conversion of old-growth forests to young forests. *Science* 247, 699–702.

Hay, S.I., Cox, J., Rogers, D.J., Randolph, S.E., Stern, D.I., Shanks, G.D., Myers, M.F. and Snow, R.W. (2002) Climate change and the resurgence of malaria in the East African highlands. *Nature* 415, 905–909.

Hayes, E.B. and Piesman, J. (2003) Current concepts – how can we prevent Lyme disease? *New England Journal of Medicine* 348, 2424–2430.

Hecht, B.P. (2003) The edge paradox: an Icelandic case study investigating impacts of multiple and novel disturbances on forest ecosystem thresholds. PhD dissertation, Yale University, New Haven, Connecticut.

Hedin, L.O., Granat, L., Likens, G.E., Buishand, T.A., Galloway, J.N., Butler, T.J. and Rohde, H. (1994) Steep declines in atmospheric base cations in regions of Europe and North America. *Nature* 367, 351–354.

Helfield, J.M. and Naiman, R.J. (2001) Effects of salmon-derived nitrogen on riparian forest growth and implications for stream productivity. *Ecology* 82, 2403–2409.

Hillel, D.J. (1991) *Out of the Earth. Civilization and the Life of the Soil*. The Free Press, New York.

Historical Research Associates (1986) *A Forest in Trust: Three-quarters of a Century of Indian Forestry 1910–1986*. Missoula, Montana.

Hodges, M. (1974) *Baldur and the Mistletoe*. Myths of the World Series, Little, Brown, Boston, Massachusetts.

Holdgate, M. (1996) *From Care to Action – Making a Sustainable World*. Earthscan, London.

Honea, J.M. (2005) Effect of salmon spawning on seasonal changes in structure and function of stream macroinvertebrate communities. Unpublished dissertation, College of Forest Resources, University of Washington, Seattle, Washington.

Hoon, P.N. (2004) Impersonal markets and personal communities? Wildlife, conservation and development in Botswana. *Journal of International Wildlife Law and Policy* 7, 143–160.

Hoots, D. and Baldwin, J. (1996) *Kudzu: The Vine to Love or Hate*. Suntop Press, Kodak, Tennessee.

Hopkin, M. (2005) Military exercises 'good for endangered species'. Firing ranges can have more wildlife than national parks. 12 August 2005. Available at: www.nature.com/ news/2005/050808/pf/050808-14_pf.html (accessed 15 August 2005).

Horton, T.R. and Bruns, T.D. (2001) The molecular revolution in ectomycorrhizal ecology: peeking into the black-box. *Molecular Ecology* 10, 1855–1871.

Houghton, R.A. (1995) Changes in the storage of terrestrial carbon since 1850. In: Lal, R., Kimble, J., Levine, E. and Stewart, B.A. (eds) *Soils and Global Change*. CRC Lewis Publishers, Boca Raton, Florida, pp. 45–65.

Houston, D.R. (1992) A host–stress–saprogen model for forest dieback–decline diseases. In: Manion, P.D. and LaChance, D. (eds) *Forest Decline Concepts*. American Phytopathological Society Press, St Paul, Minnesota, pp. 3–25.

Howe, H.F. and Richter, W.M. (1982) Effects of seed size on seedling size in *Virola surinamensis*: a within and between tree analysis. *Oecologia* 53, 347–351.

Hunter, M.L. (1990) *Wildlife, Forests and Forestry: Principles of Managing Forests for Biological Diversity*. Prentice Hall, Englewood Cliffs, New Jersey.

IFMAT (Indian Forest Management Assessment Team) (1993) *An Assessment of Indian Forests and Forest Management in the United States*. Intertribal Timber Council, Portland, Oregon.

Innes, J.L. (1988a) Forest health surveys – a critique. *Environmental Pollution* 54, 1–15.

Innes, J.L. (1988b) Forest health surveys – problems in assessing observer objectivity. *Canadian Journal of Forest Research* 18, 560–565.

Innes, J.L. (1990) *Assessment of Tree Condition*, Forestry Commission Field Book 12, HMSO, London.

Innes, J.L. (1998) Role of diagnostic studies in forest monitoring programmes. *Chemosphere* 36, 1025–1030.

Innes, J.L. and Skelly, J.M. (2001) *Field Manual for the Identification of Ozone Injury to Forest Plants.* Paul Haupt, Berne, Switzerland.

IPCC (2001) Summary for Policymakers: Climate Change 2001: The Scientific Basis. A report of Working Group I of the Intergovernmental Panel on Climate Change. Available at: www.ipcc.ch/pub/un/syreng/spm.pdf (accessed 16 November 2004).

Irving, W. (1980) *Ichabod Crane and the Headless Horseman.* Folk Tales of America, Troll Communications, New York.

IUCN (1999) Proceedings of the intergovernmental forum on forests experts meeting on protected areas. Puerto Rico, 15–19 March 1999. Available at: www.mma.gov.br/port/sbf/reuniao/doc/mechanism.pdf (accessed 2 September 2005).

Jackson, A. (1830) Second Annual Message to Congress, 6 December 1830. Available at: http://www.crinfo.org/link_frame.cfm?linkto=http%3A%2 F%2Fwww%2Esynaptic%2Ebc%2Eca%2Fejournal%2FJacksonSecondAnnualMessage%2Ehtm&parenturl (accessed 15 September 2004).

Jacobsen, T. and Adams, R.M. (1958) Salt and silt in ancient Mesopotamian agriculture. *Science* 128, 1251–1258.

Jakarta Post (2005) Villagers create forests, improve environment. Reported by Bambang M, Contributor, Wonogiri, Central Java. *Jakarta Post*, 22 February 2005 edition. Available at: www.illegal-logging.info/news.php?newsId=717 (accessed 9 September 2005).

Jenny, H. (1980) *The Soil Resource. Origin and Behavior.* Springer-Verlag, New York.

Joglar, R.L. (1998) *Los coquíes de Puerto Rico: su historia natural y conservació.* Universidad de Puerto Rico Press, San Juan, Puerto Rico.

Johnson, M.G. (1995) The role of soil management in sequestering soil carbon. In: Lal, R., Kimble, J., Levine, E. and Stewart, B.A. (eds) *Soils and Global Change.* Lewis Publishers, Chelsea, Michigan, pp. 351–363.

Josiah, S.J. (1999) *Farming the Forest for Specialty Products. Proceedings of the North American Conference on Enterprise Development through Agroforestry.* CIFOR, Jakarta, Indonesia.

Kandler, O. (1992) Historical declines and diebacks of Central European forests and present conditions. *Environmental Toxicology and Chemistry* 11, 1077–1093.

Kellert, S.R. (1993) The biological bases for human values. In: Kellert, S.R. and Wilson, E.O. (eds) *The Biophilia Hypothesis.* Island Press, Washington, DC, pp. 42–72.

Keyes, M.R. and Grier, C.C. (1981) Above- and belowground net production in 40-year-old Douglas-fir stands on low and high productivity sites. *Canadian Journal of Forest Research* 11, 599–605.

Kimmins, J.P. (1997) *Forest Ecology*, Macmillan, New York.

Klopfer, P.H. and MacArthur, R.H. (1960) Niche size and faunal diversity. *The American Naturalist* 94, 293–300.

Kobayasha, S. (1994) Effects of harvesting impacts and rehabilitation of tropical rainforest. *Journal of Plant Research* 107, 99–106.

Köerner, C. (1989) The nutritional status of plants from high altitudes: a worldwide comparison. *Oecologia* 81, 379–391.

Kovacs, M.G. (1989) *The Epic of Gilgamesh.* Stanford University Press, Stanford, California.

Kraus, F., Campbell, E.W., Allison, A. and Pratt, T. (1999) *Eleutherodactylus* frog introductions to Hawaii. *Herpetological Review* 30, 21–25.

Krohne, D.T. and Hoch, G.A. (1999) Demography of *Peromyscus leucopus* populations on habitat patches: the role of dispersal. *Canadian Journal of Zoology* 77, 1247–1253.

Lachapelle, P.R. and McCool, S.F. (2005) Exploring the concept of 'ownership' in natural resource planning. *Society and Natural Resources* 18, 1–7.

Lal, R., Kimble, J., Levine, E. and Whitman, C. (1995) World soils and greenhouse effect: an overview. In: Lal, R., Kimble, J., Levine, E. and Stewart, B.A. (eds) *Soils and Global Change*. Lewis Publishers, Chelsea, Michigan, pp. 433–436.

Landel-Mills, N. and Porras, I. (2002) Silver Bullet or Fools' Gold? A Global Review of Markets for Forest Environmental Services and Their Impact on the Poor. Available at: www.iied.org/docs/enveco/MES_prelims.pdf (accessed 20 February 2004).

Landmann, G. and Bonneau, M. (eds) (1995) *Forest Decline and Atmospheric Deposition Effects in the French Mountains*. Springer Verlag, Berlin, Germany.

Lane, R.S., Piesman, J. and Burgdorfer, W. (1991) *Lyme borreliosis*: relation of its causative agent to its vectors and hosts in North America and Europe. *Annual Review of Entomology* 36, 587–609.

Laporte, L.F. (1975) *Encounter with the Earth*. Canfield Press, San Francisco, California.

Larcher, W. (1995) *Physiological Plant Ecology*. Springer, New York.

Le Billon, P. (2002) The political ecology of transition in Cambodia 1989–1999: war, peace and forest exploitation. *Development and Change* 31, 785–805.

Lefsky, A.S., Harding, D., Cohen, W.B., Parker, G. and Shugart, H.H. (1999) Surface lidar remote sensing of basal area and biomass in deciduous forests of eastern Maryland, USA. *Remote Sensing and Environment* 67, 83–98.

Legge, A.H. and Krupa, S.V. (eds) (1986) *Air Pollutants and Their Effects on Terrestrial Ecosystems*. John Wiley and Sons, New York.

Lenz, H. (1990). *Historia del Papel en México y Cosas Relacionadas (1525–1950)*. Porrua, Mexico City, Mexico.

Leonard, L. (2003) Possible illnesses: assessing the health impacts of the Chad Pipeline Project. *Bulletin of the World Health Organization* 81, 427–433.

Likens, G.E., Driscoll, C.T., Buso, D.C., Siccama, T.G., Johnson, C.E., Lovett, G.M., Fahey, T.J., Reiners, W.A., Ryan, D.F., Martin, C.W. and Bailey, S.W. (1998) The biogeochemistry of calcium at Hubbard Brook. *Biogeochemistry* 41, 89–173.

Lindblade, K.A., Walker, E.D., Onapa, A.W., Katungu, J. and Wilson, M.L. (1999) Highland malaria in Uganda: prospective analysis of an epidemic associated with El Niño. *Transactions of the Royal Society of Tropical Medicine and Hygiene* 93, 480–487.

Lindblade, K.A., Walker, E.D., Onapa, A.W., Katungu, J. and Wilson, M.L. (2000) Land use change alters malaria transmission parameters by modifying temperature in a highland area of Uganda. *Tropical Medicine and International Health* 5, 263–274.

Linnard, W. (1982) *Welsh Woods and Forests: History and Utilization*. Amgueddfa Genedlaethol Cymru – National Museum of Wales, Caerdydd/Cardiff, UK.

Lloyd, M. (2004) Earth movers. Archeologists say Brazils rainforest, once thought to be inhospitable to humans, fostered huge ancient civilizations. The proof is in the dirt. *The Chronicle of Higher Education* 51 (3 December), A16.

Loftsson, J. (1993) Forest development in Iceland. In: Alden, J. (ed.) *Forest Development in Cold Climates*. Plenum Press, New York, pp. 435–461.

LoGiudice K., Ostefeld, R.S., Schmidt, K.A. and Keesing, F. (2003) The ecology of infectious disease: effects of host diversity and community composition on Lyme disease risk. *Proceedings of the National Academy of Sciences of the United States of America* 100, 567–571.

Lugo, A.E. (1991) Soil carbon in forested ecosystems. In: Johnson, M.G. and Kern, J.S. (eds) *Sequestering Carbon in Soils: A Workshop to Explore the Potential for Mitigating Global Climate Change*. EPA/600/3-91/031, USEPA, Environmental Research Laboratory, Corvallis, Oregon, pp. 29–30.

Luhnow, D. and Lyons, J. (2005) In Latin America, rich–poor chasm stifles growth. Many struggle to move up amid educational divide; Tehuacán's powerful clan. *The Wall Street Journal* CCXLVI (11) 18 July, A1–A4.

Lukas, S.E., Penetar, D., Berko, J., Vicens, L., Palmer, C., Mallya, G., Macklin, E.A. and Lee, D.Y.-W. (2005) An extract of the Chinese herbal root kudzu reduces alcohol drinking by heavy drinkers in a naturalistic setting. *Alcoholism: Clinical and Experimental Research* 29, 756–762.

Luoma, D., Trappe, J.M., Claridge, A.W., Jacobs, K.M. and Cazares, E. (2003) Relationships among fungi and small mammals in forested ecosystems. In: Zable, C.J. and Anthony, B. (eds) *Mammal Community Dynamics: Management and Conservation in the Coniferous Forests of Western North America*. Cambridge University Press, New York, pp. 343–373.

MacArthur, R.H. and Wilson, E.O. (1963) An equilibrium theory of insular zoogeography. *Evolution* 17, 373–387.

MacKay, F. (2002) WRM Bulletin No 62. Available at: www.wrm.org.uy/bulletin/62/AM.html (accessed 15 August 2005).

Manion, P.D. (1991) *Tree Disease Concepts*, 2nd edn. Prentice Hall, Englewood Cliffs, New Jersey.

Manion, P.D. and LaChance, D. (eds) (1992) *Forest Decline Concepts*. American Phytopathological Society Press, St Paul, Minnesota.

Mann, C.C. (2005) *1491. New Revelations of the Americas Before Columbus*. Alfred A. Knopf, New York.

Marino, B.D.V. and Odum, H.T. (1999) *Biosphere 2: Research Past and Present*. Elsevier, London and New York.

Mather, T.N., Wilson, M.L., Moore, S.I., Ribeiro, J.M.C. and Spielman, A. (1989) Comparing the relative potential of rodents as reservoirs of the Lyme disease spirochete (*Borrelia burgdorferi*). *American Journal of Epidemiology* 130, 143–150.

Mathuba, B.M. (2003) Botswana land policy. Paper presented at an International Workshop on Land Policies in Southern Africa. Botswana Ministry of Lands and Housing, Gaborone, Botswana, Africa.

Matthews, J. and Matthews, C. (1995) *British and Irish Mythology*. Diamon, London, UK.

Mayr, E. (2004) *What Makes Biology Unique? Considerations on the Autonomy of a Scientific Discipline*. Cambridge University Press, Cambridge, UK.

McCall, G. (1995) *Pacific Islands Year Book,* 17th edn. Fiji Times, Suva, Fiji.

McCool, S.F. and Cole, D.N. (1998) *Limits of Acceptable Change and Related Planning Processes: Progress and Future Directions*. Forest Service Intermountain Research Station, USDA, Ogden, Utah.

McCool, S.F., Guthrie, K. and Kapler-Smith, J. (2000) *Building Consensus: Legitimate Hope or Seductive Paradox?* Forest Service, Rocky Mountain Research Station, Report no. RMRSRP-25, USDA, Fort Collins, Colorado.

McCormick, J. (1989) *Reclaiming Paradise: The Global Environmental Movement*. Indiana University Press, Bloomington, Indiana.

McGovern, T.H., Bigelow, G., Amorosi, T. and Russell, D. (1988) Northern islands, human error, and environmental degradation: a view of social and ecological change in the medieval North Atlantic. *Human Ecology* 16, 225–270.

McIntosh, A. (1990) Home page. Available at: www.alastairmcintosh.com/articles (accessed 7 October 2005).

McShane, T.O. and Wells, M.P. (eds) (2004) *Getting Biodiversity Projects to Work. Towards More Effective Conservation and Development*. Columbia University Press, New York.

Meadows, D.H., Meadows, D.L., Randers, J. and Behrens, W.W., III. (1972) *The Limits to Growth: A Report for the Club of Rome's Project on the Predicament of Mankind*, 2nd edn. Signet, New American Library, New York.

Meadows, D.H., Meadows, D.L. and Randers, J. (1992) *Beyond the Limits: Global Collapse or a Sustainable Future.* Earthscan Publications, London, UK.

Means, J.E., Acker, S.A., Harding, D.J., Blair, J.B., Lefsky, M.A., Cohen, W.B., Harmon, M.E. and McKee, W.A. (1999) Use of large-footprint scanning lidar to estimate forest stand characteristics in the Western cascades of Oregon. *Remote Sensing and Environment* 67, 298–308.

Medina, M. (1997) Scavenging on the border: a study of the informal recycling sector in Laredo, Texas, and Nuevo Laredo, Mexico. PhD dissertation, Yale University, New Haven, Connecticut.

Medina, M. (2003) The cardboard collectors of Nuevo Laredo: how scavengers protect the environment and benefit the economy. In: Eckstein, S. and Wickham-Crowley, T. (eds) *Struggles for Social Rights in Latin America.* Routledge, New York, pp. 103–121.

Medina, M. (2004) Scavenger Cooperatives in Asia and Latin America. Available at: www.gdnet.org/pdf/medina.pdf (accessed 10 April 2003).

Medina, M. (2006) Serving the unserved: informal refuse collection in Mexico. *Waste Management and Research* 23(5), forthcoming.

Mercante, A.S. (1978) *Who's Who in Egyptian Mythology.* Clarkson N. Potter, New York.

Metafore (2005) Responsible Wood and Paper Purchasing: Experiences of the Fortune 100. Available at: www.metafore.org/downloads/fortune100_2005.pdf (accessed 17 June 2005).

Methanol Institute (2005) Homepage. Available at: www.methanol.org (accessed 15 August 2005).

Metraux, A. (1957) *Easter Island. A Stone-Age Civilization of the Pacific.* The Scientific Book Club, London, UK.

Meyerson, F. (2000) Human population growth, deforestation and protected areas management: re-thinking conservation and demographic policy for the Maya Biosphere Reserve in Guatemala. PhD dissertation, Yale University, New Haven, Connecticut.

Miller, P. and McBride, J.R. (eds) (1999) *Oxidant Air Pollutant Impacts in the Montane Forests of Southern California: A Case Study of the San Bernadino Mountains.* Springer, New York.

Minakawa, N. and Gara, R.I. (2003) Effects of chum salmon redd excavation on benthic communities in a stream in the Pacific Northwest. *Transactions of the American Fisheries Society* 132, 598–604.

Minakawa, N., Gara, R.I. and Honea, J.M. (2002) Increased individual growth rate and community biomass of stream insects associated with salmon carcasses. *Journal of the North American Benthological Society* 21, 651–659.

Minakawa, N., Munga, S., Atieli, F., Mushinzimana, E., Zhou, G., Githeko, A. and Yan, G. (2005) Spatial distribution of anopheline larval habitats in western Kenyan highlands: effects of land cover types and topography. *American Journal of Tropical Medicine and Hygiene* 73, 157–165.

Moldenke, A.R. (1999) Soil-dwelling arthropods: their diversity and functional roles. In: Meurisse, R.T., Ypsilantis, W.G. and Seybold, C. (eds) *Proceedings of the Pacific Northwest Forest and Rangeland Soil Organism Symposium.* General Technical Report PNW-GTR-461, US Department of Agriculture, Forest Service, Pacific Northwest Research Station, Portland, Oregon, pp. 33–44.

Molina, R. and Trappe, J.M. (1982) Patterns of ectomycorrhizal host specificity and potential among Pacific Northwest conifers and fungi. *Forest Science* 28, 423–458.

Molina, R., Massicotte, H.B. and Trappe, J.M. (1992) Specificity phenomena in mycorrhizal symbioses: community–ecological consequences and practical implications. In: Allen, M.F. (ed.) *Mycorrhizal Functioning: An Integrative Plant–Fungal Process.* Chapman and Hall, New York, pp. 357–423.

Montgomery, D.R. (2003) *King of Fish: the Thousand-Year Run of Salmon*. Westview Press, Boulder, Colorado.

Montgomery, D.R., Buffington, J.M., Peterson, N.P., Shuett-Hames, D. and Quinn, T.P. (1996) Stream-bed scour, egg burial depths, and the influence of salmonid spawning on bed surface mobility and embryo survival. *Canadian Journal of Fisheries and Aquatic Science* 53, 1061–1070.

Mooney, H.A. and Hobbs, R.J. (eds) (2000) *Invasive Species in a Changing World*. Island Press, Washington, DC.

Moore, J.W. and Schindler, D.E. (2004) Nutrient export from freshwater ecosystems by anadromus sockeye salmon (*Oncorhynchus nerka*). *Canadian Journal of Fisheries and Aquatic Sciences* 61, 1582–1589.

Moran, D. (1994) Debt-swaps for hot-spots: more needed. *Biodiversity Letters* 2, 63–66.

Moran, E.F. (1995) Rich and poor ecosystems of Amazonia: an approach to management. In: Nishizawa, T. and Uitto, J.I. (eds) *The Fragile Tropics of Latin America. Sustainable Management of Changing Environments*. United Nations University Press, New York, pp. 45–67.

Morishima, G.S. (1997) From paternalism to self-determination. *Journal of Forestry* 95, 4–9.

Morishima, G.S. (1998). Promises to keep: paradigms and problems with coordinated resource management in Indian country. *Evergreen* 9, 22–25.

Motanic, D. (1998) The National Indian Forest Resources Management Act: what was and what will be. *Evergreen* 9, 26–27.

Motanic, D. (2001) Intertribal Timber Council technical specialist. Phone interview by Philip Rigdon, 20 February 2001.

Mouchet, J., Manguin, S., Siroulon, J., Laventure, S., Faye, O., Onapa, A.W., Carnevale, P., Julvez, J. and Fontenille, D. (1998) Evolution of malaria in Africa for the past 40 years: impact of climatic and human factors. *Journal of American Mosquito Control Association* 14, 121–130.

MSNBC (2005) Homepage. Available at: www.msnbc.msn.com/id/6632374/print/1/displaymode/1098/ (accessed 27 September 2005).

Mueller-Dombois, D. (1992) A natural dieback theory: cohort senesence as an alternative to the decline theory. In: Manion, P.D. and LaChance, D. (eds) *Forest Decline Concepts*. American Phytopathological Society Press, St Paul, Minnesota, pp. 26–37.

Mullins, G.R, Fidzani, B. and Kolanyane, M. (2000) At the end of the day. The socioeconomic impacts of eradicating contagious bovine pleuropneumonia from Botswana. *Annals of the New York Academy of Sciences* 916, 333–344.

Munsell, J. (1980) *Chronology of the Origin and Progress of Paper and Paper-Making*. Garland Publishing, New York.

Munyekenye, O.G., Githeko, A.K., Zhou, G., Mushinzimana, E., Minakawa, N. and Yan, G. (2005) Spatial analysis of *Plasmodium falciparum* infections in western Kenya highlands. *Emerging Infectious Disease* 10, 1571–1577.

Murray, I. (2003) Environmental scientists must stop crying wolf. *Financial Times* 17 September 2003.

Myers, N. (1984) *The Primary Source: Tropical Forests and Our Future*, 1st edn. Norton, New York.

National Drug Intelligence Center (2003) Washington Drug Threat Assessment. Available at: www.usdoj.gov/ndic/pubs3/3138/meth.htm (accessed 18 October 2005).

National Forest Policy (1988) *National Forest Policy, 1988*. Available at: http://envfor.nic.in/nfap/detailed-policy.html (accessed 15 July 2005).

Neeson, E. (1991) *A History of Irish Forestry*. The Lilliput Press, Dublin, Ireland.

Newbold, P.J. (1967) *Methods of Estimating the Primary Production of Forests*. IBP. Handbook 2, Blackwell Scientific Publishers, Oxford, UK.

Newell, A. (1998) Indian forest policy rooted in federal ambivalence. *Evergreen* 9, 19–20; Milestones in shifting federal Indian policy. *Evergreen* 9, 21.

Nietschmann, B.Q. (1992) *The Interdependence of Biological and Cultural Diversity*. Occasional Paper number 21, Center for World Indigenous Studies, Kenmore, Washington.

Noss, R.F. (2002) Context matters: considerations for large-scale conservation. *Conservation in Practice* 3, 10–19.

NRC (1996) *Upstream. Salmon in the Pacific Northwest*. National Academy Press, Washington, DC.

Nupp, T.E. and Swihart, R.K. (1996) Effect of forest patch area on population attributes of white-footed mice (*Peromyscus leucopus*) in fragmented landscapes. *Canadian Journal of Zoology* 74, 467–472.

O'Connell, M.A., Hallett, J.G., West, S.D., Kelsey, K.A., Manuwal, D.A. and Pearson, S.F. (2000) *Effectiveness of Riparian Management Zones in Providing Habitat for Wildlife. Final Report*. TWF-LWAG1-00-001, Timber, Fish and Wildlife, Seattle, Washington.

Ogilvie, A.E.J. and Barlow, K.L. (2000) North Atlantic climate *c.* AD 1000: millenial reflections on the Viking discoveries of Iceland, Greenland and North America. *Weather* 55, 34–45.

O'Hara, J.L. (1999) An ecosystem approach to monitoring non-timber forest product harvest: the case study of bayleaf palm (*Sabal mauritiiformis*) in the Rio Bravo Conservation and Management Area, Belize. PhD dissertation, Yale University, New Haven, Connecticut.

O'Keefe, T.C. and Edwards, R.T. (2003) Evidence for hyperheic transfer and storage of marine-derived nutrients in sockeye streams in southwest Alaska. In: Stockner, J. (ed.) *Nutrients in Salmonid Ecosystems: Sustaining Production and Biodiversity*. American Fisheries Society, Eugene, Oregon, pp. 99–107.

Oliver, C. and Larson, B.C. (1995) *Growth and Development of Forest Stands*. Wiley and Sons, New York.

Orliac, C. (1998) Données nouvelles sur la composition de la flore de l'Île de Pâques. *Journal de la Société des Océanistes* 2, 23–31.

Orloski, K.A., Hayes, E.B., Campbell, G.L. and Dennis, D.T. (2000) Surveillance for Lyme disease – United States, 1992–1998. In: *CDC Surveillance Summaries. Monthly Morbidity and Mortality Weekly Report* 49, 1–11.

ORNL (2005) *Oak Ridge National Laboratory (US) (ORNL) and United States Forest Service (US) (USFS) and Agricultural Research Service (US) (ARS). Biomass as Feedstock for a Bioenergy and Bioproducts Industry: The Technical Feasibility of a Billion-Ton Annual Supply. A Feasibility Study, April*. Oak Ridge National Laboratory (ORNL), Oak Ridge, Tennessee.

Ostfeld, R.S. (1997) The ecology of Lyme-disease risk. *American Scientist* 85, 338–346.

Ostfeld, R.S. and Keesing, F. (2000) Biodiversity and disease risk: the case of Lyme disease. *Conservation Biology* 14, 722–728.

Padoch, C. (1992) Marketing of non-timber forest products in western Amazonia: general observations and research priorities. *Advances in Economic Botany* 9, 43–50.

Panayotou, T. and Ashton, P.S. (1992) *Not by Timber Alone. Economics and Ecology for Sustaining Tropical Forests*. Island Press, Washington, DC.

Paster, M., Pellegrino, J.L. and Carole, T.M. (2003) *Industrial Bioproducts: Today and Tomorrow*. US DOE, Office of Energy Efficiency and Renewable Energy, July 2003, Prepared by Energetics Incorporated, Columbia, Maryland.

Patel-Weynand, T. (1997) *Forests and Poverty: a View towards Sustainable Development*. Report to the United Nations Development Programme, New York.

Patel-Weynand, T. and Vogt, K.A. (1999) Certification and poverty. In: Vogt, K.A., Larson, B.C., Gordon, J.C., Vogt, D.J. and Fanzeres, A. (eds) *Forest Certification: Roots, Issues, Challenges, and Benefits*. CRC Press, Boca Raton, Florida.

Patz, J.A., Graczyk, T.K., Geller, N. and Vittor, A.Y. (2000) Effects of environmental change on emerging parasitic diseases. *International Journal for Parasitology* 30, 1395–1405.

Pearson, T., Brown, S., Petrova, S., Moore, N. and Slaymaker, D. (2005) *Application of Multispectral 3-Dimensional Aerial Digital Imagery for Estimating Carbon Stocks in a Closed Tropical Forest*. Report to the Nature Conservancy, Conservation Partnership Agreement, Winrock International, Arlington, Virginia.

Peden, W. (1955) *Notes on the State of Virginia*. University of North Carolina Press, Chapel Hill, North Carolina.

Pedigo, L.P. (1999) *Entomology and Pest Management*. Prentice-Hall, Upper Saddle River, New Jersey.

Peluso, N.L. (1991) The history of state forest management in colonial Java. *Forest and Conservation History* 35, 65–75.

Perez-Garcia, J.M., Wang, Y. and Xu, W. (1999) An economic and environmental assessment of the Asia forest sector. In: Yoshimoto, A. and Yukutake, K. (eds) *Global Concerns for Forest Resource Utilization*. Kluwer Publishing Company, Dordrecht, The Netherlands, pp. 231–242.

Perry, D.A., Molina, R. and Amaranthus, M.P. (1987) Mycorrhizae, mycorrhizosphere, and reforestation: current knowledge and research needs. *Canadian Journal of Forest Research* 17, 929–940.

Perry, D.A., Amaranthus, M.P., Borchers, J.G., Borchers, S.L. and Brainerd, R.E. (1989) Bootstrapping in ecosystems. *BioScience* 39, 230–237.

Persepsi (2005) Sertifikasi Hutan Rakyat: Pengalaman Lapangan Di Hutan Jati Jawa. Presented at the First Congress of the Indonesian Ecolabelling Institute (LEI), Jakarta, Indonesia.

Persson, T. (1980) *Structure and Function of Northern Coniferous Forests: an Ecosystem Study*. Swedish Natural Science Research Council, Stockholm, Sweden.

Peterson, J.D. (1998) Touring across America for forestry. *Evergreen* 9, 4–7; An essay by James D. Peterson. *Evergreen* 9, 9–18.

Petruncio, M. (1998) Tribal timber harvest is concentrated in the west. *Evergreen* 9, 42–43.

Pickett, S.T.A. and White, P.S. (1985) *The Ecology of Natural Disturbances and Patch Dynamics*. Academic Press, New York.

Pielou, E.C. (1991) *After the Ice Age. The Return of Life to Glaciated North America*. University of Chicago Press, Chicago, Illinois.

PIER Collaborative Report (2005) Biomass in California: challenges, opportunities, and potentials for sustainable management and development. Draft, April, Contract 500-01-016, California Energy Commission, California.

Pinedo-Vasquez, M. (1995) Human impact on varzea ecosystems in the Napo–Amazon, Peru. PhD dissertation, Yale University, New Haven, Connecticut.

Pinedo-Vasquez, M. (1999) Changes in soil formation and vegetation on silt bars and backslopes of levees following intensive production of rice and jute. In: Padoch, C., Ayres, J.M., Pinedo-Vasquez, M. and Henderson, A. (eds) *Várzea. Diversity, Development, and Conservation of Amazonia's Whitewater Floodplains*. The New York Botanical Garden Press, New York, pp. 301–311.

PNG Embassy (2005) PNG Embassy. Available at: www.pngembassy.org/ (accessed 5 October 2005).

PNG Forum (2005) Homepage. Available at: http://www.pngbd.com/forum (accessed 5 October 2005).

Pollack, M. (1978) Mummies' rags were once the rage that provided us the printed page. *Printing Impressions* 21, 60–61.

Post, W.M., Emanuel, W.R., Zinke, P.J. and Stangenberger, A.G. (1982) Soil carbon pools and world life zones. *Nature* 298, 156–159.

Preston, F.W. (1962) The canonical distribution of commonness and rarity: Parts 1 and 2. *Ecology* 43, 185–215, 410–432.

Price, S.V. (ed.) (2003) *War and Tropical Forests. Conservation in Areas of Armed Conflict*. Food Products Press, The Haworth Press, New York.

Primack, R.B. (1993) *Essentials of Conservation Biology*. Sinauer Associates, Sunderland, Massachusetts.

Prothero, R.M. (1999) Malaria, forests and people in southeast Asia. *Singapore Journal of Tropical Geography* 20, 76–85.

PSLC (2003) Polymer Science Learning Centre (PSLC) developed by the University of Southern Mississippi, University of Wisconsin Stevens Point and Chemical Heritage Foundation created a multimedia and distance learning material for polymers and described cellulose in 'Kinds of Polymers' found in 'Kids Macrogalleria' and it is in 'The Macrogalleria'. Available at: www.pslc.ws/macrog/kidsmac/cell.htm (accessed 15 September 2005).

Pyne, S.J. (1995) *World Fire – The Culture of Fire on Earth*. Henry Holt, New York.

Quazi, S.A., Ashton, M.S. and Thadani, R. (2003) Regeneration of monodominant stands of Banj oak (*Quercus leucotrichophora* A. Camus) on abandoned terraces in the Central Himalaya. *Journal of Sustainable Forestry* 17, 75–90.

Quinn, T.P. and Fresh, K. (1984) Homing and straying in chinook salmon (*Oncorhynchus tshawytscha*) from Cowlitz River hatchery, Washington. *Canadian Journal of Fisheries and Aquatic Sciences* 41, 1078–1082.

Raffles, H. (1998) Igarapé Guariba: nature, locality, and the logic of Amazonian anthropogenesis. PhD dissertation, Yale University, New Haven, Connecticut.

Raffles, H. (1999) Exploring the anthropogenic Amazon: estuarine landscape transformations in Amapá, Brazil. In: Padoch, C., Ayres, J.M., Pinedo-Vasquez, M. and Henderson, A. (eds) *Várzea Diversity, Development, and Conservation of Amazonia's Whitewater Floodplains*. The New York Botanical Garden Press, New York, pp. 355–370.

Raffles, H. (2002) *In Amazonia. A Natural History*. Princeton University Press, Princeton, New Jersey.

Ralhan, P.K. and Singh, S.P. (1987) Dynamics of nutrients and leaf mass in central Himalayan forest trees and shrubs. *Ecology* 68, 1974–1983.

Rao, M. and McGowan, P.J.K. (2002) Wild-meat use, food security, livelihoods, and conservation. *Conservation Biology* 16, 580–583.

Ratten, R. (1990) *Artisans and Entrepreneurs in the Rural Philippines: Making a Living and Gaining Wealth in Two Commercialized Crafts*. Casa Monographs, VU Press, Amsterdam.

Reynolds, J. (2001) Intertribal Timber Council – Program Manager. Phone interview by Philip Rigdon, 14 February 2001.

Richards, B.N. (1987) *The Microbiology of Terrestrial Ecosystems*. Longman Scientific and Technical/John Wiley & Sons, New York.

Rizzo, D.M. and Garbelotto, M. (2003) Sudden oak death: endangering California and Oregon forest ecosystems. *Frontiers in Ecology and the Environment* 1, 197–204.

Robinson, J. (2005) Pot farms wreak havoc in Sequoia National Park, Seattle Times, 19 August 2005. Avaliable at: http://seattletimes.newsource.com/html/nationworld/20002445-217_marijuana19.html (accessed 19 August 2005).

Robinson, J.G. and Redford, K.H. (2004) Jack of all trades, but master of none: inherent contradictions among ICD approaches. In: *Getting Biodiversity Projects to Work*. Columbia University Press, New York, pp. 10–34.

Roosevelt, A.C. (1999) Twelve thousand years of human – environment interaction in the Amazon floodplain. In: Padoch, C., Ayres, J.M., Pinedo-Vasquez, M. and Henderson, A. (eds) *Várzea. Diversity, Development, and Conservation of Amazonia's Whitewater Floodplains*. The New York Botanical Garden Press, New York, pp. 371–392.

Rosenberg, T. (2004) What the world needs now is DDT. *New York Times,* 11 April 2004.

Rosenblatt, D.L., Heske, E.J. and Nelson, S.L. (1999) Forest fragments in east-central Illinois: islands or habitat patches for mammals? *American Midland Naturalist* 141, 115–123.

Roux, G. (1992) *Ancient Iraq,* 3rd edn. Penguin Books, London, UK.

Rowe, J.S. (1983) Concepts of fire effects on plant individuals and species. In: Wein, R.W. and McLean, D.A. (eds) *The Role of Fire in Circumpolar Ecosystems*. John Wiley & Sons, New York, pp. 65–80.

Ryan, M.G., Binkley, D. and Fownes, J.H. (1997) Age-related decline in forest productivity: pattern and processes. *Advances in Ecological Research* 27, 213–266.

Salafsky, N., Margoluis, R., Redford, K. and Robinson, J. (2002) Improving the practice of conservation: a conceptual framework and research agenda for conservation science. *Conservation Biology* 16, 1469–1478.

Sanchez, M. (1993) *El Papel del Papel en Nueva España, 1740–1812.* INAH, Mexico City, Mexico.

Sanchez, P.A. (1976) *Properties and Management of Soils in the Tropics.* John Wiley & Sons, New York.

Sardjono, M.A. and Samsoedin, I. (2001) Traditional knowledge and practice of biodiversity conservation: the Benuaq Dayak community of east Kalimantan, Indonesia. In: Colfer Pierce, C.J. and Byron, Y. (eds) *People Managing Forests: the Links Between Human Well-being and Sustainability.* Resources for the Future and Center for International Forestry Research, Washington, DC, and Bogor, Indonesia, pp. 116–134.

Sargent, H.L., Jr (1972) Fishbowl planning immerses Pacific Northwest citizens in Corps projects. *Civil Engineering* 42, 54–57.

Sarkar, A.U. and Ebbs, K.L. (1992) A possible solution to tropical troubles. *Futures* 24, 653–668.

Sato, H., Yasui, K. and Byamana, K. (2000) Follow-up survey of environmental impacts of the Rwandan refugees on eastern D. R. Congo. *Ambio* 29, 122–123.

Saunders, D.A., Hobbs, R.J. and Margules, C.R. (1991) Biological consequences of ecosystem fragmentation: a review. *Conservation Biology* 5, 18–32.

Sayer, J. and Campbell, B. (2004) *The Science of Sustainable Development. Local Livelihoods and the Global Environment.* Cambridge University Press, Cambridge, UK.

Schimel, D.S., Enting, I.G., Heiman, M., Wigley, T.M.L., Raynaud, D., Alves, D. and Siegenthaler, U. (1995) CO_2 and the carbon cycle. In: Houghton, J.T., Meira Filho, L.G., Bruce, J., Lee, H., Callander, B.A., Haites, E., Harris, N. and Maskell, K. (eds) *Climatic Change 1994.* Cambridge University Press, Cambridge, UK, pp. 35–71.

Schlesinger, W.H. (1990) Evidence from chronosequence studies for a low carbon-storage potential of soils. *Nature* 348, 232–234.

Schlesinger, W.H. (1991) *Biogeochemistry: Analysis of Global Change.* Academic Press, San Diego, California.

Schlosser, T.S. (1992) *Development of Tribal Timber Resources: The Tribal Perspective – Locating the National Indian Forest Resources Management Act in the Tortured History of Indian Timber Management.* Morisset, Schlosser, Ayer & Jozwaik Attorneys At Law. MSAJ Information – Indian Law Resources. Available at: http://msaj.com/papers/timber.htm (accessed 15 September 2003).

Schmidt, K.A. and Ostfeld, R.S. (2001) Biodiversity and the dilution effect in disease ecology. *Ecology* 82, 609–619.

Schmidt, R., Berry, J.K. and Gordon, J.C. (eds) (1999) *Forests to Fight Poverty: Creating National Strategies.* Yale University Press, New Haven, Connecticut.

Schowalter, T.D. and Filip, G.M. (eds) (1993) *Beetle–pathogen Interactions in Conifer Forests.* Academic Press, New York.

Schutt, P. and Cowling, E.B. (1985) Waldsterben, a general decline of forests in Central Europe: symptoms, development and possible causes. *Plant Disease* 69, 548–549.

Seinost, G., Dykhuizen, D.E., Dattwyler, R.J., Golde, W.T., Dunn, J.J., Wang, I.-N., Gary W.P., Schriefer, M.E. and Luft, B.J. (1999) Four clones of *Borrelia burgdorferi* sensu stricto cause invasive infection in humans. *Infection and Immunity* 67, 3518–3524.

Seip, H.K. (1996) *Forestry for Human Development: A Global Imperative.* Scandinavian University Press, Copenhagen, Denmark.

Sharpe, G.W., Hendee, C.W., Sharpe, W.F. and Hendee, J.C. (1995) *Introduction to Forest and Renewable Resources,* 6th edn. WCB McGraw-Hill, Boston, Massachusetts.

Shaw, C.G., III and Kile, G.A. (1991) *Armillaria Root Disease*. Forest Service Agricultural Handbook 691, USDA, Washington, DC.

Siegenthaler, U. and Sarmiento, J.L. (1993) Atmospheric carbon dioxide and the ocean. *Nature* 365, 119–125.

Simard, S.W. and Durall, D.M. (2004) Mycorrhizal networks: a review of their extent, function, and importance. *Canadian Journal of Botany* 82, 1140–1165.

Simard, S.W., Perry, D.A., Jones, M.D., Myrold, D., Durall, D.M. and Molina, R. (1997) Net transfer of carbon between ectomycorrhizal tree species in the field. *Nature* 388, 579–582.

Sinclair, W.A., Lyon, H.H. and Johnson, W.T. (1987) *Diseases of Trees and Shrubs*. Cornell University Press, Ithaca, New York.

Singh, J.S. and Singh, S.P. (1987) Forest vegetation of the Himalaya. *Botanical Review* 53, 80–192.

Singh, J.S., Pandey, U. and Tiwari, A.K. (1984) Man and forests: a Central Himalayan case study. *Ambio* 13, 80–87.

Singh, S.P. and Singh, J.S. (1992) *Forests of the Himalaya*. Gyanodaya Prakashan, Nainital, India.

Skelly, J.M. (1989) Forest decline versus tree decline – the pathological considerations. *Environmental Monitoring and Assessment* 12, 23–27.

Skelly, J.M. (1990) On the importance of etiological accuracy during surveys to determine forest condition. *World Resources Review* 2, 250–277.

Skelly, J.M., Innes, J.L., Savage, J.E., Snyder, K.R., Van der Heyden, D., Zhang, J. and Sanz, M.J. (1999) Observation and confirmation of foliar symptoms of native plant species of Switzerland and southern Spain. *Water, Air, and Soil Pollution* 116, 227–234.

Smith, S.E. and Read, D.J. (1997) *Mycorrhizal Symbiosis*, 2nd edn. Academic Press, London, UK.

Smyth, A.H. (1905–1907) *The Writings of Benjamin Franklin*. Macmillan, New York.

Soran, B., Biro, J., Moldovan, O. and Ardelean, A. (2000) Conservation of biodiversity in Romania. *Biodiversity and Conservation* 9, 1187–1198.

Soulé, M.E. (1986) *Conservation Biology: The Science of Scarcity and Diversity*. Sinauer, Sunderland, Massachusetts.

Spiecker, H., Mielikäinen, K., Köhl, M. and Skovsgaard, J.P. (eds) (1996) *Growth Trends in European Forests*. Springer Verlag, Berlin, Germany.

Sprugel, D.G. (1991) Disturbance, equilibrium, and environmental variability: what is 'natural' vegetation in a changing environment? *Biological Conservation* 58, 1–18.

Stanek, G., Strle, F., Gray, J.S. and Wormser, G.P. (2002) History and characteristics of Lyme borreliosis. In: Gray, J.S., Kahl, O., Lane, R.S. and Stanek, G. (eds) *Lyme Borreliosis: Biology, Epidemiology and Control*. CAB International, Wallingford, UK, pp. 1–28.

Stankey, G.H., McCool, S.F. and Stokes, G.L. (1984) Limits of acceptable change: a new framework for managing the Bob Marshall Wilderness Complex. *Western Wildlands* 10, 33–37.

Stankey, G.H., Cole, D.N., Lucas, R.C., Petersen, M.E. and Frissell, S.S. (1985) *The Limits of Acceptable Change (LAC) System for Wilderness Planning*. Forest Service Intermountain Forest and Range Experiment Station, USDA, Ogden, Utah.

Steadman, D. (1989) Extinctions of birds in Eastern Polynesia: a review of the record, and comparisons with other Pacific Island groups. *Journal of Archaeological Science* 16, 177–205.

Steen, R.P. and Quinn, T.P. (1999) Egg burial depth by sockeye salmon (*Oncorhynchus nerka*): implications for survival of embryos and natural selection on female body size. *Canadian Journal of Zoology* 77, 836–841.

Stein, S.M., McRoberts, R.E., Alig, R.J., Nelson, M.D., Theobald, D.M., Eley, M., Dechter, M. and Carr, M. (2004) Forests on the edge. Housing development on America's private forests. USDA FS General Technical Report, PNW-GTR-636, May 2005, Portland, Oregon. Available at: www.fs.fed.us/projects/fote/reports/fote-6-9-05.pdf (accessed 1 September 2005).

Stevens, S. (1997) The legacy of Yellowstone. In: Stevens, S. (ed.) *Conservation Through Cultural Survival: Indigenous Peoples and Protected Areas*. Island Press, Washington, DC, pp. 13–32.

Stevenson, F.J. (1982) *Humus Chemistry: Genesis, Composition, Reactions*. Wiley-Interscience, New York.

Stewart, M.M. and Woolbright, L.L. (1996) Amphibians. In: Reagan, D.P. and Waide, R.B. (eds) *The Food Web of a Tropical Rain Forest*. University of Chicago Press, Chicago, Illinois, pp. 363–398.

Stokstad, E. (2005) The ambitious Northwest Forest Plan tried to balance desires for timber and biodiversity, but preservation trumped logging – and research. Can the plan be made as adaptable and science-friendly as intended? Learning to adapt. *Science* 309, 688–690.

Stone, R. (1995) Global warming. If the mercury soars, so may health hazards. *Science* 267, 957–958.

Sturgeon, J.C. (1997) Claiming and naming resources on the border of the state: Akha strategies in China and Thailand. *Asia Pacific Viewpoint* 38, 131–144.

Sturgeon, J.C. (2004) Postsocialist property rights for Akha in China: what is at stake? *Conservation and Society* 2, 1–17.

Sturgeon, J.C. (2005) *Border Landscapes: the Politics of Akha Land Use in China and Thailand*. University of Washington Press, Seattle, Washington.

Sundquist, E.T. (1993) The global carbon dioxide budget. *Science* 259, 934–941.

Swift, M.J., Heal, O.W. and Anderson, J.M. (1979) *Decomposition in Terrestrial Ecosystems*. University of California Press, Berkeley, California.

Taggart, J. (1999) Estimating price premiums necessary to pay for forest certification. In: Vogt, K.A., Larson, B.C., Gordon, J.C., Vogt, D.J. and Fanzeres, A. (eds) *Forest Certification: Roots, Issues, Challenges, and Benefits*. CRC Press, Boca Raton, Florida, pp. 277–285.

Tainter, F.H. and Baker, F.A. (1996) *Principles of Forest Pathology*. John Wiley & Sons, New York.

Taylor, E.K. (2003) *Some Fruits of Solitude*. Herald Press, Scottdale, Pennsylvania.

Taylor, H. (1999) What We Are Afraid of. The Harris Poll #49, 18 August 1999. Available at: www.harrisinteractive.com/harris_poll/index.asp?PID=281 (accessed 21 October 2005).

Taylor, M. (2000) Life, land and power: contesting development in northern Botswana. PhD Dissertation, University of Edinburgh, Edinburgh, UK.

Thadani, R. (1999) Disturbance, microclimate and the competitive dynamics of tree seedlings in Banj oak (*Quercus leucotrichophora*) forests of the Central Himalaya, India. PhD Dissertation, Yale University, New Haven, Connecticut.

Thomas, J.W. (1979) *Wildlife Habitats in Managed Forests: The Blue Mountains of Oregon and Washington*. Ag Handbook No. 553, USDA, Forest Service, Pacific NW Region, Portland, Oregon.

Thompson, J.D. and Tuden, A. (1987) Strategies, structures and processes of organizational decision. In: Thompson, J.D., Hammond, P.B., Hawkes, R.W. and Tuden, A. (eds) *Comparative Studies in Administration*. Garland Publishing, New York, pp 197–216.

Tilman, D. (1982) *Resource Competition and Community Structure*. Princeton University Press, Princeton, New Jersey.

TNC–WWF ALLIANCE@News (2005) First Community Certifications in Indonesia. Edition 03. March 2005. Available at: www.panda.org/downloads/forests/gdanewslettered3.pdf (accessed 17 June 2005).

Treshow, M. and Anderson, F.K. (1989) *Plant Stress from Air Pollution*. John Wiley, New York.

Turner, J. (1977) Effects of nitrogen availability on nitrogen cycling in a Douglas-fir stand. *Forest Science* 23, 307–316.

UN (2003) *United Nations List of Protected Areas 2003*. UNEP World Conservation Monitoring Centre, IUCN – The World Conservation Union, New York.

UNESCO/UNEP/FAO (1978) Tropical Forest Ecosystems: A State-of-Knowledge Report. Available at: www.fao.org/documents/show_cdr.asp?url_file=/docrep/t0550e/t0550e02.htm (accessed 28 March 2003).

UNFCCC (1992) Convention on Climate Change. UNFCCC website. Available at: www. unfccc.de/resource/conv) (accessed 4 February 2003).

UNFCCC (1997) Kyoto Protocol to the Framework Convention on Climate Change. Available at: www.unfccc.de.

US Bureau of the Census (2005) Available at: www2.census.gov/prod2/statcomp/ documents/CT1970p1-12.pdf (accessed 1 August 2005).

Usman, S., Singh, S.P. and Rawat, Y.S. (1999) Fine root productivity and turnover in two evergreen central Himalayan forests. *Annals of Botany* 84, 87–94.

Valls, O. (1980) *La Historia del Papel en España*. Empresa Nacional de Celulosas, Madrid, Spain.

van der Heijden, M. and Sanders, I.R. (2002) *Mycorrhizal Ecology*. Ecological Studies Analysis and Synthesis, Vol. 157, Springer, New York.

Viana, V.M., Ervin, J., Donovan, R.Z., Elliott, C. and Gholz, H. (1996) *Certification of Forest Products: Issues and Perspectives*. Island Press, Washington, DC and Covelo, California.

Vitousek, P.M. and Walker, L.R. (1989) Biological invasion by *Myrica faya* in Hawaii: plant demography, nitrogen fixation, and ecosystem effects. *Ecological Monograph* 59, 247–265.

Vitousek, P.M., Aber, J., Howarth, R.W., Likens, G.E., Matson, P.A., Schindler, D.W., Schlesinger, W.H. and Tilman, G.D. (1997) Human alternation of the global nitrogen cycle: causes and consequences. *Ecological Applications* 7, 737–750.

Vogt, K.A. (1991) Carbon budgets of temperate forest ecosystems. *Tree Physiology* 9, 69–86.

Vogt, K.A., Moore, E.E., Vogt, D.J., Redlin, M.R. and Edmonds, R.L. (1983) Conifer fine root and mycorrhizal root biomass within the forest floors of Douglas-fir stands of different ages and site productivities. *Canadian Journal of Forest Resources* 13, 429–437.

Vogt, K.A., Grier, C.C. and Vogt, D.J. (1986) Production, turnover, and nutrient dynamics of above- and belowground detritus of world forests. *Advances in Ecological Research* 15, 303–377.

Vogt, K.A., Vogt, D.J., Brown, S., Tilley, J.P., Edmonds, R.L., Silver, W.L. and Siccama, T.G. (1995) Dynamics of forest floor and soil organic matter accumulation in boreal, temperate and tropical forests. In: Lal, R., Kimble, J., Levine, E. and Steward, B. (eds) *Soil Management and Greenhouse Effect*. CRC Press, Lewis, Boca Raton, Florida, pp. 159–178.

Vogt, K.A., Vogt, D.J., Palmiotto, P.A., Boon, P., O'Hara, J. and Asbjornsen, H. (1996) Review of root dynamics in forest ecosystems grouped by climate, climatic forest type and species. *Plant and Soil* 187, 159–219.

Vogt, K.A, Gordon, J.C., Wargo, J.P., Vogt, D.J., Asbjornsen, H., Palmiotto, P.A., Clark, H.J., O'Hara, J.L., Keeton, W.S., Patel-Weynand, T. and Witten, E. (1997) *Ecosystems: Balancing Science with Management*. Springer-Verlag, New York.

Vogt, K.A, Larson, B.C., Gordon, J.C., Vogt, D.J. and Fanzeres, A. (1999) *Forest Certification: Roots, Issues, Challenges, and Benefits*. CRC Press, Boca Raton, Florida.

Vogt, K.A., Schmitz, O., Beard, K.H., O'Hara, J.L. and Booth, M. (2000) Conservation biology – Contemporary issues. In: Levin, S. (ed.) *Encyclopedia of Biodiversity*. Academic Press, San Diego, California, pp. 865–881.

Vogt, K.A., Grove, M., Asbjornsen, H., Maxwell, K., Vogt, D.J., Sigurdardottir, R., Larson, B.C., Schibli, L. and Dove, M. (2002) Linking social and natural science spatial scales. In: Liu, J. and Taylor, W.M. (eds) *Integrating Landscape Ecology into Natural Resource Management*. Cambridge University Press, Cambridge, UK, pp. 143–175.

Vogt, K.A., Andreu, M., Vogt, D.J., Sigurdardottir, R., Edmonds, R.L., Schiess, P. and Hodgson, K. (2005) Enhancing sustainability of forests in human landscapes by adding non-traditional values to younger forests. *Journal of Forestry* 103, 21–27.

Vtorova, V.N. and Sergeeva, T.K. (1999) On assessing environmental quality in forest ecosystems of South Vietnam. *Russian Journal of Ecology* 30, 16–21.

Walker, B.H. and Steffen, W.L. (1999) The nature of global change. In: Walker, B.H., Steffen, W.L., Canadell, J. and Ingram, J. (eds) *The Terrestrial Biosphere and Global Change*.

International Geosphere–Biosphere Programme Book Series 4, Cambridge University Press, Cambridge, UK, pp. 1–18.

Wallace, A.R. (1878) *Tropical Nature and Other Essays*. Macmillan, London, UK.

Waller L.A., Goodwin, B.J., Wilson, M.L., Ostfeld, R.S., Marshall, S. and Hayes, E.B. (in press) Exploring spatiotemporal patterns in county-level incidence and reporting of Lyme disease in the northeastern United States, 1990–2000. *Environmental and Ecological Statistics*.

Warren, W.F. (1968) Scientific Advisory Group Working Paper No. 10-68. A review of the herbicide program in South Vietnam. August 1968. Texas Tech Vietnam Centre. Available at: www.vietnam.ttu.edu/ (accessed 15 July 2003).

WCED (World Commission on Environment and Development) (1987) *Our Common Future*. United Nations, New York.

WCMC (1992) Global Biodiversity. World Conservation Monitoring Centre, Chapman and Hall, London.

Whitfield, J. (2003a) Double threat decimates apes. Hunting and the Ebola virus killing chimpanzees and gorillas. 10 April 2003. *Nature* 422, 551.

Whitfield, J. (2003b) Alaska's climate. Too hot to handle. *Nature* 425, 338–339.

Whittaker, R.H. (1975) *Communities and Ecosystems*, 2nd edn. Macmillian, New York.

Wijaya, T. (2005) Sertifikasi PHBML Antara Pesona dan Upaya Mengatasi Tantangannya. In: Hardiyanto, G. (ed.) *Label Hijau: Kompilasi Pengetahuan dan Pengalaman Sertifikasi Ekolabel di Indonesia*. Lembaga Ekolabel Indonesia (LEI), Bogor, Indonesia, pp. 169–195.

Wikipedia (2005) Papua New Guinea. Available at: http://en.wikipedia.org/wiki/Papua_New_Guinea (accessed 6 October 2005).

Wildlife and Poverty Study (2002) Prepared by the Livestock and Wildlife Advisory Group in DFID's Rural Livelihood Department. Available at: www.wildlife_poverty_study.pdf (accessed 19 December 2002).

Williams, M. (1989) *Americans and Their Forests. A Historical Geography*. Cambridge University Press, Cambridge, UK.

Willis, K.J., Gillson, L. and Brncic, T.M. (2004) How "virgin" is virgin rainforest? *Science* 304, 402–403.

Willson, M.F. and Halupka, K.C. (1995) Anadromous fish as keystone species in vertebrate communities. *Conservation Biology* 9, 489–497.

Wilshusen, P., Brechin, S., Fortwangler, C. and West, P. (2002) Reinventing a square wheel: critique of a resurgent 'Protection Paradigm' in international biodiversity conservation. *Society and Natural Resources* 15, 17–40.

Wilson, M.L. (1998) Distribution and abundance of *Ixodes scapularis* (Acari: Ixodidae) in North America: ecological processes and spatial analysis. *Journal of Medical Entomology* 35, 446–457.

Wilson, M.L., Telford III, S.R., Piesman, J. and Spielman, A. (1988) Reduced abundance of immature *Ixodes dammini* (Acari: Ixodidae) following removal of deer. *Journal of Medical Entomology* 25, 224–228.

Wilson, M.L., Ducey, A.M., Litwin, T.S., Gavin, T.A. and Spielman, A. (1990) Microgeographic distribution of immature *Ixodes dammini* ticks correlated with that of deer. *Medical and Veterinary Entomology* 4, 151–159.

Winters, R.K. (1974) *The Forest and Man*. Vantage Press, New York.

Withed, T. (2000) *Forests and Peasant Politics in Modern France*. Yale University Press, New Haven, Connecticut.

Woodham-Smith, C. (1962) *The Great Hunger, Ireland 1845–1849*. Harper and Row, New York.

Woodward, S., Stenlid, J., Karjalainen, R. and Hutterman, A. (eds) (1998) *Heterobasidion annosum: Biology, Ecology, Impact and Control*. CAB International, Wallingford, UK.

Woodwell, G.M. and Whittaker, R.H. (1968) Primary productivity in terrestrial ecosystems. *American Zoologist* 8, 19–30.

Woolbright, L.L. (1991) The impact of Hurricane Hugo on forest frogs in Puerto Rico. *Biotropica* 23, 462–467.

World Bank (1991) *The Forest Sector. A World Bank Policy Paper*. World Bank, Washington, DC. Available at: www.wrm.org.uy/actors/WB/1991policy.html (accessed 17 September 1999).

Worldwildlife (2005) World wildlife. Available at: http://www.worldwildlife.org/wildplaces/ng/index.cfm (accessed 18 September 2005).

WRI (1998–1999) Environmental Change and Human Health. Report Series: World Resources *1998–99*. A joint publication by the World Resources Institute, the United Nations Environment Programme, the United Nations Development Programme, and The World Bank, Washington, DC. Available at: http://pubs.wri.org/pubs_description.cfm?PubID=2889 (accessed 13 September 2005).

WWF (2004) Who Owns the World's Forests? Available at: www.wwflearning.co.uk/filelibrary/pdf/spr_04_main.pdf (accessed 15 September 2005).

WWF (2005) China's Wood Market, Trade and the Environment. WWF. July 2005. Available at: www.fs.fed.us/research/infocentre.html (accessed 1 September 2005).

Young, C.R. (1979) *The Royal Forests of Medieval England*. University of Pennsylvania Press, Philadelphia, Pennsylvania.

Yuniati, S. (2005) The Challenges of Developing Community-Based Forest Management in a New Indonesia. Asia–Pacific Community Forestry Newsletter 13, 1–32. February/March 2000. Regional Community Forestry Training Centre (RECOFTC). Thailand. Available at: www.recoftc.org/documents/APCF_Newsletter/13_1/APCF131.pdf (accessed 1 September 2005).

Zarin, D.J. (1999) Spatial heterogeneity and temporal variability of some Amazonian floodplain soils. In: Padoch, C., Ayres, J.M., Pinedo-Vasquez, M. and Henderson, A. (eds) *Várzea. Diversity, Development, and Conservation of Amazonia's Whitewater Floodplains*. The New York Botanical Garden Press, New York, pp. 313–321.

Zhou, G., Minakawa, N., Githeko, A.K. and Yan, G. (2004a) Association between climate variability and malaria epidemics in the East African highlands. *Proceedings of the National Academy of Sciences* 101, 2375–2380.

Zhou, G., Minakawa, N., Githeko, A.K. and Yan, G. (2004b) Spatial distribution patterns of malaria vectors and sample size determination in spatially heterogeneous environments: a case study in the west Kenyan highland. *Journal of Medical Entomology* 41, 1001–1009.

Index